Applied Biophysics of Activated Water

The Physical Properties, Biological Effects and Medical Applications of MRET Activated Water

Applied Biophysics of Activated Water

The Physical Properties, Biological Effects and Medical Applications of MRET Activated Water

Vladimir I. Vysotskii
Kiev National Shevchenko University, Ukraine

Alla A. Kornilova
Moscow State University, Russia

Igor V. Smirnov
Global Quantech, Inc., USA

World Scientific

NEW JERSEY • LONDON • SINGAPORE • BEIJING • SHANGHAI • HONG KONG • TAIPEI • CHENNAI

Published by

World Scientific Publishing Co. Pte. Ltd.
5 Toh Tuck Link, Singapore 596224
USA office: 27 Warren Street, Suite 401-402, Hackensack, NJ 07601
UK office: 57 Shelton Street, Covent Garden, London WC2H 9HE

British Library Cataloguing-in-Publication Data
A catalogue record for this book is available from the British Library.

APPLIED BIOPHYSICS OF ACTIVATED WATER
The Physical Properties, Biological Effects and Medical Applications of MRET Activated Water

ISBN-13 978-981-4271-18-9
ISBN-10 981-4271-18-7

Typeset by Stallion Press
Email: enquiries@stallionpress.com

The Authors

Vladimir I. Vysotskii
Professor
Head of Department of Theoretical Radiophysics
Radiophysical Faculty
Kiev National Shevchenko University
Vladimirskaya, St. 64
Kiev, 01033
Ukraine

Alla A. Kornilova
Ph.D.
Director of Innovation Center
Physical Faculty
Senior Researcher at Solid State Physics Department
Moscow State University
Moscow, 119899
Russia

Igor V. Smirnov
Ph.D.
President of Global Quantech, Inc.
Biotech Research company
San Marcos, CA 92078
USA
Email address: igor@gqusa.com

Preface

This book presents the results of complex experimental and theoretical studies of the characteristics of water activated by external action. The studied samples of activated water were obtained with the help of Molecular Resonance Effect Technology (MRET). We discuss the results of detailed studies of mechanical, electrodynamic, optical, and other characteristics of activated water itself and various physical phenomena associated with its application. About half of the book is devoted to the comprehensive studies of the specific features of the influence of activated water on various biological objects (plants, microorganisms, animals). These investigations were carried out from year 2004 to 2008 at several scientific institutes and universities of Ukraine, Russia, and USA, with participation from the authors of the book.

Despite the wide spectrum of the performed studies related to various branches of science, the final object of these studies was, eventually, activated water. Both in the process of these studies and in the analysis of the obtained experimental results, the authors did not attach non-distinctive or even mystic sense to the notion of activated water, and considered it always as an ordinary object of studies which has undergone the strictly controlled action of definite physical fields with definite time intervals. These time intervals (the duration of activation) were chosen from preliminary studies, and determined from the range of the most essential action.

First of all, we indicate that our investigations have no connections with extrasensorics, astrology, and other marginal fields. In connection with the existing ambiguity of the interpretation of the notion of activated water, we would like to note that the studied activated water was pure (in particular, distillated) water which had been subjected to a sufficiently long-term treatment by very weak nonionizing electromagnetic fields with definite frequency-related, spatial, and polarization-related properties. These fields were created by several identical unified generators constructed on the basis of the MRET technology. Additionally, we also refer to the use of definite strictly-controlled temperature for the storage of activated water prior to its investigation or prior to its application on biological objects.

The detailed investigations showed that, upon such electromagnetic treatment, significant quasistable changes occur in the spatial structure of water. These changes can be preserved for a very long time interval and, in turn, can lead to changes in the physical properties of activated water and to a significant modification of its biophysical and biochemical actions on living organisms. These changes can be considered as evidence of manifestation of the long-term "memory" of activated water.

The present publication is a natural and logical sequel to our preceding book (Vysotskii, Smirnov and Kornilova, 2005), where we considered some theoretical aspects of the influence of nonionizing and ionizing fields on water.

Below, we give briefly the organization of the book for the readers' convenience.

The first chapter includes the analysis of the well-known characteristics of ordinary water. In addition, we consider a number of established and new models of water structure. Here, we touch the problem of the memory of water and study some specific models of such memory. Particular attention is paid to the clathrate model of water memory, for which we performed specific calculations of the time intervals with different types of relaxation and at various temperatures (Vysotskii and Kornilova, 2004). The results of these calculations agree sufficiently well in a wide range of variation in temperature, with the data of the experiments obtained by the authors and analyzed in the subsequent chapters in detail.

The second chapter is devoted to the analysis of the main concepts of the process of water activation with the help of US patented MRET water activation technology, which is based on the effect of specific subtle, low frequency, nonionizing electromagnetic field generated by MRET activator. Written by the patent holder Dr. Igor Smirnov, this chapter describes the mechanisms of formation of volumetric fractal matrix in a polymeric compound produced in compliance with MRET technology. Dr. Smirnov also presents the brief results of scientific investigations supporting his theoretical concepts regarding the character of anomalous electromagnetic processes running in the volume of fractal matrix of MRET polymer compound following its excitation by defined external electromagnetic fields of special structure. He also analyzes the reasons for the existence of quasi-stable structural defects in the volume of water and the formation of long-range water molecular structures following the process of its activation. Then, based on the scientific investigation of the anomalous viscosity of MRET water, the author shows the compatibility of long-range polarized oriented multilayer structuring of MRET Activated Water with the analog structuring of cell water in biological systems presented in numerous theoretical studies of famous scientist such as Dr. Ling (2003) and Dr. Drost-Hansen (1991). All biological and physical studies described in this manuscript were conducted on activated water produced with the help of MRET water activator.

The third chapter of the book is related to the description of the procedure of complex studies of the physical characteristics of activated water and the analysis of the results of these studies. In the performed experiments, we show that the activation of water causes very essential changes in the electromagnetic, mechanical (including viscosity-related), and spectral properties of water. Several important experiments were conducted with participation of Prof. N. D. Gavrilova and Dr. E. Malyshkina, from Lomonosov Moscow State University, and Dr. L. N. Nikitin from Russia Academy of Science. We studied the dependence of the manifestation and the existence of these specific features of the properties of activated water on the water-activation duration, temperature, and the duration of storage after the activation. We note that some of these changes in the properties of water are so paradoxical in nature and great in magnitude that they can be undoubtedly referred to as anomalous. The dependence of changes in the properties of water on time after the completion of the activation of water is also uncommon and has no analogs in the literature. We discovered that at a low temperature, the anomalous properties of water can be preserved for many days. Some of the parameters of activated water (in particular, the hydrogen index pH) are characterized by spontaneous oscillations with a great amplitude and with the period of oscillations ranging from several hours to several days.

The fourth chapter presents the study of the influence of various types of activated water on higher plants. These investigations were conducted under the supervision of Dr. N. A. Matveeva from the Institute of Cell Biology and Gene Engineering of the National Academy of Science of Ukraine (NASU). In this chapter, we study the conditions under which activated water renders stimulating action on the growth of higher plants and, in particular, vegetable crops. We investigated the influence of the activation of water on the growth of sterile plants and callus tissue. It is shown that activated water produced in a certain mode of activation can very strongly (by hundreds of times) inhibit nonspecific growth of callus tissue. This unusual result allows one to forecast the possibility of the effective utilization of such water for the therapy of a number of diseases (in particular, psoriasis).

The fifth chapter is devoted to the study of the influence of activated water on pure microbiological cultures and on their natural associations, including a very large number of different cultures with multifunctional connections of the symbiosis type. The investigations were carried out under aerobic and anaerobic conditions. It is demonstrated that the use of water activated in certain modes increases sharply the reductase activity of microorganisms and very significantly changes the specific features of the influence of various types of antibiotics on microorganisms (enhancing such an influence or weakening it). It is also shown that, under definite conditions, activated water possesses very strong bactericidal properties and can inhibit the development of pathogenic microbiological cultures by tens and hundreds of times. These detailed investigations were conducted under the supervision of Dr. A. B. Tashyrev from Zabolotny Institute of Microbiology and Virology of NASU (Kiev). Such properties allow one to forecast the possibility of the use of a definite mode of the activation of water for its sterilization.

In the sixth chapter, we consider comprehensively the specific features of the influence of water activated in different modes on the prophylaxis and treatment of some types of oncologic diseases. In the experimental studies performed on mice, we have shown that the intake of water activated for a definite time decelerates sharply the development of tumors of two types (*Ehrlich carcinoma* and *Sarcoma 37*) in inoculated mice, and increases very significantly the lifetime of sick animals. We have found that, by the efficiency of antitumoral action, the intake of optimally activated water corresponds approximately to the methods of chemotherapy or radiation antitumoral therapy, but, contrary to them, renders no negative side action on the other organs and systems of animals. In particular, we prove that the intake of activated water increases significantly the immunity of animals, antitumoral activity of lymphocytes with natural killing properties, and the index of cytotoxic activity. Here, we have also studied the dependence of the antitumoral action of activated water on both the time and the mode of its intake (prior to the appearance of a tumor, i.e. in the mode of antitumoral prophylaxis or thereafter, which corresponds to the mode of therapy). It is demonstrated that the factor concerning the duration of the intake of water and the mode of the intake are very important circumstances which affect the treatment. These fundamental investigations were conducted under supervision and participation of Dr. Yu. V. Yanish and Dr. S. Olishevsky from Kavetsky Institute of Experimental Pathology, Oncology, and Radiology of NASU (Kiev).

In the seventh chapter, the effect of activated water on staphylococcal infection *in vivo* in animal model (on the cells of immune system) and *in vitro* on the culture of *Staphylococcus aureus* was investigated. The investigation was conducted in two steps. The experiments *in vivo* were conducted on mice infected with *Staphylococcus* culture after preventive consumption of MRET Activated Water (including the effect of activated water on the development of the local acute inflammation, on the death rate of animals in the case of intra-peritoneal staphylococcal infection, on staphylococcal infected mice, on the cellularity and the weight of lymphoid organs, and on functional activity of cells of the phagocytic system). In the *in vitro* experiments, the growth of identical staphylococcal culture was studied on meat-peptone agar treated with MRET activator. These investigations were conducted under the supervision of Prof. L. S. Kholodna from Biological Department of Kiev National Shevchenko University.

The eighth chapter is related to the analysis of some possible biophysical mechanisms of the influence of activated water on biological objects. In this chapter, we consider possible phenomena and processes, with the help of which the consumption of activated water of certain types stimulates the immunity, enhances the antitumoral activity of lymphocytes, inhibits the growth of tumors, increases the lifetime of sick animals, inhibits the growth of callus tissue, and ensures the bactericidial properties concerning the development of pathogenic cultures. All the above-mentioned consideration is carried out on the cell level of the organization of organism and is closely related to the results of the physical and biological experiments presented in this book.

In the ninth chapter, we present the total generalizing analysis of all the obtained experimental and theoretical results, some conclusions, and a number of proposals for the possible utilization of activated water in solving the applied problems of medicine, biology, biotechnology, and agriculture.

The preface, Chapters 1, 3, 4, 5, 6, 8, 9 and part of Chapter 7 were written by Prof. V. I. Vysotskii and Dr. A. A. Kornilova on the basis of their own theoretical studies and experiments carried out with their participation at the laboratories of several leading scientific institutes and universities of Ukraine and Russia. Chapter 7 was written under the supervision of Prof. L. S. Kholodna.

In the writing of the book, we used experimental results obtained over the duration of four years from the studies carried out jointly with Prof. V. I. Vysotskii and Dr. A. A. Kornilova at the Kiev National Shevchenko University (Ukraine), Lomonosov Moscow State University (Russia), Zabolotny Institute of Microbiology and Virology of the National Academy of Sciences of Ukraine (NASU), Institute of Cell Biology and Gene Engineering of NASU, Kavetsky Institute of Experimental

Pathology, Oncology, and Radiology of NASU, and Institute of Elementoorganic Compounds of Russian Academy of Science (RAN). The authors express their deep gratitude to Dr. N. A. Matveeva, Dr. A. B. Tashyrev, A. A. Tashyreva, Dr. Yu. V. Yanish, Dr. S. Olishevsky, Prof. L. S. Kholodna who performed the experiments on the study of the influence of activated water on plants, microbiological cultures, oncologic cells, animals with inoculated oncologic tumor, and on staphylococcal infection. We also sincerely thank our colleagues Prof. N. D. Gavrilova and Dr. E. Malyshkina from the Physical Department of Lomonosov Moscow State University (Russia), and Dr. L. N. Nikitin from the Institute of Elementoorganic Compounds of Russian Academy of Science who measured the viscosity, dielectric properties, and IR- and UV-spectra of activated water.

We are sincerely obliged to many colleagues and friends (especially to Dr. I. I. Samoilenko and Prof. V. D. Rusov) for useful and stimulating discussions on the above-mentioned problems.

Chapter 2 was written by Dr. I. V. Smirnov. It is substantiated by his US patented technology "Method and Device for Producing Activated Liquids and Methods of Use Thereof" (Patent number 6002479) and by numerous scientific investigations conducted in certified research institutions and universities worldwide for several years.

The authors express their deep sincere thanks to the President of the MRET Technology Corporation, Diana Suk, for her support of our studies.

Contents

Overview

The book presents the results of complex experimental and theoretical studies of the characteristics of activated water obtained under controlled action of the specific nonionizing, low-frequency, electromagnetic emission on ordinary water. This emission was produced by a special generator (activator), whose operation principle is based on the Molecular Resonance Effect Technology (MRET technology).

We discussed a number of mechanical, electrodynamic, optical, and other characteristics of activated water. It is shown that the activation of water is associated with very significant (by several and tens of times) changes in these characteristics. These changes are preserved after the completion of the activation for a long period (up to many months for the storage of activated water at a low temperature), which allows us to say about the presence of distinctive long-term memory of water.

The results of the theoretical analysis of a possible mechanism of the water memory and methods of its stimulation are given, and the comparison of the duration of existence of this memory with experimental results is made. Particular attention is paid to the clathrate model of water memory, for which specific calculations were carried out for different temperatures. It is shown that the results of the theoretical analysis and the data of physical experiments are in good agreement.

The results of specific experiments on the study of the influence of activated water on various biological objects (plants, microorganisms, animals) are throughly described and discussed.

The presented results demonstrate the significant influence of MRET Activated Water on higher plants (vegetable crops), sterile plants, and callus tissue. In particular, it is shown that activated water can very strongly (by tens and hundreds of times) inhibit nonspecific growth of callus tissue. This result allows one to forecast its use for the therapy of a number of diseases (for example, psoriasis).

We describe the results of experiments which study the influence of activated water on pure microbiological cultures and their natural associations. The studies were carried out under both aerobic and anaerobic conditions.

A significant influence of such water on the reductase activity of cultures and on the efficiency of action of various antibiotics on these cultures is discovered. It is also shown that the activation of water under definite conditions gives rise to the appearance of very strong bactericidal properties: activated water inhibits the development of pathogenic microbiological cultures by tens and hundreds of times more strongly, and this can be used for sterilization.

In this book, we present the results of studies of the use of activated water in the prophylaxis and treatment of oncologic tumors of two types (*Ehrlich carcinoma* and *Sarcoma 37*) in inoculated mice. It is shown that, in the certain mode of activation (with optimal duration) and use of this water (in particular, prophylaxis), the growth rate of tumor in inoculated mice decreases by several times, and the lifetime increases by 50–60%. By the efficiency of the antitumoral action, the optimal intake of activated water corresponds approximately to the methods of chemotherapy or radiation therapy, yet renders no negative action on other organs. It is also shown that the prophylactic intake of activated water increases significantly the immunity of animals, the antitumoral activity of lymphocytes possessing the natural killing properties, and the index of cytotoxic activity. The dependence of the antitumoral action of activated water on the time interval of its storage after the activation is demonstrated.

The effect of activated water on staphylococcal infection *in vivo* in animal model (on the cells of immune system) and *in vitro* on the culture of S*taphylococcus aureus* was investigated. In the experiments *in vitro*, the growth of identical staphylococcal culture was studied on meat-peptone agar treated with the same MRET activator. In this case, the highly-significant bacteriostatic effect of 70–100% was observed at optimal conditions of activation for different initial concentration of staphylococcal culture cells.

We also consider the possible biophysical molecular mechanisms of the direct influence of activated water on biological objects, and made a number of justified proposals for the possible use of activated water to solve the actual fundamental and applied problems of medicine, biology, biotechnology, and agriculture.

CHAPTER 1

Introduction to the Theory of Water Memory and General Principles of Water Activation

1.1. Water Structure and the Paradoxes of Water Memory

In the vital activities of any biological object, the most important chemical compound is water. It is difficult to enumerate all the functions of water in biological systems.

It is well known that water is a base of the intracellular liquid. It is the transport medium for the transfer of important chemical elements, forms the necessary states of these elements in the form of atomic and molecular ions, and ensures the normal vital activity of all systems of organism.

Water is the principal element of the very efficient system of thermoregulation and thermostabilization of all warm-blooded organisms. However, until now the mechanisms of the strikingly efficient system of thermostabilization of living organisms have not been clarified.

Intracellular water is the main participant of all radiobiological processes. It is the medium in the bulk of which the primary radiobiological processes of interaction of various types of the ionizing radiation (X-rays, gamma-quanta, fast electrons, heavy ions, and neutrons) with living objects are carried out. Water favors the formation of free-radical complexes which play decisive roles in radiation-induced damages. It is known that such indirect mechanism is responsible for more than 95% of total damages induced by radiation.

Furthermore, water is the basis of a totality of processes which underlie the phenomenon "*hormesis*", whose essence consists in the positive action of small doses of ionizing radiation on every biological object. These questions are considered comprehensively in our works (Pinchuk and Vysotskii, 2001; Vysotskii *et al.* 2002) and are generalized in the book (Vysotskii, Smirnov and Kornilova, 2005).

The spatial structure of water agrees ideally with the secondary structure of a DNA macromolecule, by guaranteeing the maximally stable existence of the double helix formed from matching pairs of nucleotides. Moreover, there are weighty arguments that the water of the primary ocean, whose composition was very close to that of the intracellular liquid, became the spatial matrix on which the first macromolecular DNA was synthesized.

At the same time, water is one of the most mysterious chemical compounds. Up to now, there is no unambiguous answer to the question of the spatial structure of water at the supermolecular level. Anomalous properties of water have for a long time become a classical example of the manifestation of characteristics of a nontrivial system. Water is the most universal solvent.

Intense doubts arise about the question on "the water memory", i.e. whether one can somehow change the properties of pure water (without changing its chemical composition) and preserve these properties for a long time. In the "near-science" and popular literature, a lot of information is available about the phenomenon that a significant change of the characteristics of water occurs under the action of various external factors. Such actions are called the process of activation of water, despite the fact that each experimenter embeds his own interpretation in this notion.

In modern scientific literature, water is considered to be activated if it (1) has undergone the action of a constant or variable electric or magnetic field, (2) is under the direct action of nonionizing SHF emission, (3) is subjected to impact or harmonic mechanical actions, or (4) is passed through heat treatment (with a decrease or increase in temperature) or a phase transformation (in particular, a freezing or a thawing transformation).

Contrary to the above, in mass media, which are far apart from science, water is said to be activated or "charged" if it is directly affected by a man who manifests strong extrasensory characteristics. Whether these methods can really influence the structure of water, or that they most likely act on the psychic state of the witnesses are questions to be solved in the future.

As one more stumbling-block, we mention experiments involving different methods of activation and utilization of water. Such experiments are numerous. They were executed by known scientists and by those who can hardly be referred to as experts or scientists. For the latter, their results were given mostly in the forms of sensational communications in newspapers, on TV, and through the internet recently, rather than in the form of scientific papers. In the majority of cases, it is very difficult or even impossible to estimate the reliability of such communications. As a rule, these sensations usually contain no information about the method of activation, about the methods and statistics of measurements, and very frequently they contradict the simplest theoretical estimates, saying nothing of more strict theoretical models. At the same time, a more involved and general theory was not developed and does not exist up to now. For this reason, the results of these experiments were often ignored by members of the scientific community, and incomprehensible results were at once and without analysis referred to as wittingly erroneous.

There exist rare communications that some of the changes induced in water during the process of activation can be preserved for a long time (hours or days). Moreover, the mass media sometimes present the information that the activated water in some way possesses unique physicochemical properties and, in this case, can render a significant positive action on vital activities of living organisms and on the course and the treatment of many diseases.

Unfortunately, scientific literature does not include the publications presenting the results of correctly executed systematic studies which demonstrate that activated water does possess particular physical properties and renders positive influence on biological systems. It is practically impossible to encounter such publications in serious journals, because the editorial boards of these journals are strongly prejudiced against such investigations.

We can list a lot of specific reasons for such a situation. However, it is obvious that this is a consequence of a deep-rooted skepticism of the question about "the water memory". Such skepticism appeared shortly after the very great successes of quantum mechanics at the first half of the 20th century in the study of relatively simple systems such as atoms, simple molecules, ideal gases and perfectly-ordered crystals. The next natural step was the transfer of the principles of the spatial organization of the ideal gas and crystals to real liquids.

At first sight, it seemed that this problem can be solved comparatively simply and rapidly. The model of water was limitedly simplified so that it was identified either with the system of a dense gas or with the ideal dynamic crystal with ordered hydrogen bonds and very great coefficient of diffusion. The duration of relaxation of any perturbations in each of these systems is very short and does not exceed several parts of one nanosecond. In the framework of such an approach, it was natural that any consideration of the long-term water memory was simply irrelevant. Moreover, those who were engaged in this field were labeled, in the best case, as "alchemists" or false scientists.

However, the structure of water turns out to be much more complicated than what was conceived in the original idea. Water has simultaneously the properties of a gas and a crystal, and its behavior frequently contradicts either of them. The comparatively simple computational methods, which allow one to determine the main characteristics of a crystal based only on the properties of a separate atom or a molecule, turned out to be clearly insufficient for the construction of a complete theory of water. (This problem will be discussed in more details in the following chapter.)

No matter how paradoxical it sounds, a certain contribution to the process of distinctive ignoration of the role of water as one of the main players in living organisms was introduced (and is introduced) by biochemistry and theoretical biology. Of course, on the systematic level, the role of water is always emphasized. But it is implicitly postulated that water plays, in fact, a passive role. Water is considered as a space for the realization of the ion transport, as a medium for removal of wastes from the organism, as a base of thermodynamics, and as the main factor of thermostabilization of organism. Water is considered as a stage, on which the great performance named "Life" is being played. Of course, there is no performance without a stage which gathers all the participants in a single collective by the principle of the unity of place and time, but the role of a stage is always passive.

But, in fact, water is a participant (possibly, a principal one) and an actor of this performance, and it possesses absolutely equal rights with others! Water directly affects all the processes in an organism, and these processes, in turn, influence water, change the properties of water, and are simultaneously subordinated to water. In this case, not only the global action of water, but also the influence of each individual molecule turn out to be important.

We give one characteristic example which confirms this assertion in full measure:

It is well known that the base model of the organization of DNA is the so-called "compressed" form A which corresponds to a maximum of the binding energy of subsequent complementary pairs of nucleotides in the absence of some external medium (in fact, in vacuum). This maximum is determined by the period $R_0 = 2.8$ Å between subsequent pairs of nucleotides in the double helix of DNA and by the angle of their mutual turn equal to $\theta = 30°$. This configuration was calculated many times, starting from the conception of the existence of two types of interaction in the regions between the successively located pairs.

The first type of interaction is the Coulomb electrostatic interaction of the ions distributed over the surface of a nucleotide. This interaction corresponds in most cases to the mutual repulsion of the pairs of nucleotides.

The second type of interaction is the dispersive van der Waals interaction, which defines the coupling between separate structural elements opposite to nucleotides (in fact, between the atoms distributed over the surface of these nucleotides). This interaction corresponds, as a rule, to the mutual attraction.

On the other hand, it is reliably known (for example, on the basis of the data of numerous X-ray diffraction studies) that only the "stretched" form B, i.e. the centered double helix form of DNA, is realized in the presence of water near DNA. In this form, the stable distance between the mean planes of any pair of nucleotides (i.e. the period of the structure of DNA) is $R_0 = 3.4$ Å, and the angle of the mutual rotation of the pairs of nucleotides $\theta = 36°$.

This fact is sufficiently paradoxical. The matter is that, both in the compressed and stretched forms of DNA, the distances between the separate pairs of nucleotides with regard to the spatial period ($R_0 = 2.8$ Å or $R_0 = 3.4$ Å) and the efficient "thickness" of each nucleotide (about 1 Å) turn out to be significantly less (the distance is about 1.8–2.4 Å) than the effective size of a water molecule which is equal to 2.76 Å. Thus, a water molecule cannot be inside the stack of complementary pairs of nucleotides. In such geometric structure, water cannot render a "direct" influence on the character of the interaction of adjacent pairs of nucleotides (on the *stacking* energy) at the expense of, for example, the screening of the field of charges or a significant modification of the dispersive interaction.

The situation looks really strange. Indeed, the effect of an increase of the period of DNA exists, and it is well-known from experiments that this effect is related to water. But no specific numerical analysis of its appearance was actually performed.

The calculations carried out in the work of Vysotsky and Hovorun (2005) showed that the answer to the mechanism of changing the stacking energy can be related to the direct influence of polar water molecules located outside the scope of the space under consideration i.e. between the nucleotides (though in close proximity to this space) on the total energy of the system. These molecules interact simultaneously and directly with different pairs of nucleotides, and create the necessary modification which ensures the turning and displacement of nucleotides from the compressed form to the normal form of DNA.

For successive analysis of such a problem, the calculation of the three-dimensional distribution of the total energy of the system, which included the interaction energy of two nearest pairs of nucleotides and the interaction energy of these pairs with a water molecule, was carried out. It was then necessary to study the position of the minimum energy as a function of the mutual orientation and distance between the mentioned pairs of nucleotides.

To study the exact dependence of possible positions of water molecules in the region of the double-strand break of DNA, the real geometric arrangement of atoms for the Watson–Crick complementary pairs of nucleotides GC (the number of atoms is 29) or AT (the number of atoms is 27) was used. All calculations were performed for the B-form of DNA, for which the propeller angle (the dihedral angle between the planes of bases) $\theta_p = 2°$, the rotation angle of the helix $\theta = 36°$, and the slope angle to the helix axis $\theta_\gamma = 5.9°$.

The total energy of the system "terminal pairs of nucleotides — a water molecule" consists of

- the interaction energy between the terminal pairs of nucleotides in the region of the double-strand break of a DNA helix;
- the total interaction energy between the atoms of each terminal pair of nucleotides in the region of the double-strand break of the DNA helix with a water molecule which is located in the region (or directly near the region) of the break.

The results of calculations of the potential energy of interactions between two neighboring pairs of nucleotides (curve 1), the interaction between these pairs and one water molecule (curve 2) and, finally, the total interaction (curve 3) are presented in Fig. 1.1.

It is seen from this figure that the presence of a single water molecule, being in the position of a stable minimum near the external of a pair of nucleotides (i.e. near external surface of DNA), shifts the position of the stable equilibrium of this pair from $R = 2.8$ Å, corresponding to a molecule of DNA in form A, to $R \approx 3.2$ Å, which is very close to the characteristic period $R_0 \approx 3.4$ Å of a macromolecule DNA in form B.

The physical reason for such a deformation of the DNA helix in the presence of a water molecule is related to that, in the scope of the strongly interacting system "a water molecule + a pair of nucleotides", all its components are revealed as participants with quite equal rights, and the very strained state of DNA is a result of the distinctive compromise between two tendencies:

- on the one hand, the interaction between two pairs of nucleotides at the distance $R \approx 3.3$ Å corresponds to the appearance of a force

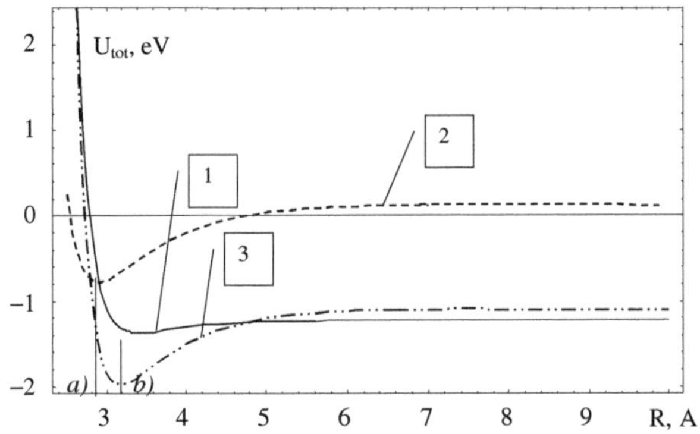

Figure 1.1. Influence of a single water molecule on the structure of DNA. The dependence of the total energy of the system "a water molecule — terminal nucleotides" (curve 3), the total interaction energy between the terminal pairs of nucleotides (curve 2), and the total interaction energy between two neighboring pairs of nucleotides and the water molecule (curve 1) on the distance between nucleotides R. In all the cases, the water molecule is at the point near the external surface of nucleotides which corresponds to the absolute minimum of the energy of its interaction with nucleotides. The points with coordinates *a*) and *b*) determine the positions of the minimum of the potential energy in the absence of water and in the presence of one molecule of H_2O.

$F = -\mathrm{d}U/\mathrm{d}r$ which tends to compress a DNA helix to the period $R_0 \approx 2.8$ Å (this is a position of the minimum interaction energy of the pairs of nucleotides in vacuum);

- on the other hand, the interaction of a water molecule with the same two pairs of nucleotides causes the appearance of another force which tends to pull a water molecule as far as possible in the region between nucleotides and, hence, separate these pairs and increase the distance between them to $R_0 \approx 3.5$ Å (this distance corresponds to a minimum of the interaction energy of a water molecule with a pair of nucleotides).

It is seen from the results of calculations that the inclusion of even one water molecule causes a very strong deformation of DNA. This result presents a sufficiently grounded explanation for the physicomolecular mechanism of the deformation of DNA when it comes into contact with water.

It should be noted that taking into account other water molecules which are near the external surface of DNA will allow one to calculate more exactly the structure of the stretched form of DNA.

This example demonstrates the obvious (but, nevertheless, very frequently ignored) argument that water is really an active player, rather than a passive one. One more important conclusion which follows directly from the performed analysis consists in that a molecule of the bound water near the surface of DNA is the integral component of DNA like other atoms which belong to the composition of nucleotides. This result was always perceived as intuitively obvious, but it was not confirmed by the results of direct calculations.

It is noted that whereas the problem concerning the structure and the properties of water was the object of numerous studies, the applied aspects of the influence of activated water on biological systems are presented by a collection of uncoordinated rare results which are frequently in poor agreement between themselves and sometimes mutually contradictory.

Below, we present some data from the literature which characterize changes in the properties of water and the specific features of its action.

For example, the main characteristics of water change after it passes through the region with a constant magnetic field (Klassen, 1973). In particular, if the intensity of this field varies from 1900 to 5700 Oe, the pH of chemically pure water (bidistillate) changes by 5.1–9.1%, and the surface tension changes by 2.2–7.3%. Such magnetic treatment changes the spectrum of infrared absorption of water. In water which passed through a region with a sufficiently strong magnetic field, the efficiency of the processes of dehydration of dissolved diamagnetic ions decreases and, on the other hand, the efficiency of the dehydration of paramagnetic ions increases.

For the same water, its ability to wet surfaces is also significantly changed. In this case, the effect turns out to be somewhat ambiguous: if the surface contains Si, then the wettability grows, and it decreases, as a rule, if Si is absent.

The magnetic treatment of water causes a very significant change in the dissolving rate for many salts. For example, the dissolving rate of magnesium sulfate increases by 120 times under the action of a strong magnetic field, which is realized in the mode of a sharp change in the direction of this field on water.

In a number of works, the derived results testify that a significant change in the refraction index of aqueous solutions of the proteins of blood plasma occurs under the action of a weak microwave emission. The refining experiments showed that the main contribution to this effect was given by a change in the refraction index of water itself.

The enumerated anomalous properties of water which has undergone the action of a magnetic field preserve for many hours and days.

It follows from experiments that such activated water possesses the changed physicochemical properties and, in some cases, can render a specific influence on biological objects (including a beneficial action on the treatment of some diseases).

One more aspect concerning the problem of water memory is associated with the possibility to preserve the information. This is related to dissolved chemical compounds with a great degree of dissolution in water. In fact, such a problem corresponds, by all canons, to classical homeopathy with its fundamental principle of manifold dissolution in water.

Among the works which caused great resonance at that time, we focus on the work (Davenas, 1988) published in the authoritative journal *Nature*. In this work, it was found that water preserves the information of trace amounts of some biologically active substances (i.e. that water was, in fact, chemically activated). This piece of information was not lost after extremely strong dilution, when the molecules of a dissolved substance were almost completely absent in water. This work describes the studies on classical immunology performed by a group of researchers led by the French biochemist J. Benveniste. They studied the specific influence of protein molecules on blood cells which are named basophils. These molecules induce their specific response called degranulation. According to the concepts of biochemistry, the greater the concentration of such proteins, the higher the rate of degranulation. Also, the rate decreases as the concentration decreases. Contrary to this idea, it was found in experiments that the clearly pronounced effect of degranulation of basophils was observed even at the extremely great dilution of the solution of protein molecules (antipolyglobulins), where their relative concentration was at most 10^{-30} (which corresponds to only one molecule per 70 liters of water!). Since the volume of a cuvette, where these studies were carried out, was much less than 70 liters, this means that none of protein molecules was present in the volume of water after the manifold dilution.

The experiments which were executed in this work demonstrated that this type of information can be stored for a long time in unperturbed water at sufficiently low temperatures, but it is efficiently "erased" under classical actions such as ultrasound, strong heating, or the phase transition such as the freezing of water and the thawing of ice.

We note that the medical aspect of the action of activated water is poorly studied, though this action has been confirmed by many experiments.

As an example of the influence of activated water on the vital activity of simplest microorganisms, we point out the results of studies performed on the Faculty of Biology of Lomonosov Moscow State University (these results can be found in the unpublished doctoral dissertation by Zenin, 1999).

The activation of water was realized by subjecting it to a variable magnetic field created by a standard magnetic mixer between 7 min and 15 min. Such water undergoes a number of physicochemical changes (including a change of its conductivity). In particular, the preliminary studies showed that, for distillated water undergone processing for 7 min, its conductivity increased at least by 10–15 times. The growth of the conductivity obeyed an almost linear law. After the cessation of the action of the variable magnetic field, a slow relaxation of the conductivity which decreases to the initial value was observed for 20–25 min. The other situation occurred at a more prolonged action. At the initial stage (during the action of a variable magnetic field), an analogous increase of the conductivity of water was observed. However, after the cessation of the action of the magnetic field, the effect of a spontaneous additional increase of the conductivity of water by 2.5–3 times for 15–20 min was observed instead of its decrease. Such result shows that the system responsible for water memory is characterized by the threshold time interval of irreducible activation from 7 till 15 min.

This activated water was used immediately after the completion of the activation to study the action on *Spirostoma*. If water was subjected to the short-term subthreshold action of a magnetic field, the motive activity of infusoria was inhibited for a short time interval of 30–40 min. If water has undergone the long-term (above-threshold) activation, *spirostoma* were irreversibly paralyzed with the full suppression of all signs of activity.

There are many such facts demonstrating the unusual behavior of water subjected to certain physical actions. Concluding the brief and incomplete analysis, we mention the following circumstances:

On the one hand, the popular literature and the mass media present a lot of information (frequently, with obvious advertising characteristics) on the beneficial effect of “charged” and activated water. Such water was advertised very frequently as a distinctive panacea for any diseases without any substantiation. It is natural that the scientific value of such communications is close to zero.

On the other hand, we failed to find the description of at least one cycle of studies which are executed with sufficient completeness by the commonly-used rules of scientific studies and which present the reliable results concerning not only the action of activated water of the same type on biological objects, but also the properties of this water. We received impression that the study of all specific features of the influence of activated water on biological systems turns out to be outside the field of interests of classical biology.

It is necessary to note that we had no preliminary "barrier-type separation", at which all similar results would be rejected *a priori* without analysis. We think that water, being life's cradle, renders very strong influence on the character of vital activity. Many genetic and somatic diseases, like problems in the mechanisms of division of cells, reproduction, and mutations, are integrally related to water.

Prior to the detailed investigation of these questions, we consider the specific features of the structure and the problem of the memory of ordinary and activated water in more details.

1.2. The Clathrate Model and a Water Memory Cell

There exists a great number of various theories and models explaining the structure and properties of water. In each of them, the basic position is the idea of hydrogen bonds as the main factor defining the formation of structurized agglomerates. For this reason, water is a cooperative system, and it contains the chain formations of hydrogen bonds.

Water possesses a number of unique properties, among which a particular place is occupied by its long-term "memory". Numerous experiments, some of which were presented above, have confirmed the existence of water memory, which is activated under the action of some physical fields (for example, a magnetic field, impact mechanical action, sharp change in the temperature or pressure) and can store the information about this action for many hours and days.

The above-presented facts painted one visible side of the problem. Just this side arouses the greatest interest, but simultaneously raises the greatest number of objections. The other side of the problem is based on the explanation of these effects and on the clarification of their mechanisms. We would consider it in more details.

At first sight, it seems that water, as a specific physicomolecular object, cannot have long-term memory. This follows from the following simple evaluations:

For a long time, the continual (quasicrystalline) model of water was dominant. In the framework of this model, the spatial structure of the potential energy for each of the molecules H_2O is the almost periodic three-dimensional system of potential wells and barriers. This relief is a result of the self-consistent motion of all water molecules which combines two independent processes: the oscillatory motion in each of the potential wells and a random (fluctuation-related) hop to the neighboring well. The mean frequency of oscillations in potential wells is approximately the same as the Debye frequency in solids ($\omega_D \approx 10^{13}\,s^{-1}$). The mean duration of a hop to the neighboring potential well $\tau_0 \approx 10^{-13}$ s. The mean duration of the stay in a well,

$$\langle\tau\rangle = \tau_0 e^{\Delta W/k_B T} \approx 10^{-9} - 10^{-10}\,s, \tag{1.1}$$

is determined by the temperature T of water and the activation energy, $\Delta W \approx 0.2$ eV, of the process of diffusion (by the height of a barrier between the neighboring wells). Staying in the framework of this model, it is easy to conclude that the water memory must be preserved not longer than the value of $\langle\tau\rangle$ which is by many orders less than the values given by numerous experiments.

The continuously increasing number of reliable experiments indicates that the continual model describes inadequately the structure of water.

The presence of a spatial structure in the bulk of water was first proved by Bernal in 1933.

The calculations on the basis of quantum chemistry showed that water molecules participate in the formation of molecular ensembles and can form various types of associated molecules: hydrol H_2O, dihydrol $(H_2O)_2$, trihydrol $(H_2O)_3$, etc. Further studies showed that much greater associations (clusters) of water molecules can be formed in water, the structure of which resembles small pieces of ice. As a rule, these clusters are unstable and spontaneously disappear. The dynamics of such associations underlies the cluster model of water (Nemethy, 1962). In the framework of such a model, one may consider that water is a two-phase system — a crystalline liquid with the intense processes of crystal-forming and strong intermolecular bonds (hydrogen bridges), with the ability to form agglomerates of hundreds of molecules and generate the infinite number of forms of the liquid-crystalline phase in water which is named a complex lattice-like structure. Such a structure is characterized by the great number of eigenfrequencies.

The more detailed studies (for example, Samoilov, 1957) showed that the so-called "clathrate" model is most close to the reality. In the final form, this model was developed by Pauling (1959). The Pauling clathrate model is

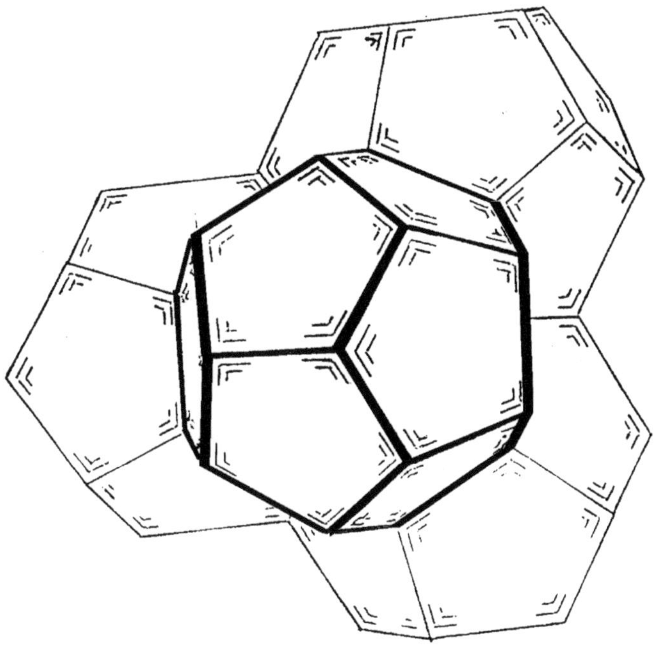

Figure 1.2. System of clathrate hydrates in water.

based on the idea that the union of atoms of oxygen and hydrogen is able to create spatial flexible tetrahedral frames. The spatial structure of the frame is given in Fig. 1.2.

The formation of tetrahedral frame is promoted by the circumstance that the natural spatial angle between the OH-bonds in a free molecule of water H_2O is equal to 104.5°, which is sufficiently close to the exact value of the tetrahedral angle of 108°. For the additional bending of this bond by an angle of 3.5°, a small energy is required, and the very presence of an additional bending significantly enhances the stiffness of the crystalline frame (a similar situation occurs, for example, in such purely structural element as a prestressed reinforced concrete).

At nodes of the crystalline frame, there are very large (on the scale of a water molecule) microcavities (microvoids) with rigid atomic walls. The main elements of this structure are regular polyhedrons i.e. dodecahedrons coupled with one another. Such systems are called "clathrate hydrates". This frame structure is held by hydrogen bonds. They firmly fasten the system of pentagonal dodecahedral polyhedrons of ions of oxygen and hydrogen which form the walls of microcavities. Each polyhedron can be characterized by an inscribed sphere with the radius $R_c \approx 2.6$ Å. Each polyhedron has 12 pentagonal faces, 30 edges joining these faces, and 20 vertices. At each vertex, three edges come together. At the vertices of each polyhedron, 20 molecules of H_2O are situated, and each molecule of H_2O has three hydrogen bonds. By the data in Zenin (1999), three polyhedrons can be joined in stable associates containing 57 water molecules. From these 57 molecules, 17 ones have completely saturated hydrogen bonds and form a tetrahedral hydrophobic central frame. Moreover, the surface of each of four dodecahedrons contains 10 centers of the formation of a hydrogen bond (O-H or O).

Outside of this frame, there are quasifree molecules of "ordinary" isotropic water, whose properties and structure approximately correspond to the continual model. Microcavities of the frame are joined with the external space by windows of about 2.5 Å in diameter, which is not much less than the width of a water molecule ($2R \approx 2.76$ Å). Finally, each of the microcavities is separated from the "external" amorphous quasifree water by a ring-like potential barrier of about $\Delta R \approx 0.13$ Å in width which borders a window. The relative amount of molecules of "frame" water at room temperature is 20–30% and increases as the temperature decreases. In the volume of a microcavity, one of the molecules of H_2O, CH_4, O_2, or N_2, for example, can be freely arranged.

Due to the presence of a strong and symmetric (relative to the centers of microcavities) electrostatic field, there exists a certain ban on the formation of hydrogen bonds of water molecules in microcavities with their walls. In this case, there occurs such nontrivial phenomenon as the repulsion of free water molecules from the walls of the frame formed also of water molecules (i.e. water molecules in the volume of water become hydrophobic)! The mean density of the clathrate frame (without the filling by water molecules) is 0.80 g/cm^3, i.e. microcavities occupy 20% of the entire volume of the structurized frame of water. If the microcavities are filled by water molecules, then the density of water is close to 1 g/cm^3.

The results of direct measurements (Zenin, 1999) showed that the optical properties of structurized and amorphous water at the same temperature differ very strongly. In particular, the difference of the refractive indices of the clathrate frame and amorphous water reaches 4–5% in some cases, which testifies to the spatial ordering of the clathrate frame.

The Pauling clathrate model explains very well all the properties of water (including its anomalous compressibility). The structure of DNA perfectly corresponds to the spatial structure of such frame water. In this case, every macromolecule of DNA puts water in order at the distance to 300–500 Å from its surface. The possibility to join the Pauling clathrate model with the cluster model was considered in many works. In this case, the separate elements of clathrate frames can be joined from time to time by hydrogen bonds and form groups possessing an ordered structure (i.e. clusters). Since there exists a very strong interconnection between the neighboring hydrogen bonds, the appearance and elimination of hydrogen bonds occur in a correlated manner and are synchronized in time. Such a character of the bond allows us to suppose that the "flickering clusters" arise and disappear in water. The lifetime of clusters is of the order of 10^{-10} s, i.e. of the order of 1000 molecular oscillations.

The considered specific features of the bulk water structure indicate that water molecules are always distributed between two systems weakly coupled with each other: the quasiamorphous nonstructurized water and the quasicrystalline structurized system of clathrate hydrates. During the external action on water (i.e. on the activation of water), there occurs a significant change of its structure and parameters. In view of the scale and mechanism of activation, two different hierarchical levels of organization of the water structure (a macrolevel and a microlevel) exist.

The first hierarchical level of the water structure (the macrolevel of the structure) corresponds to the global spatial structure of water and defines the form and the position of its spatial frame. This level is characterized by the presence of the system of clathrate hydrates which form stable dodecahedral polyhedrons of ions of oxygen and hydrogen. Inside the volume of each of these polyhedrons, there exist the empty microcavities with rigid hydrophobic walls. The dodecahedral polyhedrons with the help of stable hydrogen bonds are joined in binary, triple, and more complicated associates which can be further combined in very large associates (macroclusters). The space between macroclusters is filled with the quasiamorphous water.

Thus, the macrolevel of the structural organization of water corresponds to the equilibrium distribution between the phase of amorphous water and the phase of water, represented as a system of macroclusters with a complicated hierarchy. Under the action of the external factors, this distribution can be changed. For example, the volume of macroclusters increases with decrease in temperature, whereas the volume of quasiamorphous water decreases. As the temperature grows, the volumes of macroclusters decrease, and, in addition, each macrocluster can be divided into several smaller parts. It is natural that the volume of quasiamorphous water increases in this case. The same changes can occur under other types of action, e.g. under the action of ultrasound on the aqueous medium. Due to the strong dependence on external actions, the macrolevel of water structure has no sufficient efficiency in order to organize a water memory system that is stable to the external destructive actions. At the same time, it is obvious that, in the absence of very strong destructive actions, the process of recording of the information in the form of the system of regularly-joined clathrate cells is quite possible. Simply saying, separate polyhedrons of the clathrate frame can be combined by several alternative means. In this case, of importance is that the realization of a certain orientation of a specific pair of polyhedrons leads automatically so that the subsequent polyhedrons will be joined to this "bare" associate in exactly the same way, which induces the appearance of ordered macroclusters. Such a system has obvious long-range order, which can explain the possibility of a global structurization of great volumes of water. This question was sufficiently considered in Zenin's work (1999).

The second hierarchical level of the water structure (microlevel) corresponds to the processes of motion and distribution of separate molecules of H_2O between microcavities of the three-dimensional clathrate frame of water and the quasiamorphous nonstructurized water. This microlevel defines the nonstationary evolution of H_2O molecules. The process of evolution is determined by two possible directions: molecules can leave the volume of quasiamorphous water, enter into the volume of these microcavities, and be there in the hydrophobic form for a long time or, on the contrary, can pass from microcavities into the volume of the quasiamorphous water.

It is quite obvious that the microlevel of the water structure is distinguished by a much greater stability relative to the action of external destructive factors than the macrolevel. Under all external transformations of the clathrate frame which are characteristic of the macrolevel, hydrophobic H_2O molecules remain in the stable state in the volume of microcavities. Such a stability makes the microlevel of water structure to be an efficient object for the organization of a water memory system. Earlier, nobody has considered such a memory system in detail.

We now show how the presence of the clathrate frame of water can lead to the formation of the long-term memory in water and to the possibility of recording and using the information (Vysotskii and Kornilova, 2004; Vysotskii, 2005).

Consider the initial water which is in the state of thermodynamic equilibrium and is characterized by the definite temperature T. This state corresponds to the maximum of the entropy. Such water can be produced as a result of the long-term boiling and the slow cooling or a very long standing. In this case, the number of microcavities in the system of clathrate hydrates which are filled by water corresponds to the Boltzmann distribution with regard for the statistical weights of the states of H_2O molecules in microcavities and in amorphous water. This is an equilibrium water or an ordinary one.

In the frame, 18% of microcavities are filled by H_2O molecules at a temperature of 4°C, 38% at the normal temperature of a man (36.6°C), and about 50% at 55°C.

Such a law of distribution is related to several circumstances:

- the Boltzmann distribution at the given temperature,
- the degeneration multiplicity of the initial and final states of a H_2O molecules in the composition of amorphous water near an entry window to the volume of a microcavity and in the volume of this microcavity, and
- the ratio between the volume of all the amorphous water and the volume of the clathrate frame.

All three quantities vary with changes in the temperature, which hampers the execution of an exact calculation of the dynamics of the population of microcavities. At the same time, it is obvious that the binding energy of water molecules is close to zero in the volume of clathrate microcavities (due to the hydrophobic character of the interaction with walls), and the state of a H_2O molecule in the bulk of quasiamorphous water is determined by the depth of the potential well conditioned by the interaction with other water molecules. The depth of this well corresponds to the energy of activation $\Delta W \approx 0.2\,\text{eV}$ under the diffusion, which in fact decreases the energy level of a H_2O molecule relative to that of the same molecule in the clathrate frame by ΔW. Based on this consideration, it becomes obvious that the necessary energies of activation for the entry to a microcavity, ΔE_M, and for the output from it, $\Delta E_M - \Delta W$, will be different (Fig. 1.3).

According to this fact, the time for an "excess" water molecule to be present in a microcavity and the time of existence of an "excess" vacancy in an empty microcavity will also be different.

Upon the breaking of thermodynamic equilibrium, the H_2O molecules are redistributed between amorphous water and microcavities to a new equilibrium state. We now show that the spontaneous transition between these

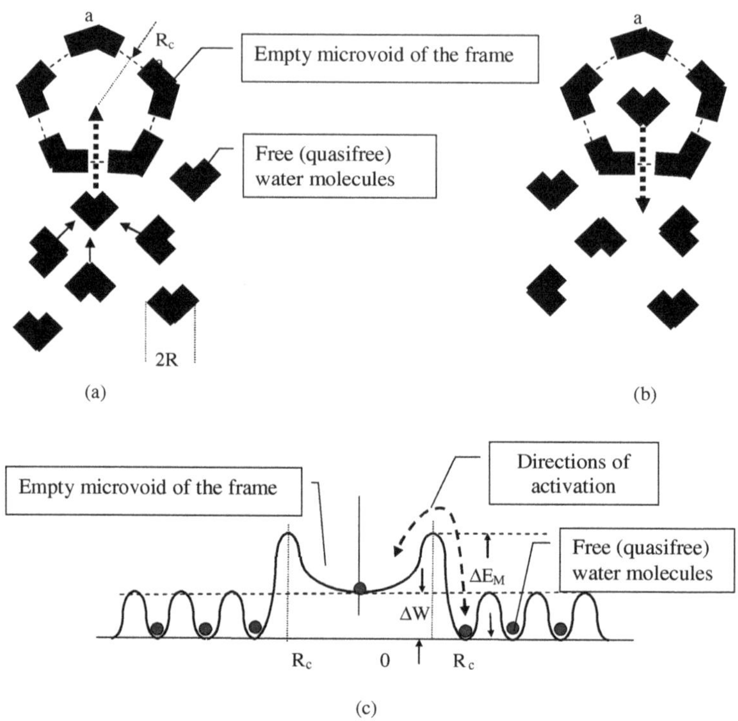

Figure 1.3. Process of (a) thermostimulated activation and (b) deactivation of microvoids of the clathrate frame of water under the increase or decrease of temperature; (c) structure of the potential energy of molecules of amorphous and bound water in the microvoid volume and near its boundary.

states is strongly inhibited due to the very small probability of the tunneling of H_2O molecules through "narrow" windows, and the duration of existence of each state turns out to be very great. Let us determine the duration of relaxation under such a redistribution.

Such a relaxation corresponds to the transition of water molecules in two possible directions: (a) from the state of amorphous water into the volume of microcavities (if the initial amount of water molecules in microcavities was less than a value conditioned by the Boltzmann distribution, which can happen under fast heating of the whole water); and (b) from the state of "excess" water in microcavities into amorphous water (if the amount of water molecules in microcavities was greater than the equilibrium value, which corresponds, for example, to the fast cooling of the whole water).

The process of relaxation of each of these transitions depends on the thermodynamic probability

$$W = e^{-\Delta E_M / k_B T} \tag{1.2}$$

for one of the water molecules to get the energy ΔE_M as a result of many random interactions with other molecules. This energy should be sufficient for a short-term deformation of a water molecule (associated with the increase of the interaction energy between a proton and the ion of oxygen) resulting in the short-term decrease of its size to that of the window of a microcavity and, respectively, the entry of this molecule inside the microcavity.

Since the frequency of collisions of any of the water molecules with the surface of any structural object in water is equal to the frequency of oscillations of molecules near a local equilibrium position $\omega_D \approx 1/\tau_0 \approx 10^{13}$ s, the total probability of capture of a molecule by an empty microcavity per unit time is $F = W/\tau_0$. This formula allows us to determine the average duration of existence of a nonequilibrium (empty) state of a microcavity in the volume of the spatial tetrahedral frame of water (the duration of relaxation of a vacancy in microcavities):

$$T_{1W} = 1/F_1 = \tau_0 e^{\Delta E_M / k_B T}. \tag{1.3}$$

It is obvious that this duration will determine the duration of existence of the water memory on the filling of this microcavity (for example, under the heating of water).

For the determination of the duration of relaxation [Eq. (1.3)], it is necessary to find the quantity ΔE_M which characterizes the height of the potential barrier separating the amorphous water from the empty space in a clathrate. To this end, it is necessary to consider the process of deformation of a water molecule in more details.

The vibrational motion of a proton in a H_2O molecule in the direction perpendicular to the bond line OH corresponds to a harmonic oscillator (Fig. 1.4).

The potential energy corresponding to a displacement of the hydrogen ion by a value r relative to the equilibrium position can be written in the form of the energy of a harmonic oscillator

$$V(r) = \frac{M_H \omega_H^2 r^2}{2}. \tag{1.4}$$

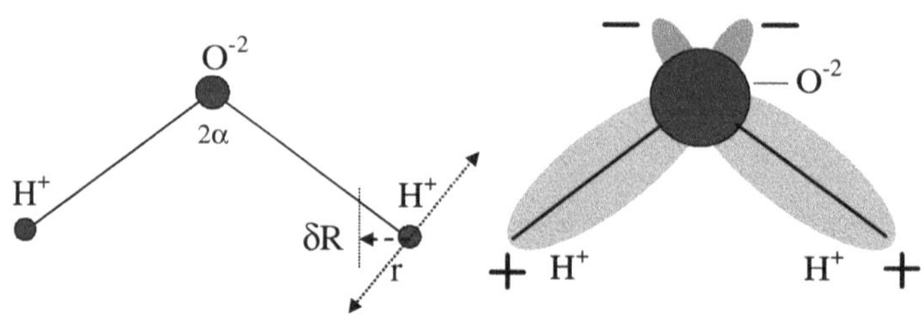

Figure 1.4. Normal oscillations of the proton in a water molecule. Two arrows directed oppositely show the direction of a normal oscillation of the proton.

Here, M_H is the reduced mass of a hydrogen atom, and $\omega_H \approx 3.10^{14}\,\mathrm{s}^{-1}$ is the frequency of normal oscillations of a proton in a water molecule in the direction perpendicular to the bond line OH (Zatsepina, 1999).

With regard for the fact that the angle between the bond lines of protons with the nucleus of oxygen $2\alpha \approx 104°27' \approx 104.5°$, we find that, in order to deform the external size of a water molecule by a value $\delta R \approx 0.26\,\text{Å}$ sufficient for the passage of a molecule into a microcavity, the deformation energy

$$V(\delta R) = \frac{M_{\rm H}\omega_{\rm H}^2(\delta R)^2}{2\cos^2\alpha} \approx 1.1\,\mathrm{eV} \tag{1.5}$$

is required. This value determines the threshold energy $\Delta E_M = V(\delta R)$ defining the process of relaxation of water. This threshold exceeds strongly the thermal energy of water molecules equal to $k_B T \approx 0.025\,\mathrm{eV}$ at room temperature. It is seen that the duration of relaxation T_{1W} depends very strongly on the threshold value of the deformation energy of a water molecule ΔE_M and its temperature T. A great value of ΔE_M leads to a small probability to overcome the barrier in the region of the input window of a microcavity. Finally, the probability of spontaneous deactivation of water is very small, which corresponds to a very great time interval of the storage of the information.

We now execute some numerical evaluations. At a temperature of water $T = 293\,\mathrm{K}$ (20°C), the duration of relaxation (the duration of existence of "the water memory") $T_{1W} \approx 10$ days. As the temperature of water increases or decreases, the duration of relaxation sharply decreases or increases, respectively (see Table 1.1).

For the alternative direction of the relaxation (the transition of a water molecule from the volume of a microcavity into the volume of amorphous water), the duration of relaxation T_{2W} is also determined by a relation

Table 1.1. Dependence of the duration of relaxation of water (the duration of "the water memory") on its temperature.

T, °C	1	10	20	30	36.6	40	50	60	70	90
T_{1W}	300 days	49 days	10 days	58 h	24 h	15 h	4.4 h	1.3 h	27 min	3 min
T_{2W}	30 min	14 min	4 min	1.5 min	45 s	30 s	12 s	4 s	1.5 s	0.3 s

[Eq. (1.3)], in which the energy of activation is changed ($\Delta E_M - \Delta W \approx 0.9$ eV instead of $\Delta E_M \approx 1.1$ eV). In addition, it is necessary to take into account the following: because the internal size of microcavities exceeds significantly that of the potential well for each molecule in the volume of quasiamorphous water, the effective frequency of collisions of a water molecule with walls in microcavities $\omega_D \approx 1/\tau_0$ will be less (and the period τ_0, respectively, will be greater) than that in the volume of water. The results of calculations of the duration of relaxation T_{2W} on the reverse transition of H_2O molecules from the volume of microcavities into the quasiamorphous water are presented in Table 1.1.

We note that, in order to calculate the quantities T_{1W} and T_{2W}, it is necessary to know the exact values of the activation energy and the height $V(\delta R)$ of the potential barrier [Eq. (1.5)] regulating the entry into microcavities of the clathrate frame.

We determined these parameters based on the model calculations. The main error can be related to the approximate value, $\delta R \approx 0.26$ Å, of the width of the potential barrier bordering a window being the input to the volume of microcavities in the clathrate. Since δR is present in the exponential component defining the duration of relaxation [Eq. (1.3)] of the nonequilibrium population of quantum states in the volume of microcavities, even small changes in $V(\delta R)$ can very significantly influence both the durations of relaxation. Upon correction of these parameters, the corresponding values of T_{1W} and T_{2W} can be significantly changed.

While varying a value of ΔE_M by $\pm 5\%$ relative to the above-given value of $\Delta E_M \approx 1.1$ eV, the calculated duration of "the water memory" is changed by several orders. These data are presented in Table 1.2.

The obtained values of T_{1W} and T_{2W} correspond to the relaxation of water from the nonequilibrium state to an equilibrium one corresponding to its temperature.

We note that such activated water has other electromagnetic and mechanical properties. Because part of water molecules can be isolated

Table 1.2. Dependence of the duration of relaxation of water on temperature for various values of the height of the potential barrier at the input to the volume of microcavities of the clathrate frame. The quantities $T_{1,2W}$ (5%) and $T_{1,2W}$ (−5%) correspond to the durations of relaxation on the increase or decrease of the height barrier by ±5%.

T, °C	1	10	20	30	36.6	40	50	60	70	90
T_{1W} (5%)	7.2 years	550 days	160 days	23 days	9 days	5.5 days	35 h	10 h	3 h	20 min
T_{1W} (−5%)	24 days	5.7 days	31 h	8 h	3.4 h	2.2 h	45 min	12 min	4 min	36 s
T_{2W} (5%)	8 h	2.2 min	34 min	10 min	4.5 min	3.1 min	1 min	22 s	8.5 s	1.5 s
T_{1W} (−5%)	10 min	3.2 min	57 s	18.5 s	9 s	6.4 s	2.4 s	0.9 s	0.4 s	0.07 s

in the volume of clathrate microcavities, the coefficient of absorption of water in the region of frequencies of the microwave range can be significantly changed. This effect is related to the presence of saturated hydrogen bonds in the condensed water and their absence for quasifree H_2O molecules localized in microcavities. In a certain sense, the totality of water molecules in various microcavities is analogous to water vapor, of which each molecule is positioned in its own potential well. We recall that just such an effect of the growth of the coefficient of absorption of water vapor in the microwave range was discovered in Carion's work (1978) devoted to the study of the difference of the absorptions of water and vapor.

If this process is analyzed from the viewpoint of the theory of information, then the formation of a stable nonequilibrium distribution of the population of various clathrate microcavities can be considered by the process of recording of the information in the volume of water. Such water can be named as activated.

The very great duration of relaxation T_{1W} allows us to consider water as a two-level (two-band, to be more exact) bistable system with the great lifetime in each of two states. Such a system allows one to realize the recording and storage of the information (in the form of the ratio of the numbers of filled and unfilled microcavities) and to efficiently utilize this information at the expense of a change in the properties of water on the transition of a great number of H_2O molecules and other atoms, molecules, and ions dissolved in water from the state of amorphous water into the volume of closed microcavities and *vice versa* (Fig. 1.3).

Though the duration of inverse relaxation T_{2W} on the output of water molecules from the volume of microcavities turns out significantly less than the duration of direct relaxation on the input into these microcavities, it exceeds, in any case by many orders, the typical duration of relaxation [Eq. (1.1)] $\langle\tau\rangle \approx 10^{-9} - 10^{-10}$ s owing to the fluctuations of a hydrogen bond in the volume of amorphous water.

There exists one more circumstance which can lead to a sharp increase of the duration of the water memory. It is related to the possibility to populate the microcavities by molecules other than H_2O.

As known, water is a weak electrolyte, and the probability of the equilibrium thermal fluctuation-related dissociation

$$H_2O \leftrightarrow H^+ + OH^- \tag{1.6}$$

is always nonzero. The equilibrium relative concentration η_{H^+} of the ions of hydrogen is determined by the hydrogen index $\mathrm{pH} = -lg\eta_{\mathrm{H}^+}$. In a normal (neutral) distillated water at room temperature, pH $\approx$ 7, which corresponds to $\eta_{H^+} \approx 10^{-7}$, and the total concentration of ions of each sign $n_{H^+} = n_{OH^-} \approx 6 \cdot 10^{15}\,cm^{-3}$.

On the heating of water to the boiling temperature 100°C, pH decreases almost by δ(pH) $\approx$1. In this case, the values of $n_{H^+} = n_{OH^-}$ grow approximately by one order in magnitude.

After a number of transformations (including the neutralization of the ions H^+ and OH^-, their decay, and the subsequent formation of a number of molecular products), the equilibrium distribution of main products of thermolysis (H, H_2, OH, and H_2O_2) is formed in water. The typical ratios of the relative concentrations of these products are as follows:

$$\eta_{OH}/\eta_H \approx 4.3, \quad \eta_{H_2O_2}/\eta_H \approx 1.25, \quad \eta_{H_2}/\eta_H \approx 0.75. \tag{1.7}$$

Thus, every cm^3 of water at room temperature contains about 3×10^{-15} to 6×10^{-15} atoms and molecules of H, H_2, OH, and H_2O_2. Molecules and atoms of H, H_2, and OH have sizes significantly less than those of water molecules, and can easily enter into and leave the volume of microcavities in the volume of the clathrate frame. For this reason, such molecules and atoms by themselves cannot be carriers of the information.

The situation will be changed significantly, if the volume of a microcavity can contain simultaneously two particles which can form stable molecules of two types:

$$\begin{gathered} H + OH \rightarrow H_2O, \\ OH + OH \rightarrow H_2O_2. \end{gathered} \tag{1.8}$$

We note that the probability of reactions of the second type [Eq. (1.8)] is significantly greater. This is conditioned by a larger concentration, η_{OH}, of OH molecules as compared to η_H of atoms H.

The size of a H_2O_2 molecule is greater than that of a molecule H_2O by 9%. This implies that this molecule must be deformed by $\delta R \approx 0.5$ Å in order to leave a microcavity. By assuming that the frequency of normal oscillations of a proton, ω_H, in a H_2O_2 molecule coincides with the analogous frequency of oscillations of a proton in a water molecule, we can find the threshold energy of the deformation $V(\delta R) \approx 4$ eV from formula (1.5).

This value determines the energy $\Delta E_M = V(\delta R)$ related to the process of relaxation of molecules of hydrogen peroxide in water. A molecule of H_2O_2, which is positioned in a microcavity as a result of reactions, cannot leave it and will be "locked" there for a long time. This time exceeds the duration of relaxation of water molecules presented in Tables 1.1 and 1.2 by many orders.

We also notice that the activation of water can be realized in various ways, e.g. the process of heating or cooling, as well as magnetic fields or ultrasound.

Such periodic coherent actions can stimulate the formation of quasistable clusters, each of which joins several mutually ordered clathrate frames. In such a system, the behavior of isolated water molecules in periodically positioned microcavities is similar to the motion of hydrogen in palladium, where a very high degree of saturation of the lattice is ensured. Periodic actions can also affect the parameters of the clathrate frame of water by changing, for example, the transparence of the potential barrier in the windows of microcavities (it is the problem of tunneling of a H_2O molecule through a nonstationary barrier). In addition, a strong periodic magnetic field can stimulate transitions between the energy levels which characterize the state of H_2O molecules in microcavities and in amorphous water (for example, at the expense of multiphotonic nonlinear processes upon interaction with magnetic moments), which leads to the nonequilibrium population of water molecules in microcavities and corresponds to the activation of water.

Such external action can induce nonequilibrium population of microcavities which cannot be attained by changing the temperature. A similar activation of water can also be reached by uniform compression.

On the basis of such a system of long-term memory, we can interpret many effects leading to the activation and manifestation of anomalous properties of water. In conclusion, we note that the considered phenomena refer to pure water and do not concern the influence of dissolved admixtures (including ions and microparticles of Fe), the very presence of which can induce other effects (for example, in the presence of an external constant magnetic field).

The process of spontaneous and decaying luminescence of water is an indirect confirmation of the fact that the activation of water can be related to a change in the population of microcavities in the clathrate frame and the transition of water molecules from the bound state in the volume of microcavities in amorphous water. This effect was observed many times by many researchers. The reason of such luminescence can be related to the release of local activation energy for a water molecule after its passage of the potential barrier regulating the input or output from microcavities in the clathrate frame. This excessive energy can be directly lighted up after the completion of such overbarrier transition or can stimulate a number of physicochemical transformations leading to luminescence (i.e. can be its catalyst).

For example, Dr. Voeikov of Lomonosov Moscow State University observed the effect of luminescence using methods of chemiluminescent analysis with addition of salts of bivalent iron and luminol as a fluorophor in water. In particular, such decaying luminescence was observed in artesian water and in water which was in a closed bottle for a sufficiently long time. It is of interest to note that the temporal dependence of the intensity of luminescence depended significantly on the bottle material (glass, ceramic, or plastic). The duration of existence of such luminescence at room temperature was 5–7 days, which well agrees with the data presented in Tables 1.1 and 1.2. In water which was subjected to a preliminary heating to a high temperature and then to a gradual long-term cooling, luminescence was not registered. Luminescence was also not observed in water which was stored for a long time at a constant temperature.

Let us consider some of the experiments on activation of water (Gapochka, 1994) which indirectly confirm, in our opinion, the considered clathrate mechanism of water memory. In these experiments, the influence of a sufficiently powerful microwave emission on water was studied. The experiments were carried out with the use of several modes of action which differed by the frequency, power, type of modulation, and duration of action.

The differential spectra of the optical density of water preliminarily undergone the SHF irradiation were studied in the range $\lambda = 190$–900 nm on a spectrophotometer "Hitachi 557". It was found that the SHF irradiation did not lead to any considerable change in the optical density in the range $\lambda = 350$–900 nm, but gave rise to its significant increase in the range $\lambda = 190$–350 nm.

In Fig. 1.5, we show a change in the optical density of distillated and bidistillated water irradiated by a sequence of powerful pulses of the SHF emission relative to the density of analogous unirradiated distillated water. For activation, a cuvette with water was positioned in a waveguide connected with a SHF generator. The parameters of emission are as follows: the emission frequency ω is 2.71 GHz; the power, -800 kW; the duration of pulses, -1 μs; the repetition frequency of pulses, -230 Hz; the total time of irradiation, -5 s.

The measurements were carried out in 24 h after the action of the SHF emission on water. It was found that the SHF irradiation induces no

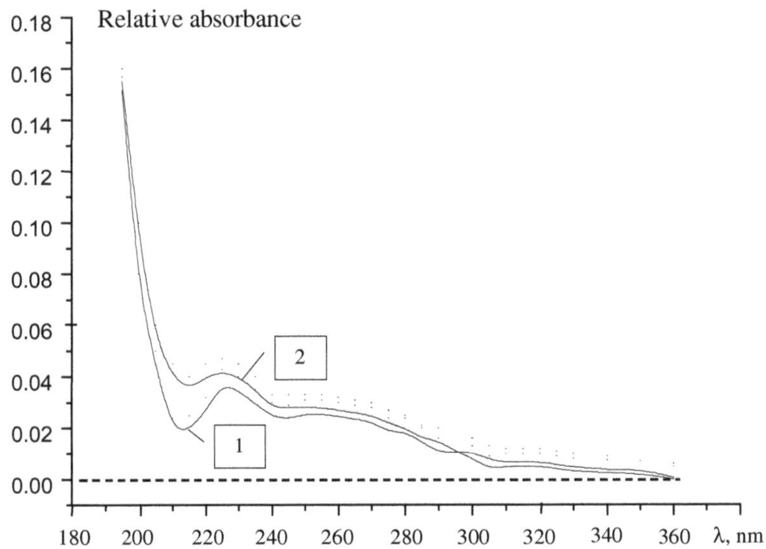

Figure 1.5. Change of the optical density of water of different types irradiated by powerful pulse SHF emission: 1 — distillated water, 2 — bidistillated water.

considerable changes in the optical density in the range 350–900 nm, but leads to its significant increase in the range $\lambda = 190$–350 nm. It is seen from Fig. 1.5 that the changes are more significant in bidistillated water on the activation than in distillated water. This testifies that the effect of activation is related namely to water, rather than to dissolved salts. For both types of water, two peaks of the absorption at $\lambda \approx 225$ nm and $\lambda \approx 255$ nm are clearly seen. In addition, we see a very great additional increase in the absorption at $\lambda \approx 190$ nm characteristic of any water.

In Fig. 1.6, we present the results characterizing the dependence of the optical density of the identical distillated water on the frequency of the SHF emission. In this case, a generator of continuous emission was used. Its parameters are as follows: the emission frequencies $\omega = 40.00$ GHz, 45.55 GHz, and 53.55 GHz; the power $= -10$ mW; the duration of irradiation $= -20$ min. The measurements of the parameters of water were executed in 24 h after the SHF emission.

It is seen from the figure that a very significant change in the optical characteristics of activated water occurs also under the action of a comparatively low-intensity (but continuous) emission. This testifies that the

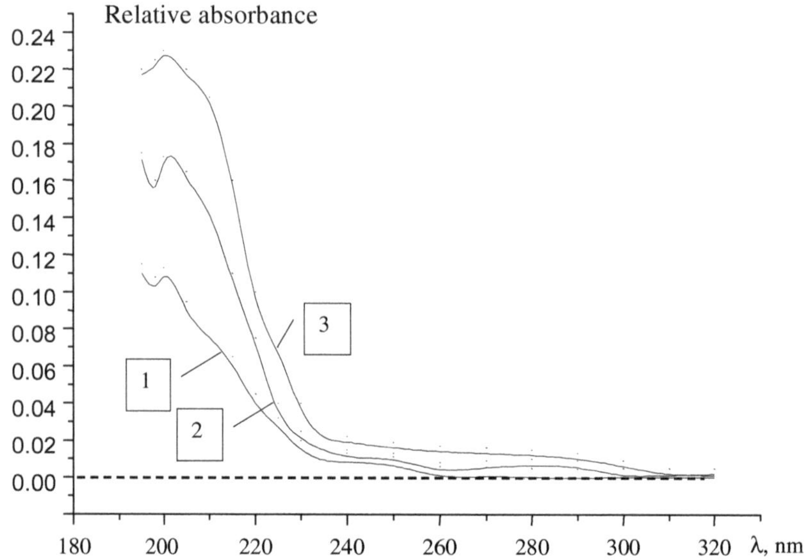

Figure 1.6. Changes in the optical density of distillated water irradiated by a continuous low-intensity SHF emission with frequencies: $\omega = 40.00$ GHz (1), $\omega = 45.55$ GHz (2), and $\omega = 53.55$ GHz (3).

process of activation is accumulative and depends not only on the excitation intensity, but also on the excitation duration.

It is of interest to note that though the total energy of the pulse SHF field, which passes through a cuvette, corresponds to 1 kJ and is two orders greater than that of the continuous irradiation, the resulting change in the optical density of activated water is much greater in the latter case. This result emphasizes the strong dependence of the effect of activation on the frequency of the action.

One more peculiarity of the results of these measurements consists in that the change in the optical density non-monotonously depends on the irradiation frequency: while the frequency grows from 40.00 GHz to 53.55 GHz, the optical density firstly decreases, and then increases to the maximum.

Gapochka's work in 1994 presents the study of the influence of intense SHF irradiation of water on the parameters of its NMR spectrum. The shift of the center of a NMR line was studied on a spectrograph "Tesla-BS-497" in 24 h after the completion of the activation of water. It was found that, for all three used methods of activation (continuous emission with frequency

$\omega = 2.45$ GHz, power of 450 W, and energy of 2.25 kJ; pulse emission with frequency $\omega = 2.71$ GHz, power of 800 kW, and energy of 0.92 kJ; pulse emission with frequency $\omega = 0.9$ GHz, power of 1000 kW and energy of 16 kJ), the center of a NMR line was shifted, respectively, by 32 ± 2, 38 ± 2, and 34 ± 2 Hz relative to the position of the control line for nonactivated distillated water. This displacement is related to an increase in the electron density in the region of a proton in a H_2O molecule.

We recall the optical density D connected with the coefficient of absorption $k''(\omega)$, the thickness of the medium L, and its dielectric permittivity $\varepsilon(\omega)$ by the relations

$$D = 2k''(\omega)L, \quad \text{and}$$
$$k''(\omega) = \frac{\omega}{c} Im\sqrt{\varepsilon'(\omega) + i\varepsilon''(\omega)} \approx \frac{\omega\varepsilon''(\omega)}{2c\sqrt{\varepsilon'(\omega)}}. \qquad (1.9)$$

It is obvious that the increase in the optical density of water in the same cuvette of constant thickness L upon irradiation of the SHF field is related to both the increase in the imaginary part and the decrease in the real part of the dielectric permittivity of water $\varepsilon(\omega)$ in the UV region of the spectrum.

It is difficult to substantiate the presented results, if we do not account for the above-considered clathrate model of the mechanism of activation of water. The matter is that the change in the optical density of water in the UV range can be related only to stable changes in the electron configuration of water molecules. The experiments show that such changes preserve for at least 24 h.

It is known that any stable structural changes of the configuration of hydrogen bonds can only lead to a change of the specific features of the microwave spectrum and do not influence the spectrum of transitions in the UV range. At the same time, these effects can be rather easily substantiated, if we assume that the redistribution of the populations of isolated water molecules in the volume of clathrate microcavities occurs in the process of activation. If we take into account that the spectrum of the UV absorption of these quasifree molecules significantly differs from the spectrum of the UV absorption of bound molecules, then the change in the populations leads to a change of the absorption. In addition, it is possible that the above-considered anomalies in the absorption of activated water in the UV range are related to molecules of hydrogen peroxide which can be localized in microcavities of the clathrate frame.

1.3. Program, Equipment, and Research Techniques for the Investigation of Activated Water

The performed brief analysis of the existent models of water structure clearly demostrates that water is an object which is much more complicated than a crystal or a gas. The above-considered results of numerous experiments testify that activated water has a number of specific physical properties. Many other experiments indicate that water activated in some way has distinctive memory and can render a strong influence on biological objects in a great time interval after the completion of the activation.

All these undoubtedly interesting results are uncoordinated, refer to completely different methods and modes of activation, and do not form a single logically connected system.

A significant drawback consists also in that the available studies have a highly specialized character. If, for example, the physical properties of activated water were studied, then the specific features of its action on biological objects were not considered. In particular, it is not known how the specific anomalous physical properties act on plants, microorganisms, animals, and men. On the contrary, if the specificity of action of activated water on a specific biological object was analyzed, its physical properties were not investigated.

Moreover, the registration of a definite character of the action of such water on biological objects of some type (for example, on a specific plant) does not allow one to forecast the result of the action on animals or plants of the other types.

Is the factor of the influence of activated water on living objects single and universal? Are there different mechanisms which adapt only to specific objects? How does the duration of activation of water affect the specificity of its action on different biological objects? To what degree do the anomalies of physical properties of activated water and the specific features of its action on a biological object correlate with one another? How do the duration and the mode of the storage of activated water affect the specificity of its action on biological objects?

We can pose infinite number of such questions but note that there are no answers to them in the literature. If the answers to similar questions are absent and the basic principles and the mechanism of action of activated water are not clear, the consequences of the use of activated water cannot be reliably forecasted.

As the above-considered general problems of activation of water, specific mechanisms and models of the long-term memory of water, and the analysis of specific features of the action of activated water on biological objects are very important and actual, we made the decision, jointly with our colleagues, to carry out a successive cycle of physical and biological studies of water activated in a single manner. We believed that this would allow us to perform not only the qualitative, but quantitative analysis of both the properties of activated water and the specific features of its influence on different biological objects and on different stress-involved situations (including, in particular, the presence of oncologic diseases).

Especially critical was the problem of the choice of a method of activation of water and the choice and the use of a specific device which would realize such an activation for the whole cycle of studies. First of all, we abandoned those methods of activation which can be conditionally named "force".

Regarding "force" methods, we refer to those methods of activation which are related to the high-energy action on water (for example, the action of high-power emission or the action by ionizing fields). It is clear that, on such "force" action, there appear many side factors (for example, secondary radicals) not related to the very process of activation of water which is understood as the distinctive recording of the information in the bulk water without change of its charge or chemical state.

We also discarded the application of such methods of activation, with which it is difficult to get the reproducible characteristics of activated water. We considered that, in order to carry on the complex studies, it is worth using the standard seriously-produced device, whose characteristics are invariable and whose effect on water is the same for any number of acts of activation of various samples of identical water.

Based on such considerations, we performed a cycle of complex studies, by using the activator of water which was earlier studied by us and which allowed us to obtain a number of interesting, reliably reproducible and sufficiently convincing results in the whole spectrum of physical and biological applications.

Activation of water was realized with the help of a device whose principle of action is based on the so-called Molecular Resonance Effect Technology (MRET). It is described in a patent (Smirnov, 2000), and its author is one of the authors of this book. The preliminary information about this device is given in the book by Vysotskii, Smirnov and Kornilova (2005). This activator is the source of a low-intensity low-frequency complex field which

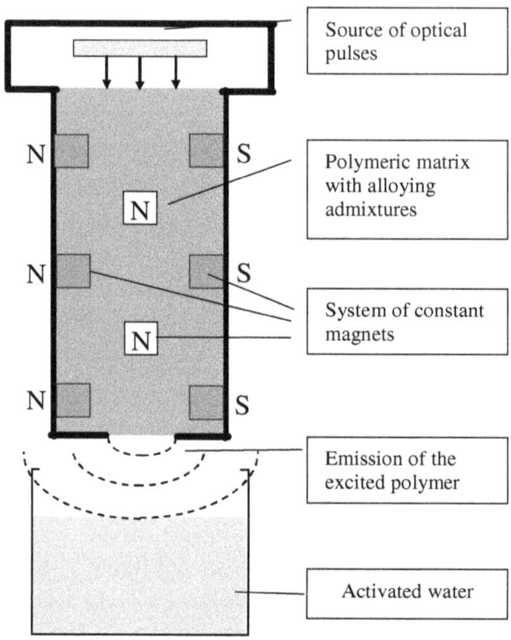

Figure 1.7. Scheme of a device for the activation of water due to the irradiation by a weak variable electromagnetic field.

has both electric and magnetic components and is registered only in the near zone of the working unit of the activator. We note that nobody carried out systematic successive physicotechnical studies of water activated with this device.

The scheme of the device used for the activation of water is presented in Fig. 1.7. The structure of the device allows one to obtain a great amount of activated water with the identical characteristics.

The device includes the working body consisting of a polymeric matrix formed by oriented threads of a linear polymer (like the epoxy polymer) with the alloying admixtures of various chemical elements and compounds (including those in the form of an admixture of magnetic materials). The polymer itself, according to the patent, has both ferroelectric and piezoelectric properties. This polymer is surrounded by the system of mutually perpendicular pairs of oriented constant magnets, the field intensity of each magnet being 4000 Gs.

Above the upper part of the polymer, there is a matrix of light-emitting diodes, to which a pulse voltage with a repetition frequency of about 8 Hz (in the range 7.2–8.2 Hz) is applied. These light-emitting diodes emit a sequence of optical pulses in the range of wavelengths of about 0.6 μm. Optical pulses acting on the oriented polymeric system alloyed by admixtures create a redistribution of electric charges and a small deformation which leads to a periodic displacement of the electric charges and the alloying magnetic admixtures. The motion of the electric charges and the magnetic moments of atoms in the strong circular field of the constant magnets generates a variable low-intense electromagnetic field with complicated spatial structure. This field oscillates with the frequency of the repetition of optical exciting pulses and includes both the electric and magnetic components. The detailed description of MRET activator is presented in Chap. 2 and a detailed illustration is given in Fig. 2.7.

The generated electromagnetic field acts on water positioned near the open end of the polymeric activator and changes its structure, which is reflected in the long-term change of the properties of water. It is that the screening of the field by nonmagnetic materials (for example, with the help of a piece of thin glass, plastic, or even several layers of paper) significantly weakens the intensity of the action on water. Since the magnetic field is not practically changed at such a screening, it is obvious that the electric component of the variable electromagnetic field with a complicated configuration plays the significant role in the activation of water (for example, by means of the influence on the dipole moments of molecules and clusters in the bulk and on the surface of water). The influence of the magnetic component of the field and the general analysis of the action of a specific activator on the structure and properties of water will be given later.

The general view of two used activators with the identical polymer matrix is presented in Fig. 1.8.

The program of the study of the properties of activated water includes the execution of systematic complex studies in several fields of physics and biology:

- study of the physicomolecular characteristics of activated water (dielectric permittivity, conductance, refractive index, viscosity, efficiency of the Raman scattering, parameters of molecular motion, a value of the hydrogen index pH, etc.);
- study of the influence of activated water on "pure" microbiological cultures;
- study of the influence of activated water on the associations (cenoses and supercenoses) of microbiological cultures;

Figure 1.8. The general view of two used activators.

- study of the influence of activated water on higher plants;
- study of the influence of activated water on nonspecialized rapidly growing cell systems (like a callus tissue in botany or psoriasis in medicine);
- study of the influence of various fractions of activated water on the prophylaxis and the treatment of oncologic diseases, including the studies *in vivo* (mice) and *in vitro* (cells of various oncologic cultures); and
- study of the influence of various fractions of activated water on staphylococcal infection *in vivo* in animal model and *in vitro* on *Staphylococcus* culture.

The program foresaw the study of the properties of activated water at different durations of activation, different temperatures of the storage of water, and different durations of its storage after the activation.

We studied the influence of activated water on growth of microbiological cultures, on the efficiency of action of various antibiotics on these cultures, on basic biochemical reactions, on the formation of colonies, on the sterilization of various media from pathogenic cultures, etc. These studies were carried out under both aerobic and anaerobic conditions.

Since water of any type, being in the region where cells are dividing, renders direct influence on the character and direction of this process, one of our principal aims was to investigate the influence of activated water on the processes of growth and division of anomalous cells (in the first instance, of oncologic tumoral cells). In addition, taking into account that activated water introduced into any real biological system will be inevitably mixed with ordinary nonactivated water already present in the system, we also planned to execute a number of experiments aimed at the study of the influence of such natural dilution on the efficiency of the action of activated water.

All these studies were realized fully, and their results are presented. In the course of the studies, we employed modern laboratory measuring facilities and up-to-date methods used in comprehensive systematic studies in various fields of physics, chemistry, microbiology, botany, biochemistry, and genetics.

The studies were carried out from 2004 to 2008 with the attraction of highly skilled experts from Kiev and Moscow. The list of the institutions which took part in our experiments is given in the Preface.

CHAPTER 2

Molecular Resonance Effect Technology as the Basic Method for Activation of Liquid Substances

2.1. Introduction to the Theory of Fractal Matrix

The basic principles of formation of complex structural systems in nature are based on the concept of structural resonance interactions. These principles govern the life activities of biological systems on this planet as well as the physical processes in the universe. They constitute the foundation of existence and stability of any complex system. There are several basic principles: fractalization, complementarity, and formation of the lattice of "barrier" membranes.

The principle of fractalization is realized through the iterative algorithm of formation of complex structural systems based on the existence of the initial prototype matrix which governs the formation of the object. In this case, the iterative formation of the final system consists of the successive reflection of initial prototype matrix on the final structure of the whole system. As a result, the final multilevel fractal structure has long distance correlations in the arrangement of particles. Any small fragment of fractal system reproduces the structure of the whole system under the increasing scale. This principle clearly describes the hierarchy organization of fractal system. This principle can be seen in the formation of crystalline lattice of mono-crystals, development and growth of biological systems where genetic prototype is developed through the certain algorithm of replication from single cell to the organism, where every cell has a unique basic matrix in the form of DNA structure.

Another principle that governs the formation of fractal system is the principle of complementarity. The main criterion of the integrity of fractal system is minimization of tendencies leading to spontaneous formation of "inside" conflicts and contradictions in the system. It states that in order to achieve stability of any complex system, the level of inside "contradictions" of this system should be directed to null. This statement is correct for any three-dimensional system as well as any volumetric system that has the infinite number of different kinds of structural vectors. The basis of formation of stable complex system should be the structural module which has precise, balanced matrix structure and can clone self projections in the surrounded environment. The fractal cloning of structures considers the formation of self-similar replications of the initial basic module with specific coefficient of similarity. The object which is formed as a result of fractal cloning process has dimensions that are proportional to the dimensions of initial basic module.

The next basic principle that governs the formation of fractal system in nature provides the idea of existence of the lattice of "barrier" membranes. Any fractal system is separated by barrier membranes relative to the central zone of the system. These membranes play the roles of transformers or converters of the previously existing algorithm, signal into another algorithm, or signal which level is more adequate for the present system. In this case the transmission of signal from the central zone of the fractal system to the peripheral zone of the same system and *vice versa* is related with the step-by-step adaptation of this signal. This principle can be interpreted as a process of quantum transformations of the entropy of the object, such that each barrier membrane of the system is considered to be some kind of a fractal "space–wave" filter which modifies previously existing algorithm or provides signals to enter into new algorithm. This concept leads to the conclusion that fractal matrix encountered with any physical process or agent has the ability to affect this process or agent in a way obviously characterized by the matrix's structure.

These main principles govern the development of all dynamic systems in nature, such as:

- the self-similar process of the growth of fruits and vegetables,
- the growth of living cellular structures,
- the growth of crystals, and
- the formation of polymer materials during polymerization process.

There is an assumed relationship between structural transformations of lattices of an implemented physical space and the lattice space of arithmetical progression bodies. This assumption can explain the phenomenon of the formation of highly organized periodical fractal structures in crystals and polymer materials. For the development of new composite materials in nanotechnology, it seems a matter of prime importance. To find a way of making a strict mathematical description of an optimum packing of the spaces based on fractal properties of both spatial structures and numerical continuum structures.

An atom has a complicated structure consisting of a massive positive nucleus and negatively-charged electron shells, and in general, the structure is electrically neutral. From the standpoint of the general postulate of the quantum mechanics, the Heisenberg principle of uncertainty, and the de Broglie wave dualism, any electron can be completely described by a wave function (Ψ), and therefore has a purely wave structure. Quantum theory is based on Schrödinger's equation:

$$\widehat{H}\Psi = E\Psi, \tag{2.1}$$

where $\widehat{H}$ is Hamiltonian operator, E is eigenvalue and in which electrons are considered as wave-like particles whose "waviness" is mathematically represented by a set of wave functions (Ψ) obtained by solving Schrödinger's equation.

A powerful branch of nanotechnology, the tunnel-probe technology, is based on the wave properties of the electron, and so there are no reasons to doubt the wave nature of the electron. A linear dimension of an atom (i.e. of its electron shell) is 10^{-8} cm, and correspondingly, the volume of the atom is approximately 10^{-24} cm^3. At the same time, a linear dimension of a nucleus is approximately 2×10^{-13} to 5×10^{-13} cm, and the volume occupied by the nucleus is 10^{-37}–10^{-38} cm^3. The difference between the nucleus volume and the volume of the whole atom is 15 orders, i.e. it can be stated that $99.9 \cdots 9\%$ of the atom consist of wave structures represented by electrons, and only $0.0 \cdots 1\%$ of the atom is a corpuscular component. A balance of attractive and repulsive forces between the involved atoms is needed for a solid-state structure to exist. Modern science considers four types of solid-state structures — molecular, ionic, covalent, and metalic structures — which are formed with various bond types. All of them are formed due to the interaction of the external valence electrons of the atom, i.e. due to the interaction of merely wave structures. However, the external valence electrons are also responsible for the effect of repulsive forces. According to one of the main postulates of quantum mechanics, i.e. Pauli's principle, two structures having the same set of quantum numbers cannot exist within the same space volume. Since the atoms and electrons of the same elements are statistically indistinguishable, when atoms come closer and external electronic shells overlap due to Pauli's principle, the electronic levels split, which is adequate to the effect of repulsive forces. The occurrence of a balance between the attractive and repulsive forces results in the fact that the atom itself behaves as a harmonic oscillator, and the entire solid-state structure can be represented as a system of harmonic oscillators, i.e. a system having a wave oscillation spectrum other than the wave structures involved in the process.

Thus, due to interaction between the wave structures of valence electrons, processes of self-organization of a solid body's atoms into ordered structures take place. It is the wave structures which tend toward self-organization processes and certain kinds of long-range interaction, for example, interference and diffraction. If we view a fractal structure as a complex hierarchical system based on self-similarity principles, then any solid-state structure can be considered as a fractal structure. Since any solid body is a wave structure, it is reasonable to make use of the resonance phenomenon to affect this body by means of an appropriate physical agent. This universal physical agent is the electromagnetic field (Serov, 2003).

2.2. The Fractal Matrix Characteristics of MRET Polymer Material

In our case, the epoxy polymer material is a good example presenting all qualities of volumetric fractal matrix mentioned above (Fig. 2.1).

The epoxy polymer samples were studied with the help of small-angle X-ray scattering (SAXS). The analysis of the entire scattering curve of an epoxy compound suggests a fractal behavior of the internal surface on a scale between 100 nm and 10 nm and, in the tail end of the SAXS curves, reveals maxima corresponding to those of two regular spheres with radii of the order of 7 nm and 14 nm. The analysis of the beginning of the curves yields one or two correlation lengths close to 100 nm and 20 nm. These results are consistent with the general model of IPN (interpenetrating polymer network) structures as revealed by other physico-chemical techniques (Sobry *et al.*, 1991).

Most of polar polymers possess comparatively high values of relative permittivity (dielectric constant), which means that both bonding and non-bonding electrons in the molecular structure of these polymers can be easily displaced by external electromagnetic force. While many polymers are highly flexible and form an amorphous solid upon the process of

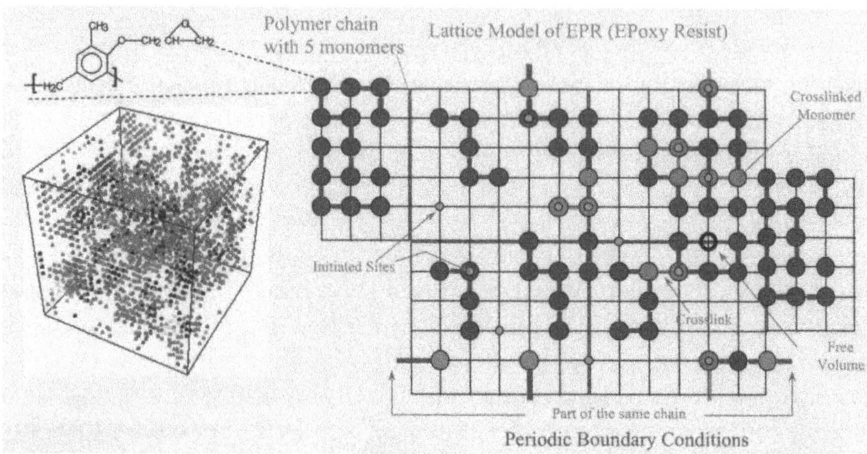

Figure 2.1. Lattice model of Epoxy polymer fractal structure (Patsis and Glezos, 1999).

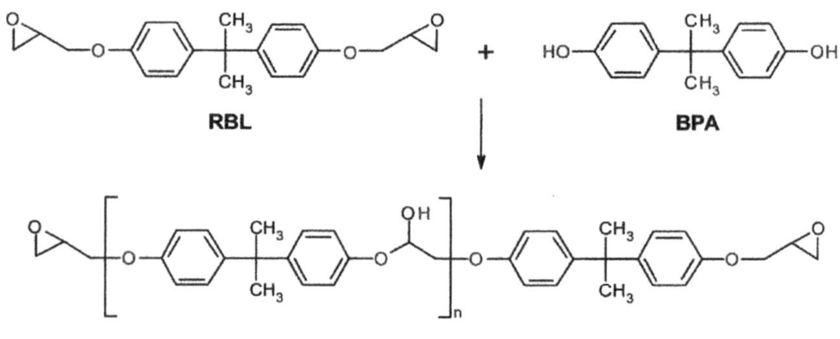

Figure 2.2. Epoxy polymer typically contains highly polar hydroxyls and amines. Once all the amine sites have reacted with the epoxy sites, a three-dimensional network is achieved.

polymerization, a large number of polymers such as epoxy actually form partially crystalline structures. Epoxy is formed by mixing Bisphenol A with low-molecular weight liquid resin that contains epoxy groups. The principal reaction of epoxy groups with phenolic hydroxyl functions leads to linear polymer chains formation (nM — Mn, where $n > 38$) (Fig. 2.2).

In the case of epoxy polymer, the kinetics to a large extent determines the final crystalline structure of the polymer. The introduction of foreign agents (substances) in the parent lattice of epoxy polymer leads to the effect of superimposed periodicity, and as a result develops modulated crystalline structures with specific fractal microstructure, phase transition, network topology, and polarity. It is the main patented process of formation of the specific MRET fractal matrix structures in epoxy polymer which belongs to the patent holder of Molecular Resonance Effect Technology (MRET). A number of studies show that external electromagnetic field can affect local orientations and phase transitions in polymer crystalline systems of longitudinal chains. The longitudinal polymer crystalline system is a macromolecule of consecutively copolymerized liquid crystals and flexible polymer sequences. The external electromagnetic field can seriously modify the local orientation order of the system and affect phase transition parameters and dielectric properties of the polymer compound. A simple molecular mechanism exists since the polar parts of the molecule in epoxy are rigidly attached to the chain backbone. The orientation of the polar groups in electromagnetic field affects the backbone orientation. The extension of the local orientation of crystalline structure of epoxy introduced to electromagnetic field has been determined with an anisotropy parameter, based on the Ultrasound Critical-Angle Reflectometry.

The external electromagnetic field generates an excitation in the crystalline structures of polymer compound. The existence of orientations and phase transitions in crystalline systems of epoxy polymer introduced to external electromagnetic field leads to the origination of subsequent relaxation and strain phases in macromolecular structures that induces the phenomenon of piezoelectricity. This is the electrical response of a material to the change of pressure in molecular structures of polymer compound. Piezoelectricity can only be observed in materials having a noncentrosymmetrical structure and elastic properties. Both properties can be found in polar polymer compounds. Several investigations conducted on polymers with cholesteric elastomer structures indicated that uniaxial compression parallel to the helicoidal axis of the cholesteric structure leads to a compression of the helix. Simultaneously, an electrical charge at the surface of the elastomer is observed. Actually, there exists a linear correlation between deformation of the sample and the electric voltage that resembles piezoelectricity. According to the theory of Brand, the following Eq. (2.2) gives the expansion of the free energy of a cholesteric elastomer. Only the terms dealing with deformations and electric field effects are written down explicitly; all other contributions are summarized in F_0:

$$F = F_0 + 1/2Ce^2 + \varepsilon E^2 + q_0\xi Ee. \qquad (2.2)$$

Here C is Young's modulus, e the deformation, ε the dielectric permittivity, E the electrical field, $q_0 = 2\prod/p$ the cholesteric reciprocal pitch, and ξ the coupling coefficient between E and e. Minimizing the free energy with respect to the electric field yields a relation between deformation and the electric field:

$$E = -(q_0\xi/2\varepsilon)e, \tag{2.3}$$

where the term $q_0\xi/2\varepsilon$ defines the piezoelectric coefficient h. Here, it has to be noted that h is inversely related to the pitch of cholesteric elastomer. According to Eq. (2.3), the piezoelectric coefficient is directly proportional to the reciprocal pitch of the cholesteric phase. There is an excellent linear relationship with respect to the pitch dependence of the piezoelectric effect (Fig. 2.3).

The correlation between the piezoelectric coefficient and the order parameter reflects a coupling, and shows that the piezoelectric effect of polymer compounds directly depends on the state of order of the liquid crystalline phase structures (Meier and Finkelmann, 1993).

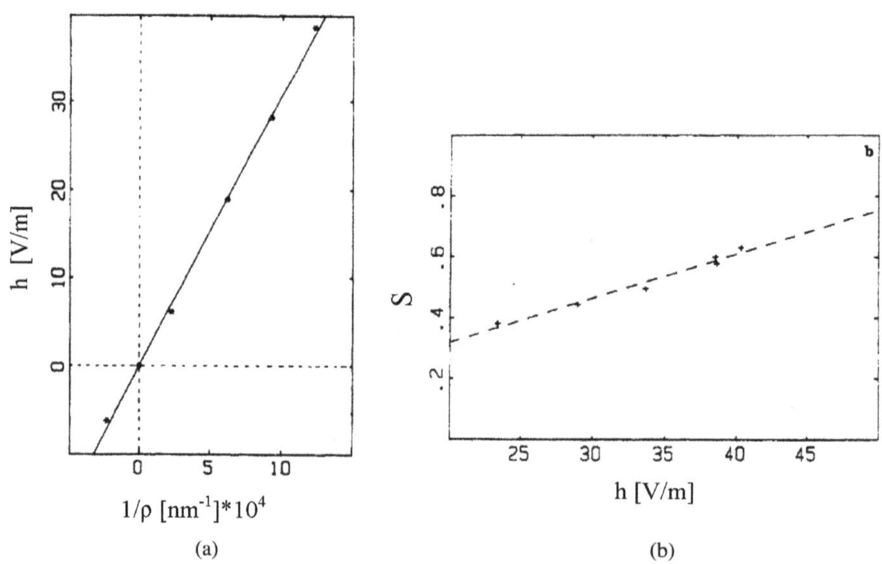

Figure 2.3. (a) Piezoelectric coefficient (h) versus the reciprocal pitch (1/p) of the elastomer. (b) Order parameter (S) of the cholesteric phase versus piezoelectric coefficient (h) (Meier and Finkelmann, 1993).

It means that there is a direct correlation between the topology of polymer molecular structures and intensity of piezoelectric phenomenon. Since the fractal structures of the polymer have the complex organization state, and due to the phenomenon of piezoelectricity, the epoxy polymer compound generates modified subtle electromagnetic signals of the random nature that can affect electrodynamic properties of water. A coherent resonance interaction, including both a spatial resonance and a resonance of the oscillating frequency of microscopic orbital currents of protons in water-molecular hexagons, leads to the process of deviation from the stochiometric composition of water and the reorganization of water clathrate structures with minimum input of energy.

The electromagnetic nature of the physical field generated by volumetric fractal structure of epoxy polymer compound developed in compliance with MRET requirements and its resonance interaction with water molecular structures were proved by investigation conducted at the National University of Singapore under supervision of Prof. Ong Khim Chye, Gary (2004). This investigation provides evidence that MRET Activated Water mixed with cement has the ability to enhance compressive strength of concrete due to the special electrodynamic and other physical properties of MRET Activated Water. It was also observed that the anomalous electrodynamic and physical properties of MRET Activated Water were lost after the contact of water with metallic surface. It is well-known fact that water molecular structures acquire dipole moment after introduction to external electromagnetic field. These dipole moment characteristics are relatively stable when water is kept in container made of dielectric material. In metallic container, the water molecular structures go through the process of rapid relaxation and lose their acquired dipole moment. This fact provides evidence that MRET Activated Water acquires the anomalous physical properties due to the interaction with a physical field of electromagnetic nature. This subtle electromagnetic field is generated by volumetric fractal matrix structure of MRET polymer compound.

The study at the National University of Singapore was conducted in order to establish the significance of the beneficial effect of MRET Activated Water on the compressive strength of mortar and concrete. In total, three batches of specimens (two mortars and one concrete) with different mix proportions were cast and tested. In each batch, two groups of cube samples were prepared. One group, acting as the control, was cast using normal tap water. The other group was prepared using MRET Activated Water in place of normal tap water. The mortar samples were 50 mm cubes while the concrete samples were 100 mm cubes. The two batches of mortar specimens were designed to investigate the effect of MRET Activated Water on the development of compressive strength with age up to 28 days, while the batch with concrete specimens was designed to explore whether the contact with the metal mixer and conventional steel cube moulds has an effect on the compressive strength of the specimens tested. The used cement was Ordinary Portland Cement (OPC) Type I SS: 2000. The physical properties and size gradations of the coarse and fine aggregates (sand) were tested in accordance to ASTM C136. Normal tap water used to cast the control specimens was supplied by the Public Utilities Board (PUB) of Singapore. Its quality is shown in Table 2.1.

MRET Activated Water was prepared by using the MRET apparatus. In accordance to their guidelines, the optimum duration of activation is 30 min. Each time, the apparatus is capable of activating 1 litre of water. After which, the activated water was stored in plastic containers.

- Mortar specimens

Two batches of mortar specimens were prepared. In Batch 1, a total of 24 samples of 50 mm cubes were cast (12 cubes cast using normal tap water acted as the control, while the other 12 cubes used the activated water). In Batch 2, a total of 32 samples of 50 mm cubes were cast (16 cubes cast using normal tap water acted as the control, while the other 16 cubes used the activated water). Ordinary steel moulds were used to cast the control samples while plastic moulds were used to cast the specimens containing MRET Activated Water. This is to prevent metallic surfaces from coming into contact with the activated water used in the specimens during casting.

Table 2.1. Quality of tap water.

Test item (in mg/L where applicable)	Result	WHO guideline
pH value	7.0–9.0	—
Turbidity (NTU)	<5	5
Sulphate (as SO_4)	30–60	250
Phosphate (as PO_4)	< 0.1	—
Silica (as SiO_2)	1–10	—
Total dissolved solids	200–350	1000
Total alkalinity (as $CaCO_3$)	20–50	—
Residual chlorine (as chloramines or free chlorine)	<2.0	5

• Concrete specimens

One batch of concrete specimens was prepared. A total of 48 samples of 100 mm cubies were cast (24 cubes cast using normal tap water acted as the control, while the other 24 cubes used the activated water). Ordinary steel moulds were used to cast both the control samples as well as the MRET Activated Water samples. This was to investigate the effect that contact with metallic surfaces might have on the compressive strength of the test specimens cast using MRET Activated Water. The preparation process for both the control and MRET Activated Water samples were the same.

2.2.1. *Results and discussions*

• Testing of specimens

All specimens were tested in the Denison Compressive Test Machine for their compressive strength. The Mortar Batch 1 specimens were tested for their 7-, 14- and 28-day compressive strength. The Mortar Batch 2 specimens were tested for their 3-, 7-, 14- and 28-day compressive strength. The loading rate for all mortar specimens is 12.5 kN/min. The concrete specimens were tested for their 3-, 7-, 14- and 28-day compressive strength at a loading rate of 200 kN/min.

• Batch 2 mortar results

Figure 2.4 shows that the increase in compressive strength of specimens cast using MRET Activated Water was evident from the age of 3 days. Figure 2.5 shows that the 3-day compressive strength of the specimens cast with activated water is about 15% higher than that of the specimens cast with normal water. This confirms the acceleration effect that activated water has on concrete cast using activated water, as reported by Chau (1996).

• Concrete specimens

Figure 2.6 shows that the compressive strength of the concrete samples cast using MRET Activated Water is very similar to those cast using normal water. The fresh concrete mix cast using MRET Activated Water came into contact with metallic surfaces at three stages during casting. After coming into contact with metallic surfaces during the casting process, the concrete specimens cast using activated water did not register an increase in compressive strength. The beneficial effect of the MRET Activated Water

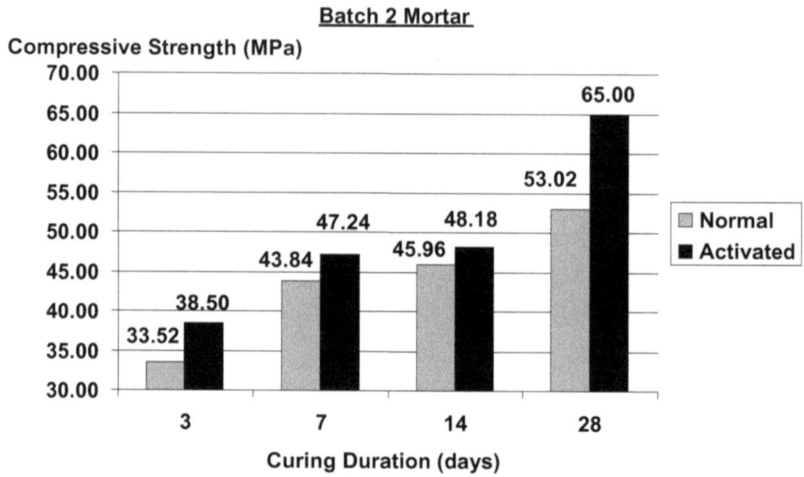

Figure 2.4. Compressive strength of mortar cubes cast using normal and MRET Activated Water.

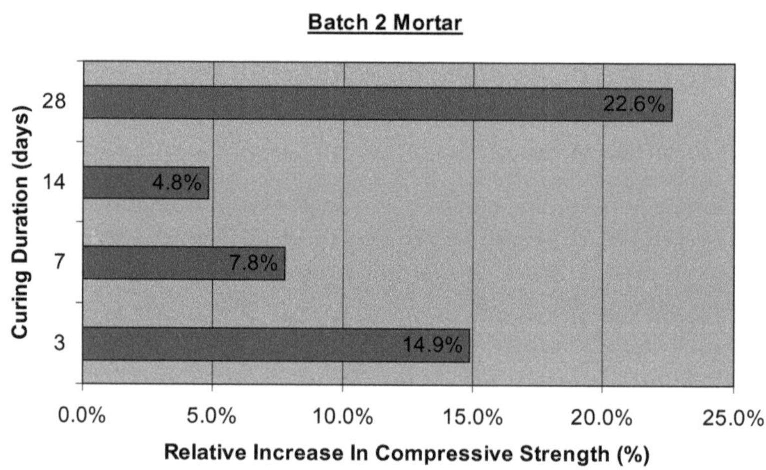

Figure 2.5. Percentage increase in compressive strength of mortar cubes cast using MRET Activated Water.

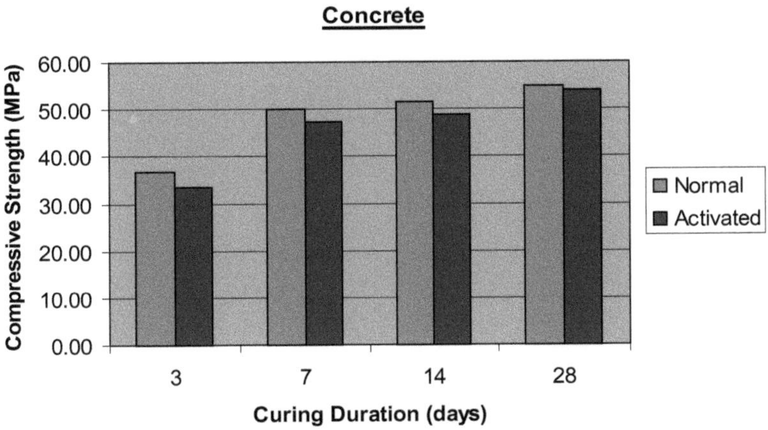

Figure 2.6. Compressive strength of concrete cubes cast using normal and MRET Activated Water that came into contact with metallic surfaces at three stages during casting procedure.

appears to have been neutralized by the metallic surfaces coming into contact with fresh concrete.

2.3. Method and Device for the Production of Activated Liquids

Water plays a critical role in most chemical and biological processes. It is well-known that water quality can have a significant effect upon those processes. Therefore, considerable time and effort has been spent to purify water from various sources. Such purification processes, while useful, merely remove most of the dissolved and suspended foreign particles in water, but do not alter the nature of the water itself. While this is of advantage in reducing the opportunities for the foreign materials to adversely affect the chemical and biological processes, such purification techniques do not overcome the fundamental limitation that the water itself imposes on the process. Molecular Resonance Effect Technology provides process which can alter the water itself, so that enhanced properties of altered water can advantageously be used to improve the basic functions of the chemical and biological processes (Smirnov, 2003).

MRET activation device includes a polymeric body and container for liquids (Fig. 2.7). This polymeric body forms volumetric fractal matrix

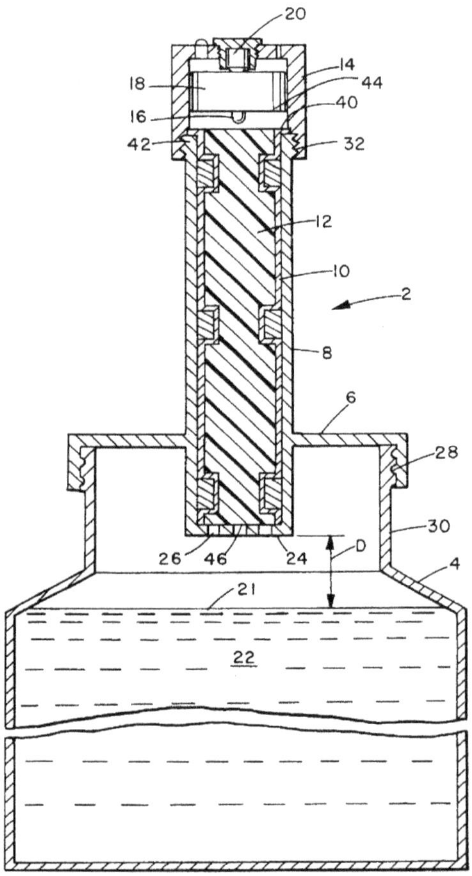

Figure 2.7. MRET activation device.

of specific topology and molecular orientation when small quantities of inorganic and organic substances are incorporated into it (Smirnov, 2000).

The device (2) is made up of a liquid reservoir (4), a cap (6) that includes a vertical housing (8) in which the column (10) containing MRET polymer compound (12) is encased. There is a chamber (14) which is attached to the housing that contains an electrical circuit (18) to activate polymer compound. The lower end (24) of housing (8) contains openings (26) to allow the polymeric body to emit low frequency oscillations toward the water (22). The housing is positioned with its lower end toward the water surface and disposed such that the distal end (46) of MRET polymeric body is spaced apart from the water surface by a distance D of at least about 2–3 cm. The cap (6) is removably mounted to the top neck (30) of reservoir so that it can be easily removed to fill or empty the reservoir. The volumetric fractal matrix structure of MRET polymer compound is exposed to an electromagnetic radiation of the visible light with a wavelength range of 400–800 nm and a frequency range of 7.2–8.2 Hz emitted by the diode (16). Recently, a number of interesting optical phenomena, including localization of optical excitations and dramatically enhanced optical nonlinearities, have been observed for fractal objects. The localization in fractals results from their scale-invariant morphology, which does not support the propagating waves typical for translational invariant systems. It has been observed that hot-spot localizations are very sensitive to even small changes in linear polarization plane and frequency of the applied external electromagnetic field. Another agent that produces an excitation in MRET polymeric body is external magnetic field formed by several pairs of magnets (34) in order to generate low frequency electromagnetic signals. The magnet pairs are disposed so that the north-south orientation of poles are reversed from pair to pair and form heterogeneous symmetric magnetic field in the polymeric body. In order to investigate the nature of the magnetic field that provides the excitation in crystalline structures of the polymer compound, the measurements of magnetic flux density were performed with the help of AlphaLab, Inc. DC Magnetometer (USA). This magnetometer is certified to display magnetic flux density in one axis with a scaling accuracy of +/− 2% over temperature range of 30° to 110°F in the dynamic range of 0 to +/− 20 kg. Accuracy of absolute zero field level is determined by the user when setting the "OFFSET" control.

2.3.1. *Testing of device for production of activated liquids*

Tables 2.2 and 2.3 as well as Figs. 2.8–2.11 describe the magnetic field structure observed during the measurement procedure.

The analysis of data received during the measurement of external magnetic field generated by the pairs of magnets placed around the polymeric body leads to the conclusion that the said magnetic field has heterogenic, symmetric structure. It also provides evidence that the characteristics of magnetic volumetric structure in the middle of the polymeric body differ from the characteristics of magnetic structure observed under the activator during the function of MRET activator. The analysis of the characteristics of magnetic field under the activator provides evidence that magnetic flux

Table 2.2. Results of measurements of magnetic flux density in the middle horizontal layer of polymer compound (Gs).

X/Y cm	0	1	2	3	4	5	6
0	17.2	16.3	13.1	12.5	13.1	16.3	17.2
1	15.2	8.5	7.6	7.2	7.6	8.5	15.2
2	12.3	7.0	4.1	3.8	4.1	7.0	12.3
3	0	0	0	0	0	0	0
4	−12.3	−7.0	−4.1	−3.8	−4.1	−7.0	−12.3
5	−15.2	−8.5	−7.8	−7.2	−7.6	−8.5	−15.2
6	−17.2	−16.3	−13.1	−12.5	−13.1	−16.3	−17.2

Table 2.3. Results of measurements of magnetic flux density under MRET activator (Gs).

X/Y cm	0	1	2	3	4	5	6	7	8	9	10	11
0	1.5	3.0	4.0	4.0	0.1	−4.0	4.0	0.1	−4.0	−4.0	−3.0	−1.5
1	1.5	2.0	2.6	2.6	1.0	−0.1	0.1	−1.0	−2.6	−2.6	−2.0	−1.5
2	1.0	1.5	1.8	1.8	0.8	0.1	−0.1	−0.8	−1.8	−1.8	−1.5	−1.0
3	0.5	0.6	0.7	0.7	0.6	0.05	−0.05	−0.6	−0.7	−0.7	−0.6	−0.5
4	0.3	0.4	0.5	0.5	0.4	0.02	−0.02	−0.4	−0.5	−0.5	−0.4	−0.3
5	0.03	0.04	0.06	0.05	0.04	0.01	−0.01	−0.04	−0.05	−0.05	−0.04	−0.03

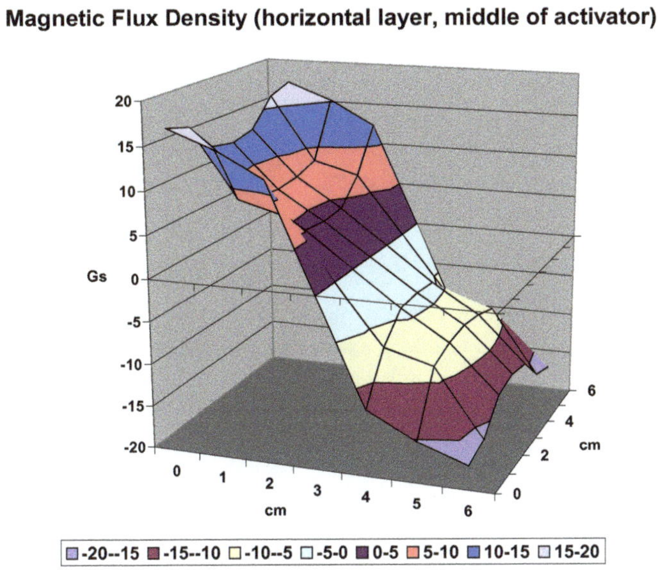

Figure 2.8. The magnetic field volumetric structure in the middle horizontal layer of polymeric body of MRET activator.

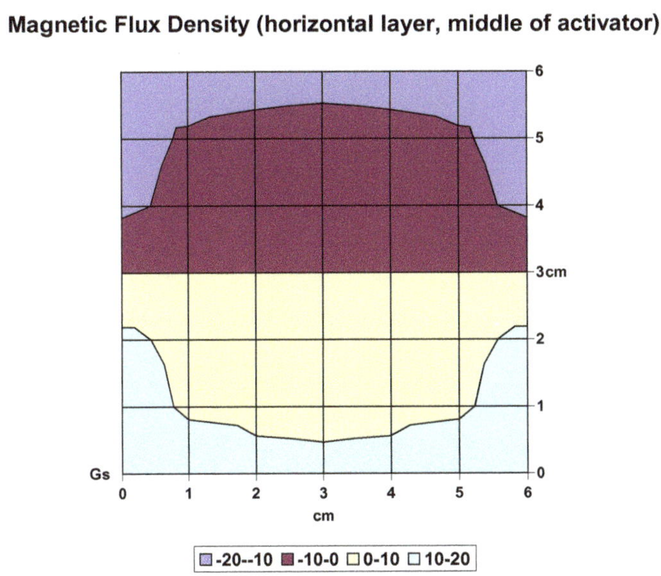

Figure 2.9. The horizontal projection of magnetic field that occurred in the middle horizontal layer of polymeric body during the function of MRET activator (units: Gs).

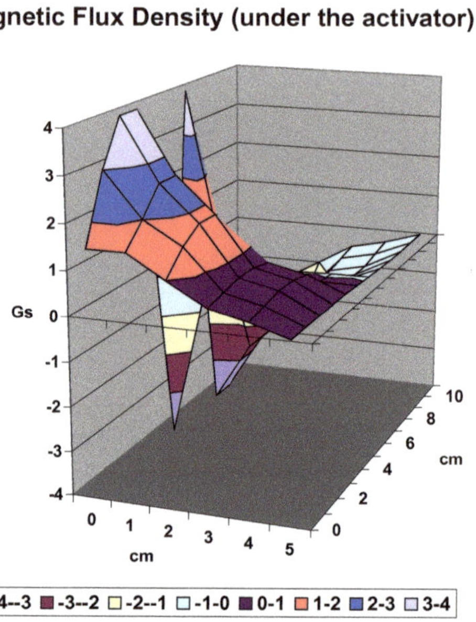

Figure 2.10. The magnetic field volumetric structure observed under the activator during the function of MRET activator (Gs).

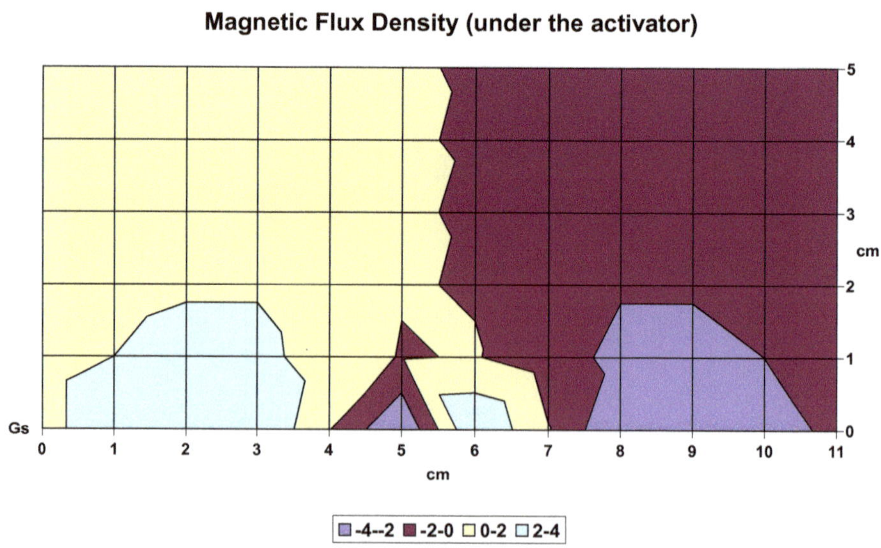

Figure 2.11. The horizontal projection of magnetic field observed under the activator during the function of MRET activator (Gs).

density reduces to zero value at the distance of 2–3 cm from the low edge of the activator. The activation of water molecular structure is induced by subtle low-frequency oscillations generated by the activator. In order to understand the nature of electromagnetic oscillations generated by the volumetric fractal matrix of MRET polymer compound, it is reasonable to consider this polymeric body as a fractal aggregate composed of metal nanoparticles, since a number of metallic salts are incorporated in MRET polymeric structure.

Recent studies of the metal-dielectric compounds exposed to external electromagnetic fields reveal their remarkable optical and magnetic properties. It was discovered that the structures of these compounds usually consist of small nanorings which self-assemble into larger nanorings of about 10 nm, which in turn self-assemble into even larger nanorings in the range of 100–1000 nm in size. For example, a close examination of Co-thiol-polymer composite by transmission electron microscopy (TEM) in the HAADF mode shows formation of nanoring structures with typical fractal topology (Fig. 2.12).

The resulting materials present the anomalous optical properties such as absorption and emission in the range of visible zone of the electromagnetic spectrum, and a strong nonlinear optical behavior. The nonlinear optical behavior is due to the presence of the metallic clusters inside the polymer matrix. It was found that these properties can be modified with the concentration of incorporated metallic nanoparticles in polymer structure. For example, Fig. 2.13 shows a few samples of different colors that can be obtained by changing the concentration of Co^{2+} in metal-dielectric compounds and their corresponding UV-visible spectra.

A well-defined absorption peak was found in the region near 500 nm, and the most intense emission peak was found in the region near 560 nm. Optical excitation in fractal structures is not uniformly distributed but rather concentrated in hot spots much smaller in size than the size of the fractal cluster and often the wavelength as well. The strong electromagnetic fields in these hot spots can result in large enhancement of optical nonlinearities and other effects that require intense fields. In fractal aggregates composed of metal nanoparticles, the optical excitations are associated with plasmon modes, which can have high resonance quality factors and thus provide particularly strong enhancement in the hot spots (Drachev *et al.*, 2001).

The mechanism of light absorption of these composites is related to the light localization inside the cluster. The reason for the localization is

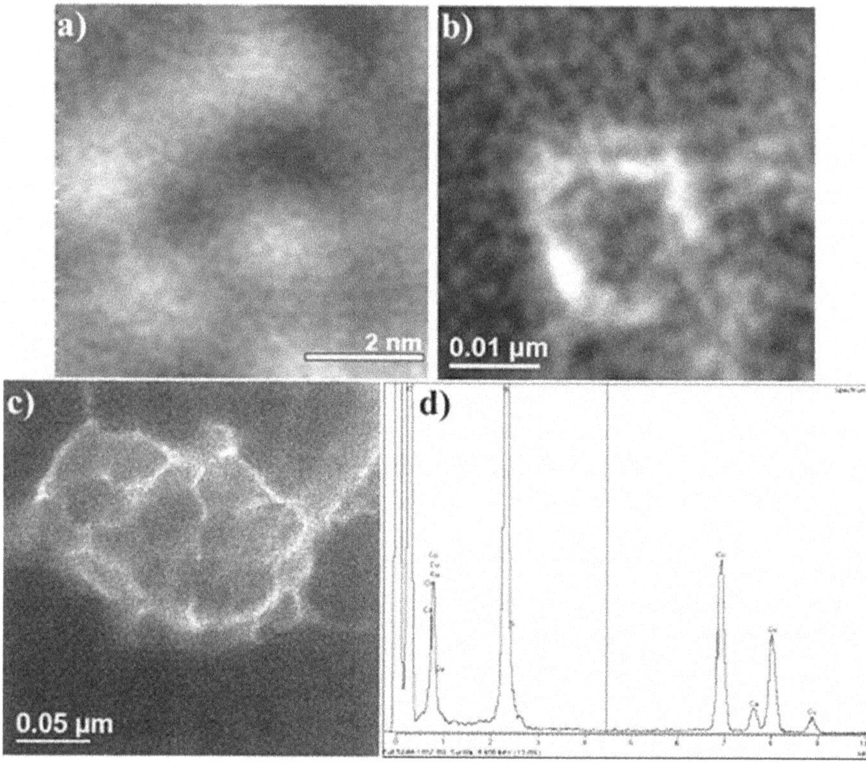

Figure 2.12. (a) TEM–HAADF images of a 2 nm ring structure in a Co-thiol sample where the arrangement of the clusters can be observed. (b) A larger nanoring of 10 nm formed by units smaller than the ones shown in (a). (c) A larger structure of rings observed inside the polymer. (d) EDS pattern showing the presence of Co and S in the region shown in (c) (Marin-Almazo *et al.*, 2004).

a specific behavior of the photon trajectory inside the scatterers of fractal system. Due to multiple rescattering, this trajectory acquires the features of Antoine's set having zero topological dimensions. Respectively, the photon trajectory acquires a "mechanical rigidity" due to a singularity of energy density on it. The interlaced sections (due to multiple scattering) of the photon trajectory are the reason for the self detention of photon inside the fractal structure. The localization cross-section is a critical function of the fractal dimension and depends on the specific frequency range. Among other peculiarities are the induced emission of photons localized

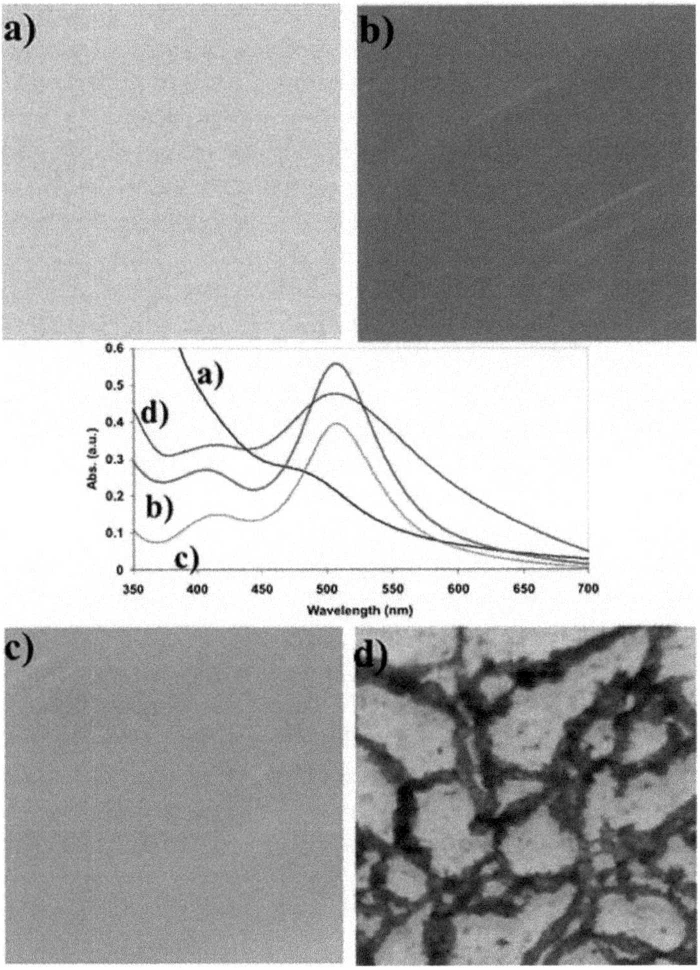

Figure 2.13. Image of four different composites polymer $+ Co^{2+}$-thiol $+$ Co systems and their corresponding UV-vis spectra. The concentration of Co and Co^{2+}-thiol was changed to produce the different absorption peaks. The polymer in (d) contains an excess of Co^{2+}-thiol (Marin-Almazo *et al.*, 2004).

inside cluster and also a possibility of free photon transmission through the fractal system (Maksimenko, 1999). The similar mechanisms may explain the origin of electromagnetic processes in the volumetric matrix of MRET polymer composite under the effect of visible light with a wavelength range of 400–800 nm emitted by the diode.

It was also discovered that such composites exhibit a superparamagnetic effect in the wide range of temperature values. In addition to superparamagnetic behavior, it was found that nanoring structures of these compounds act as the giant aromatic macromolecules, and an electrical current circulates through them when they are exposed to external electromagnetic field. This generated current opposes the external field and gives rise to a negative contribution to the magnetization, which would have to be added to the negative magnetization arising from the polymeric matrix. Thus, there are two contributions to the negative part of the magnetization: the intrinsic diamagnetism from the polymeric matrix and the diamagnetism from the nanorings, which increases as the applied external field increases.

Due to the complexity of all possible interactions between all the magnetic species present in the volumetric fractal matrix of MRET polymer compound and their spatial arrangement, more experimental and theoretical work is required to achieve a full understanding of these interactions and their effect on the magnetic properties observed. It is reasonable to conclude that the origin of subtle low frequency oscillations that alter water molecules arises from the complex electromagnetic processes that take place in the volumetric fractal matrix of MRET polymer composite. These processes include the combination of anomalous strong electrical activity in nanorings structures, modified magnetic properties, and enhanced piezoelectric properties of MRET polymer material.

The mechanism that explains the effect of electromagnetic fields on water is related to the existence of defects in molecular structure of water. The stable structural changes in water were detected in experiments by the UV luminescence spectrophotometer. They have been attributed to different water structural defects that include specific centers of luminescence. The nuclear proton spins were considered to be primary targets of external magnetic fields, since proton lattice of water molecules is unstable and asymmetric. The structural metastability of water was associated with microscopic orbital currents of protons in water-molecular hexagons and deviation from the stoichiometric composition of water. The effects of memory of water interacting with electromagnetic fields were supposed to originate from the oscillations of water-molecular hexagons. The orientation of nuclear proton spins may influence biochemical processes in biological systems. This is a result of associations and disintegrations of above mentioned structural defects of water, since ionic structural defects are chemically active. The recombination of water defects is classified within both classical and quantum types of description. The nuclear proton spins exposed to external resonance magnetic field are losing their dynamic correlation that leads to recombination of water defects and deviation of stoichiometric composition of water. The recombination of water defects in magnetic field is a result of proton spin orientations that initiate the quantum transition of proton from one potential position to another potential position in the lattice of hydrogen bonding in water. In the case of resonance magnetic field, nuclear proton spins have the tendency to orient themselves along the lines of magnetic field. In this case, proton spins are oriented in predominant directions for a long period of time and their precession slows down or even stops. This situation may lead to formation of stable structural defects in water.

The stable water molecular clusters were discussed based on observed low-frequency spectra of the water electric conductivity in a number of experiments. The clusters were assumed to memorize an electromagnetic activation in water molecular structure.

The research regarding the physical parameters of water confirmed that MRET treatment of distilled water led to substantial modification of basic physical-molecular properties of distilled water. The level of modification of properties of MRET water depends on the duration of the process of activation. The results also confirmed the ability of MRET Activated Water to keep its anomalous characteristics for several hours or days at room temperature and especially at low temperature (known in physics as the "long-term water memory" phenomenon) (Vysotskii, 2004, 2005).

The experiment conducted on MRET Activated Water subjected to tangent pressure revealed that at very low velocity of motion of water (tangent pressure in the range of 0.004–0.005 Pa, temperature 20°C) the viscosity of water activated for 60 min decreased about 200–250 times compare to non-activated water from the same source. The most significant phenomenon of anomalous low viscosity of activated water, the decrease about 300–500 times, was observed for water activated for 30 min (Vysotskii, Olishevsky and Kornilova, 2006; Vysotskii, Tashyrev and Kornilova, 2006).

These results confirm the hypothesis regarding the modification of molecular structure in MRET Activated Water. Particularly, the anomalously low viscosity of MRET Activated Water in the area of very low tangent pressure confirms the polarized-oriented multilayer molecular structuring of MRET water: the high level of long-range molecular coupling (hydrogen bonding) inside the "layer" and very low level of molecular coupling between the "layers".

The significant modification of electrodynamic characteristics of distilled water subjected to applied electromagnetic field in the range of low frequencies was observed after MRET activation. The electrical conductivity of MRET Activated Water at 20°C in the range of frequencies of 0.1 Hz–100 kHz decreased by 80–90% in 30-min activated water, and by 66–70% in 60-min activated water respectively compared to non-activated distilled water. The dielectric permittivity in the very low frequencies range of 0.1–1000 Hz decreased by 80–90%, and in the range of frequency of 1–100 kHz, it decreased by 18% in 30-min activated water; the decrease by 70–85% was observed in the range of 0.1–1000 Hz in 60-min activated water compared to non-activated water from the same source (Vysotskii, Olishevsky and Kornilova, 2006; Vysotskii, Tashyrev and Kornilova, 2006). It is reasonable to admit that in the range of very low frequencies 0.1–1000 Hz, the long-range multilayer molecular structures of MRET water (with high level of molecular coupling inside the "layers" and extremely low level of hydrogen bonding between the "layers") create lower level of resistance of water dipoles to the alignment and, as a result, the dielectric permittivity of MRET water is substantially lowered by 70–90% compared to non-activated water.

These experimental and theoretical results were obtained by Prof. V. I. Vysotskii and Dr. A. A. Kornilova with their colleagues in 2004 to 2008 and will be discussed in Chap. 3.

The substantial decrease of dielectric permittivity also confirms the direct correlation between viscosity and dielectric permittivity of water in the range of low frequencies of applied EMF (Chaplin, 2005).

Thus, the anomalous viscosity of MRET water (subjected to very low tangent pressure) and electrodynamic characteristics of MRET water (subjected to applied electromagnetic field of low frequency range) confirmed the high level of long-range dynamic structuring of water molecules in polarized-oriented multilayer formations in activated water produced with the help of MRET activation process. The fundamental biophysical theories revealed the scientific paradigm regarding polarized-oriented multilayer (PM) structuring of cell water in biological systems (Ling, 2001, 2003; Drost-Hansen, 1971, 1991). According to the PM theory, there is the assumption that the formation of distinctive dynamic structure by the cell water results from its interaction with some intracellular proteins. More specifically, the dynamic structure of cell water results from its direct or indirect interaction with the positively-charged CO groups (P sites) and negatively-charged NH groups (N sites) on the "backbones" of a pervasive

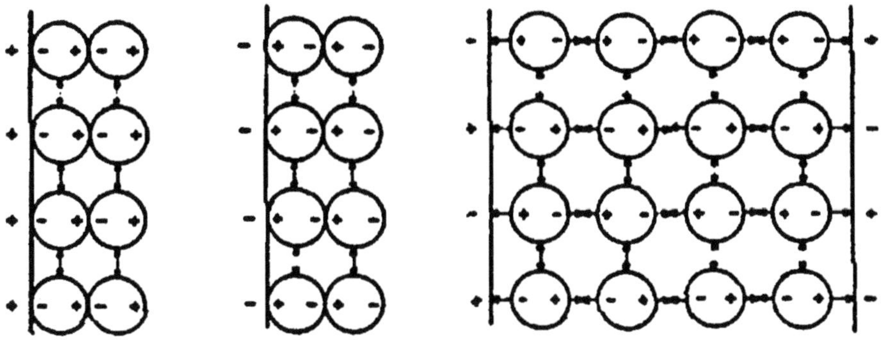

(a) (b) (c)

Figure 2.14. The illustration of the way that individual ions and checkerboards of (a) evenly distributed positively charged P sites alone, or (b) negatively charged N sites alone, polarize and orient water molecules in immediate contact and further away. Emphasis was, however, on uniformly distanced bipolar surfaces containing alternative positive (P) and negative (N) sites called an NP surface. (c) When two juxtaposed NP surfaces face one another, the system is called an NP–NP system (Ling, 2003).

matrix of fully-extended proteins. These P- and N-site-bearing proteins and the water molecules with which they interact constitute what is called a NP–NP system. Electrical polarization and directional orientation of multiple layers of water molecules may occur under the influence of one or two (juxtaposed) checkerboard(s) of alternative positive and negative sites (Fig. 2.14).

Parenthetically, water molecules may also be polarized and oriented in "layers" by a NO system or a PO system, in which electrically-neutral O sites replace properly-spaced electrically-charged P or N sites of a classic NP system respectively. The aggregate physical impacts of the NP sites on the bulk-phase water may be somewhat arbitrarily divided into three components: to enhance the average water-to-water interaction of (all) the water molecules in the system (Component 1); to reduce the translational as well as rotational motional freedom of the water molecules (Component 2); and to prolong the stay or *residence time* of each water molecule at a specific preferred location (Component 3) (Ling, 2003).

The interaction of water dipoles with pervasive matrix of fully-extended proteins constitutes the basis for the cellular transduction mechanism. Based on this scientific approach, the similarity of molecular formations of cell water and MRET Activated Water can contribute to their compatibility, easy bio-availability and assimilation of MRET Activated Water, as well as to the enhancement of cellular functions in biological systems.

The anomalous electrodynamic characteristics and viscosity of MRET Activated Water provide some evidence regarding the possible effect of MRET water on the proper function of cells in biological systems. It is well-known that cellular processes in biological systems are driven by the low energy of bio-chemical reactions inside and outside the cells and cellular structures. Consequently, such processes create subtle low frequency electromagnetic field and low tangent pressures along water surfaces and the membranes between the cells. The anomalously low viscosity, dielectric permittivity and electrical conductivity of MRET water in the range of very low frequencies that exist in biological systems can contribute to the enhancement of the cellular transduction mechanism, and result in improved intracellular/extracellular water exchange and the proper function of cells in biological systems.

Thus, specially modified magnetic field can induce the formation of metastable structural composition of water that is related to the intensity of protons recombination in the lattice of hydrogen bonding in water. One of the primary magneto-biological mechanisms associates the effects of subtle magnetic fields with modified states of liquid water in biological systems. The structural changes in water that result from the influence of external magnetic field are further transmitted to the biological level, since water takes part in a variety of metabolic reactions. The concept of MRET activation effect on water molecules and resulted physical and biological effects were confirmed by a number of experiments that are discussed in the following chapters of this book.

CHAPTER 3

Study of the Physical Properties of MRET Activated Water

3.1. Methods and Equipment to Study the Dielectric Permittivity and the Conductivity of Activated Water

In this section, we consider the main electrodynamic characteristics of activated water. It is necessary to note that the electrodynamic properties of water play very important roles in intracellular processes in biological systems. Below, we give several arguments to clarify this role:

(1) Water is the natural background in the scope of which all biochemical processes are running. In nature, only four types of interactions (strong, weak, electromagnetic, and gravitational) are known. Two of the interactions are purely nuclear, and the gravitational one reveals itself only on the cosmic scale. Therefore, it is clear that only the electromagnetic interaction is essential in the scope of any biological system. For the sake of simplicity, we notice that the whole specificity of any biological process is eventually reduced to certain electromagnetic interactions. Just for this reason, the electromagnetic properties of water, which play the decisive role in its self-organization and in its influence on other objects, must be comprehensively investigated. These properties are revealed in all, without exception, biochemical and biophysical processes.
(2) Water is the most active object of radiobiological processes, by transforming (as a rule, through the generation of free radicals) the primary ionizing radiation into the collection of factors inducing mutation and degradation of biological molecules. The decisive role in these processes is played by water, and its electrodynamic properties are the factor of its action.
(3) Specific features of the electrodynamic characteristics of water [first of all, a very great value of its dielectric permittivity $\varepsilon(\omega)$] are the reason

for the natural dissociation of molecules of many chemical compounds and the formation of the necessary ion composition of vitally important microelements. Otherwise, the normal operation of many systems of a living organism (in particular, the operation of selective membranes) would be impossible.

(4) The change in the dispersive properties of water can render very strong influence (by means of modification of the electrostatic forces between separated charges and forces of the van der Waals type defining the interaction of the systems of neutral atoms and molecules) on the long-range interaction of basic elements of living systems such as cells, viruses, biological macromolecules, enzymes, etc.

(5) Viscosity, surface tension, and other mechanical properties of water which play a very important role in the processes of vital activity (in particular, in ion transport and cell division) turn out to be directly related to the electrodynamic characteristics of water.

This list is far from being complete and can be continued.

Firstly, we consider the dispersive characteristics of the dielectric permittivity of activated water. Of importance are both the real $\varepsilon'(\omega)$ and imaginary parts $\varepsilon''(\omega)$ of the complex-valued dielectric permittivity $\varepsilon'(\omega)$. These two characteristics determine both the refractive index of the electromagnetic field and the conductivity of water. They affect, in a decisive manner, long-range interactions of biological microsystems (in particular, of viruses) and biological macromolecules (van der Waals interactions), as well as the ion transport.

For the determination of the complex-valued dielectric permittivity in a wide range of frequencies, we used the method of measurement of the complex-valued resistance (impedance) under the AC circuit. In view of the importance and the nontriviality of the obtained results, we will consider the essence of physical processes, which characterize the very process of measurement, in more details.

In the present study, we use a Novocontrol installation including an analyzer of the impedance (a dielectric analyzer ALPHA operating in the real time scale in the range of frequencies from 3×10^{-6} Hz to 3×10^{7} Hz), a measuring cell containing water under study, a thermostating system for the measuring cell (temperature control — QUATRO Cryosystem) providing the heating and the heat setting in a very wide interval of temperatures from $-160°$ to $400°$C, and the system for the automatic accumulation and processing of data.

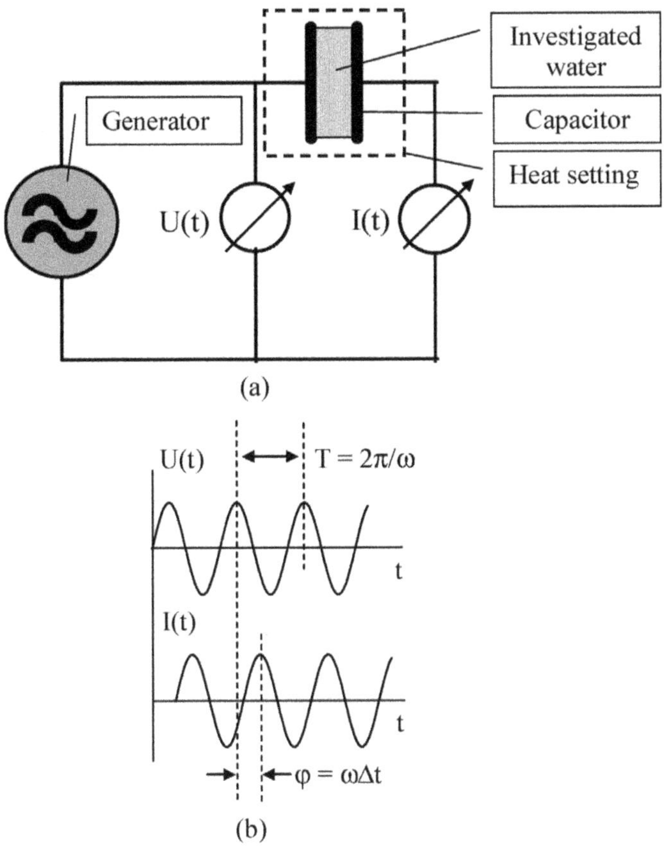

Figure 3.1. Scheme of a radioelectronic circuit for the measurement of (a) the electrodynamic characteristics of water and (b) the diagrams of voltage $U(t)$ and current $I(t)$ in various sections of this circuit.

The liquid under study [activated or nonactivated (control) water] was positioned without gaps between two plane electrodes in the volume of the measuring cell, thus forming the system of a capacitor which includes the resistance $R = Z'(\omega)$ and the capacitance $C = Z''(\omega)$ joined in parallel.

The scheme of measurements and the main specific features of changes of the voltage and the current are given in Fig. 3.1.

In a simplified form, the scheme of measurements includes a stabilized source of alternating voltage (a generator), a cuvette under study with activated water (capacitor), the measuring system allowing one to determine the voltage on the cuvette, the current through it, and the difference of phases between the current and the voltage.

We apply the voltage U_0 with a constant frequency ω to a capacitor. This voltage U_0 induces the current I_0 of the same frequency which passes through a capacitor under study. Due to the presence of a complex-valued resistance in the circuit, there exists the phase shift named the phase angle φ between the current and the voltage [Fig. 3.1(b)].

The relation between $U(t)$ and $I(t)$ and the phase angle φ are determined by the electric properties of the material of a sample [the conductivity $\sigma(\omega)$ and dielectric permittivity $\varepsilon(\omega)$] and its linear sizes. For the sake of simplicity, the formulas are presented usually in terms of complex-valued quantities:

$$U(t) = U_0 \sin(\omega t) = \mathrm{Re}(U^* e^{i\omega t}),$$
$$I(t) = I_0 \sin(\omega t + \varphi) = \mathrm{Re}(I^* e^{i\omega t}), \tag{3.1}$$

where

$$U^* = U_0, \quad I^* = I' + iI'', \quad I_0 = \sqrt{I'^2 + I''^2}, \quad tg\varphi = \frac{I''}{I'}.$$

For any material medium, the measured impedance of a capacitor containing a sample,

$$Z^* = Z' + iZ'' = \frac{U^*}{I^*}, \tag{3.2}$$

is connected with the dielectric permittivity of the medium $\varepsilon(\omega)$ by the relation

$$\varepsilon^*(\omega) = \varepsilon'(\omega) - i\varepsilon''(\omega) = \frac{-i}{\omega C_0 Z^*(\omega)}, \tag{3.3}$$

where C_0 is the capacitance of a capacitor of the same geometry as that of the liquid under study.

The complex-valued conductivity is expressed through the same impedance as

$$\sigma^*(\omega) = \sigma'(\omega) - i\sigma''(\omega) = \frac{1}{Z^*(\omega)}\frac{d}{S}, \tag{3.4}$$

where d is the distance between the capacitor plates (the sample thickness or, in the case under study, the thickness of the layer of studied water between the electrodes), and S is the surface area of the electrodes.

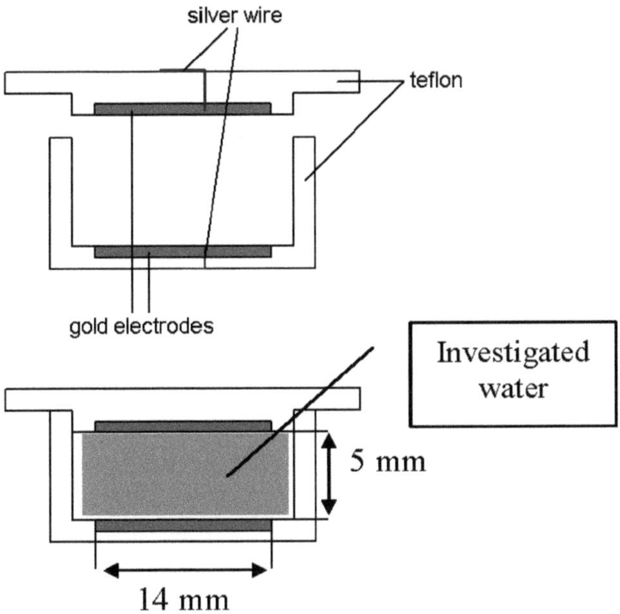

Figure 3.2. Construction of a dismountable cell for the measurement of electrodynamic characteristics of ordinary or activated water.

The important quantity in the analysis of the properties of water is the dielectric loss tangent $tg\delta(\omega) = \frac{\varepsilon''(\omega)}{\varepsilon'(\omega)}$.

In Fig. 3.2, we present the construction of a measuring cell. The cell is a dismountable cylindrical chamber, whose height $d = 5$ mm, manufactured of Teflon. On the chamber bottom is positioned the lower electrode specially produced of pure gold. The analogous electrode is on a removable lid of the chamber. The use of gold as the material for electrodes is due to the necessity to minimize the influence of the process of measurement on a state of the measuring cell. In particular, it is desirable to completely eliminate the chemical reactions running on the interaction of the products of electrolysis of water with electrodes (for example, to eliminate the oxidation of electrodes), to decrease the diffusion of the metal ions of electrodes in the water under study, etc. Such factors of errors have special significance in the case of studies of distillated activated water, in which the initial presence of extraneous ions is insignificant. In the opposite case (in the presence of the diffusion of ions), the ion composition of water is distorted, and the oxidation of electrodes yields the increase of the resistance of a measuring cell. Such processes can lead to significant distortion of the measured electrodynamic characteristics.

One more source of errors can be related to the influence of the external medium on the process of measurements. As an example of such an influence, we mention the passage of a current along the external circuit bypassing the studied cuvette, e.g. through the air surrounding the cell.

To prevent such an influence, the measuring cell with all current-carrying electrodes was positioned at a constant temperature in an air-line and was continuously, throughout the measurement, blown off with pure gaseous nitrogen. As known, gaseous nitrogen is a very good dielectric, and the disposition of the measuring cell in the atmosphere of nitrogen sharply decreases the influence of the environment. Besides protection from the external influence, we solved simultaneously the other problem, namely to stabilize the temperature of a cell with the studied liquid on the prescribed level for the entire time interval of measurements. We note that the duration of one batch of measurements (with regard for the fact that the lower boundary of the range of the studied frequencies is very small and is equal to $\omega_{\min} = 0.1$ Hz) took a sufficiently great time. In particular, the duration Δt of the measurement for each specific frequency near the lower boundary must exceed the value $\Delta t_0 \equiv 1/\omega_{\min}$ by several orders.

3.2. Anomalous Electrodynamic Characteristics of Activated Water

In our studies, we determined the electrodynamic properties of activated water as functions of three basic parameters:

- the water activation duration,
- the water storage duration after the activation, and
- the temperature of the storage of activated water.

The influence of some of these parameters was investigated comprehensively (for example, the influence of the storage duration and the storage temperature after the activation), and that of the water activation duration — only for two values which correspond to the optimal parameters for the use of activated water in medicobiological experiments.

The scheme and the procedure of studies were as follows:

As the base quantity, the water storage temperature was fixed at 20°C. At this temperature, we studied the basic parameters for the initial water (nonactivated distillated water) and different fractions of activated water (the durations of activation were 30 min and 60 min) for different durations of its storage after the activation, and till the time point of the measurement (immediately after the activation, i.e. for the storage duration $t = 0$ and for several time intervals during storage prior to the measurements: 0.5 h, 1 h, 2 h, 5 h, and 24 h).

For the fraction of activated water which corresponds to the strongest change in the physical properties (it turned out that this occurred for the activation duration equal to 30 min), we furthered the study at different storage temperatures i.e. 4°C, 20°C, 40°C, and 72°C.

For all of the above, the activation was realized at the optimal temperature for the given laboratory, 20°C.

To clarify the question about the time interval during which the activated water preserves its properties (in fact, the question about the duration of "the memory of water"), we studied the influence of the *storage duration* at different temperatures on the properties of water. To this end, activated water was positioned in a cooler at a temperature of about 4°C, was stored in the laboratory at a fixed temperature of 20°C, or was positioned in a thermostat at a temperature of 40°C.

After the planned storage of a specific sample was terminated, activated water was positioned in the measuring cell which was then placed immediately into the device. The device was kept at the stabilized temperature, which is equal to the storage temperature, throughout the measurements. In the case where water was studied immediately after the completion of its activation (such an activation was carried out in all the cases at a fixed temperature of 20°C), the treated water was placed into the measuring cell which was rapidly cooled or heated to the necessary temperature, and only then its study was performed. A typical cooling rate for water was 3°C/min, while the heating rate for water prior to its study was 10°C/min.

After cooling or heating water to the necessary temperature, we carried out additional stabilization of temperature for two minutes in each sample of activated water. This time interval is necessary for all transient processes to be completed and for the maximum homogeneity of temperature to be attained in the whole volume of the cell under study.

Only then we carried out the measurements of the parameters of activated water. The process of measurement took approximately five minutes.

One of the analyses was realized with water heated in the process of measurements up to 72°C. Such studies were performed without considering the influence of the storage duration at such high temperature. This is conditioned by the fact that no anomalous electrodynamic properties of activated water were registered at such a temperature in the process of measurements performed directly after the activation, whereas these anomalies were well manifested during the storage and the measurement at lower temperatures. This is caused by the very short duration of the relaxation of anomalous electrodynamic properties of water at high temperatures. It is obvious that, with regard for these circumstances, it became senseless to study the influence of additional storage at such high temperature.

Below, we present some results of the studies.

In Fig. 3.3, we present the results of the base studies of the electrodynamic characteristics of initial distillated water completely analogous to water, which was then subjected to the procedure of activation. The studies were carried out at a temperature of 20°C. It is worth noting that the characteristics of such water coincided qualitatively with the standard dependences which were obtained by the producers of the Novocontrol system during the factory tests. This correspondence confirms that the parameters of the experimental equipment in use meet the starting requirements.

It follows from these results that the initial distillated water has a low value of conductivity, $\sigma \approx 3.10^{-5}$ S/cm^2, and the dielectric permittivity was close to the "standard" one, $\varepsilon'(\omega) \approx 90$, at the frequency of alternating electric current in the range of 10^3–10^7 Hz. This value of the dielectric permittivity well agrees with the Debye mechanism of orientational polarizability of independent molecules of water. The maximum value of the dielectric loss tangent $(tg\delta)_{\max} \approx 120$ and this corresponds to a frequency of 1522 Hz.

With the frequency decreased to $\omega < 10^2$ Hz, a sharp decrease of the conductivity and a very sharp monotonous increase of the dielectric permittivity were registered. In the studied region of superlow frequencies (at $\omega \leq 10^{-1}$ Hz), the effective dielectric permittivity reaches a very great value $\varepsilon'(\omega) \approx 10^8$.

It is necessary to say several words about the physical sense of such sharp increase in $\varepsilon'(\omega)$. In the region of low frequencies, the total polarization is determined by the orientational polarization of molecules of water (according to the Debye mechanism). This orientational mechanism is principal up to the frequencies $\omega_D \approx 10^{10}$–10^{11} Hz. Above this frequency, the polarization is determined by vibrational states of molecules in the IR range (at $\omega > \omega_i$) and by electron transitions in atoms of water in the UV range (at $\omega > \omega_e$).

We will show that the registered anomalously great values of $\varepsilon'(\omega)$ can be attained only under condition that the studied distillated water contains

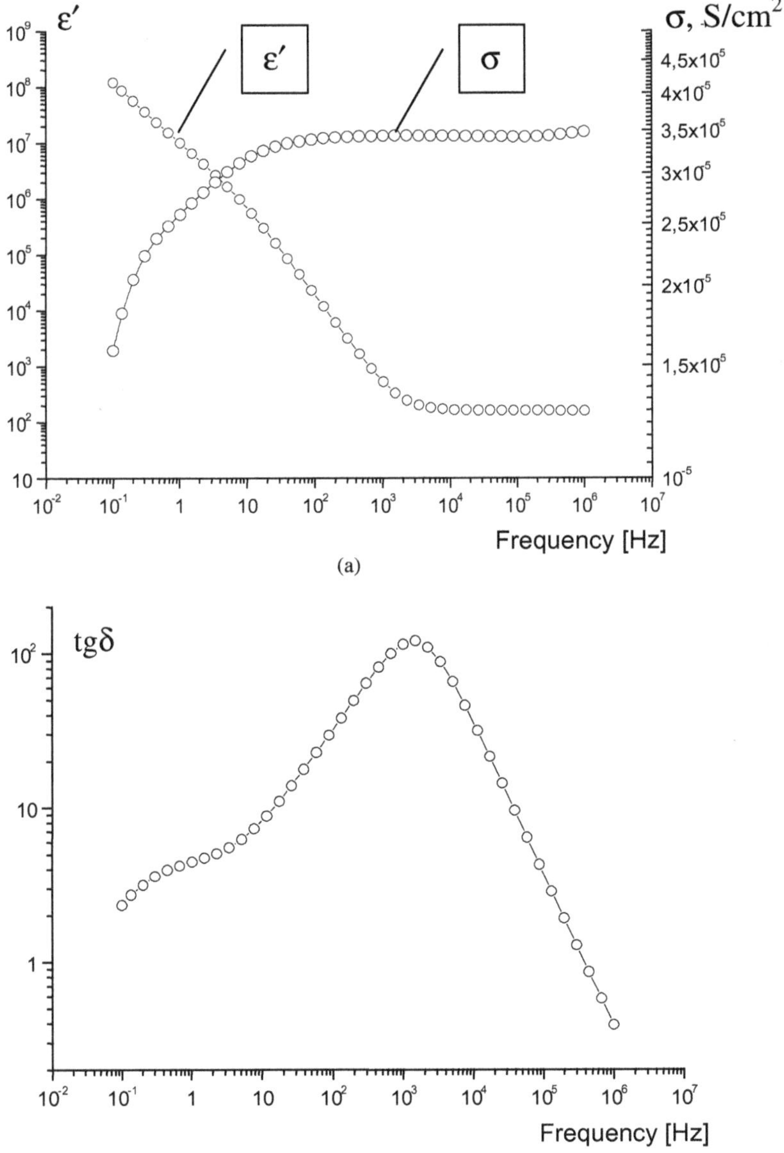

Figure 3.3. (a) Dielectric permittivity ε' and conductivity σ and (b) dielectric loss tangent $tg\delta$ of the initial nonactivated distilled water at 20°C versus the frequency.

sufficiently great structurized objects consisting of molecules of water which are moving as a single formation. The measurement of $\varepsilon'(\omega)$ in the region of superlow frequencies allows us to indirectly evaluate the size and the mass of large supermolecular objects, whose existence is related to the structure of water.

As it is well known, the dielectric permittivity of ordinary water in the region of low frequencies is described by the formula

$$\varepsilon(\omega) = 1 + \frac{4\pi n_w \alpha_1}{(1 - i\omega\tau_1)} \equiv 1 + \frac{\varepsilon_1' - 1}{(1 - i\omega\tau_1)} = \varepsilon'(\omega) + i\varepsilon''(\omega). \qquad (3.5)$$

Here, n_w is a concentration of water molecules, $\alpha_1 = p_1^2/3k_BT$ is orientational polarizability of one molecule of water, $p_1 = |\vec{p}|$ is the dipole moment of one molecule of water, $\tau_1 \equiv 1/\omega_1 = 4\pi\eta R^3/k_BT$ is the time of relaxation of orientational polarization of one water molecule with averaged radius R, η is a coefficient of viscosity of water, and $\varepsilon_1' \equiv \varepsilon'(\omega = 0) = 4\pi n_w \alpha_1$ is the reference expression for a real part of dielectric permittivity of water at low frequencies $\omega \ll 1/\tau_1 \approx 10^{10}\,\mathrm{c}^{-1}$.

In addition, the imaginary part of the dielectric permittivity is related to the conductivity

$$\varepsilon''(\omega) = 4\pi\sigma(\omega)/\omega.$$

From these formulas, it is easy to receive explicit expressions for the imaginary and real parts of dielectric permittivity and conductivity

$$\varepsilon'(\omega) = 1 + \frac{4\pi n_w \alpha_1}{1 + (\omega/\omega_1)^2} = 1 + \frac{(\varepsilon_1' - 1)}{1 + (\omega/\omega_1)^2},$$

$$\varepsilon''(\omega) = 1 + \frac{4\pi n_w \alpha_1 (\omega/\omega_1)}{1 + (\omega/\omega_1)^2} = \frac{(\varepsilon_1' - 1)(\omega/\omega_1)}{1 + (\omega/\omega_1)^2}, \quad \text{and} \qquad (3.6)$$

$$\sigma(\omega) = \frac{(\varepsilon_1' - 1)(\omega^2/\omega_1)}{4\pi[1 + (\omega/\omega_1)^2]}.$$

It is obvious that formula (3.5) characterizes the properties of water only in the region of low frequencies, where the contribution of the considered mechanism of the Debye orientational polarization turns out to be more significant than the influence of electron levels.

In the case of collective interaction of water molecules and with presence of large clusters in water, formula (3.5) should be changed.

Let's consider the following model system. Water consists of both independent molecules and clusters of water molecules. Each cluster consists of $N \gg 1$ molecules. If in a unit volume of water, K clusters are present, the relative part of clusterized water is $g = NK/n_w < 1$. The maximum dipole moment of each rigidly structured cluster is $p_N^{(\max)} = Np_1$, $\alpha_N^{(\max)} = N^2\alpha_1$. For the simplicity of the analysis, we will consider only this case.

In such system, particles of two different types — usual molecules of water and clusters — exist in water.

The expression for the dielectric permittivity at low frequency for such two-component model has the following modified form

$$\varepsilon(\omega) = 1 + \frac{4\pi(1-g)n_w\alpha_1}{(1-i\omega\tau_1)} + \frac{4\pi g n_w \alpha_N}{(1-i\omega\tau_N)} = \varepsilon'(\omega) + i\varepsilon''(\omega). \qquad (3.5a)$$

Here, $\alpha_N = N^2\alpha_1$ is the orientational polarizability of one cluster, $\tau_N \equiv 1/\omega_N \approx 4\pi\eta NR^3/k_BT = N\tau_1$ is the time of relaxation of orientational polarization of one cluster that consist of N oriented water molecules.

From the formula (3.5a) for $\varepsilon(\omega)$, the expressions for imaginary and real parts of the dielectric permittivity and conductivity of such two-component water are as follows:

$$\begin{aligned}
\varepsilon'(\omega) &= 1 + \frac{4\pi(1-g)n_w\alpha_1}{1+(\omega/\omega_1)^2} + \frac{4\pi g N n_w \alpha_1}{1+(\omega/\omega_N)^2} \\
&= 1 + \frac{(1-g)(\varepsilon_1'-1)}{1+(\omega/\omega_1)^2} + \frac{gN(\varepsilon_1'-1)}{1+(\omega/\omega_N)^2}, \\
\varepsilon''(\omega) &= 1 + \frac{4\pi(1-g)n_w\alpha_1(\omega/\omega_1)}{1+(\omega/\omega_1)^2} + \frac{4\pi g N n_w \alpha_1(\omega/\omega_N)}{1+(\omega/\omega_N)^2} \qquad (3.6a) \\
&= \frac{(1-g)(\varepsilon_1'-1)(\omega/\omega_1)}{1+(\omega/\omega_1)^2} + \frac{gN(\varepsilon_1'-1)(\omega/\omega_N)}{1+(\omega/\omega_N)^2}, \quad \text{and} \\
\sigma(\omega) &= \frac{(1-g)(\varepsilon_1'-1)(\omega^2/\omega_1)}{4\pi[1+(\omega/\omega_1)^2]} + \frac{gN(\varepsilon_1'-1)(\omega^2/\omega_N)}{4\pi[1+(\omega/\omega_N)^2]}.
\end{aligned}$$

Let's consider different special cases.

In the area of relatively low frequencies $\omega_N \ll \omega \ll \omega_i$, ω_e and at $\omega_1 \approx \omega$, the formula (3.6a) for two-component water takes the form:

$$\varepsilon'(\omega) \approx 1 + \frac{(1-g)(\varepsilon_1'-1)}{1+(\omega/\omega_1)^2} + \frac{g(\varepsilon_1'-1)(\omega_1/\omega)^2}{N} \approx 1 + \frac{(1-g)\varepsilon_1'}{1+(\omega/\omega_1)^2},$$
$$\varepsilon''(\omega) \approx \frac{(1-g)(\varepsilon_1'-1)(\omega/\omega_1)}{1+(\omega/\omega_1)^2} + \frac{g(\varepsilon_1'-1)(\omega_1/\omega)}{N} \approx \frac{(1-g)\varepsilon_1'(\omega/\omega_1)}{1+(\omega/\omega_1)^2}, \tag{3.7}$$

and

$$\sigma(\omega) \approx \frac{(1-g)(\varepsilon_1'-1)(\omega^2/\omega_1)}{4\pi[1+(\omega/\omega_1)^2]} + \frac{g\varepsilon_1'\omega_1}{4\pi} \approx \frac{(1-g)\varepsilon_1'(\omega^2/\omega_1)}{4\pi[1+(\omega/\omega_1)^2]}.$$

The expressions in (3.7) differ from (3.5) for one-component water on a trivial factor $(1-g)$. In particular, from Eq. (3.7) it follows that in the case of absence of the second (cluster) component, we came to the reference asymptotic formulas

$$\varepsilon'(\omega) \to \varepsilon_1',$$
$$\varepsilon''(\omega) \to \varepsilon_1'(\omega^2/\omega_1) \to 0, \quad \text{and} \tag{3.7a}$$
$$\sigma \to \varepsilon_1'(\omega^2/\omega_1)/4\pi \to 0.$$

Other situation takes place for lower frequencies $\omega_N \ll \omega \ll \omega_1$. For this frequency area, we have:

$$\varepsilon'(\omega) \approx 1 + (1-g)(\varepsilon_1'-1) + \frac{g(\varepsilon_1'-1)(\omega_1/\omega)^2}{N} \approx \frac{g\varepsilon_1'(\omega_1/\omega)^2}{N},$$
$$\varepsilon''(\omega) \approx (1-g)(\varepsilon_1'-1)(\omega/\omega_1) + g(\varepsilon_1'-1)(\omega_1/\omega) \approx g\varepsilon_1'\omega_1/\omega, \tag{3.8}$$
$$\sigma(\omega) = \frac{(1-g)(\varepsilon_1'-1)(\omega^2/\omega_1)}{4\pi} + \frac{g(\varepsilon_1'-1)\omega_1}{4\pi}.$$

At last, in the case of extremely low frequencies $\omega \ll \omega_1$, ω_N from the Eq. (3.6a), we have:

$$\varepsilon'(\omega \to 0) = (1-g)(\varepsilon_1'-1) + gN(\varepsilon_1'-1);$$
$$\varepsilon''(\omega \to 0) = (1-g)(\varepsilon_1'-1)(\omega/\omega_1) + gN^2(\varepsilon_1'-1)(\omega/\omega_1) \to 0; \tag{3.9}$$
$$\sigma(\omega \to 0) = (1-g)(\varepsilon_1'-1)(\omega^2/\omega_1)/4\pi + gN^2(\varepsilon_1'-1)(\omega^2/\omega_N)/4\pi \to 0.$$

The obtained outcomes are well correlated with the experimental data.

From Eq. (3.8), it follows that at $\omega \ll \omega_1$, a sharp increase of dielectric permittivity $\varepsilon'(\omega) \sim 1/\omega^2$ and decrease of conductivity $\sigma(\omega)$ take place. We have received the same results in all experiments. From Eq. (3.9), the limiting values of these characteristics are as follows:

$$\varepsilon'(\omega \to 0) \to gN\varepsilon_1'; \quad \sigma(\omega \to 0) \to 0.$$

From comparison of experimental result $\varepsilon'(\omega)_{\max} \geq 10^8$ [Fig. 3.3(a)] and the result of calculation of $\varepsilon'(\omega)$ [Eq. (3.9)], it is possible to receive estimation for the value of $N : N > 10^8$. These N clusters can be identified with elements of the clathrate frame.

This result further confirms the presence of elements of the stiff structure in the water bulk.

From the other hand, it is obvious that such idealized model explains only qualitative properties of $\varepsilon'(\omega)$ and $\sigma(\omega)$ behavior. The multicomponent model is more actual when it is necessary to take into account presence of clusters with different sizes and with a different degree of polarization in water ($p_N < Np_1$).

There are several additional remarks. The part of the discussed effect in the region of low and superlow frequencies can be related to the influence of free ions which are products of the natural dissociation of molecules of water. The process of dissociation is represented as

$$H_2O \leftrightarrow H^+ + OH^-.$$

Protons (H^+) cannot be in the free state for a long time. They are either transformed into hydrogen atoms H, or form the ions of oxonium ($H_2O + H^+ \leftrightarrow H_3O^+$) which are also unstable complexes. In some cases, the ion-molecular complexes $H_5O_2^+$ can also be formed.

In the natural state of pure water, the relative concentration η_{H^+} of the ions of hydrogen is determined by the hydrogen index $\text{pH} = -\lg\eta_{H^+}$. In normal (neutral) distillated water at room temperature, the hydrogen index $\text{pH} \approx 7$, which corresponds to the relative concentration $\eta_{\text{H}^+} = \eta_{\text{H}^-} \approx 10^{-7}$ and the total concentration $n_{\text{H}^+} = n_{\text{OH}^-} \approx 6 \times 10^{15}\,\text{cm}^{-3}$ of ions of each sign.

In this case, about 5×10^{15} pairs of ions can be present in the volume of the measuring cell. At a relatively high frequency $\omega > 10^3$ Hz of the electric field applied to the capacitor (the water understudy is between its plates), ions with different signs have no time to displace in the direction to different electrodes, and their presence does not affect the medium polarization.

At a low frequency of the electric field (at $\omega < 10^2$–10^3 Hz), these heavy ions have time to react to a change in the vector of the field intensity, and move synchronously with the field by periodically forming two oppositely charged layers on the opposite surfaces of electrodes, of which the sign and the value of charges change synchronously with the field frequency. Such a separation of charges causes the appearance of a very great additional electric dipole moment $\Delta\vec{p}(t)$, which is equivalent to a sharp change of both dielectric permittivity $\Delta\varepsilon'$ and conductivity $\Delta\sigma$. It is quite obvious that the frequency, at which the separation of charges and the formation of two surface charged layers occur, depends on both the distance between the capacitor plates and the drift velocity of ions. Taking this circumstance into account, we notice that the value of the total dielectric permittivity $\varepsilon'(\omega)$ in such a system is somewhat different from the traditional value which characterizes the infinite medium. At the same time, the former is a very convenient characteristic of the liquid medium for a constant value of the distance between the plates, and allows one to register the structural changes which can appear on, for example, the activation of water. In particular, the measurement of $\varepsilon'(\omega)$ in the region of superlow frequencies allows us to indirectly evaluate the size and the mass of large supermolecular objects, whose existence is related to the structure of water.

In Fig. 3.4, we present the specific features of the electrodynamic characteristics of water activated for 30 min and stored prior to the time point of measurement at a temperature of 5°C. The dielectric permittivity $\varepsilon'(\omega)$ of this water in the region of frequencies $\omega < 10^3$ Hz at the initial time point (immediately after the completion of the activation) decreases approximately by five times as compared to the initial (nonactivated) water. The maximum value of the conductivity of water also decreases by the same number of times. With increase in the storage duration of activated water, the very slow recovery (relaxation) of these quantities occurs. In particular, for five hours of the storage, the maximum value of the conductivity grew only by 25% (from $\sigma_{\max} \approx 7.2 \times 10^{-6}$ S/sm^2 to $\sigma_{\max} \approx 9 \times 10^{-6}$ S/sm^2). The dielectric loss tangent was not practically changed for this time. Its maximum value $tg\delta_{\max} \approx 65$ and corresponds to the frequency $\omega \approx 451$ Hz.

The estimates show that the full relaxation of the electrodynamic parameters occurs for a very long time. Upon storage of water at a low temperature, its relaxation time can attain several days or even weeks.

In Fig. 3.5, we present the specific features of the electrodynamic characteristics of water activated for 30 min and stored up to the time point of measurement at a temperature of 20°C. It is seen that this water preserves

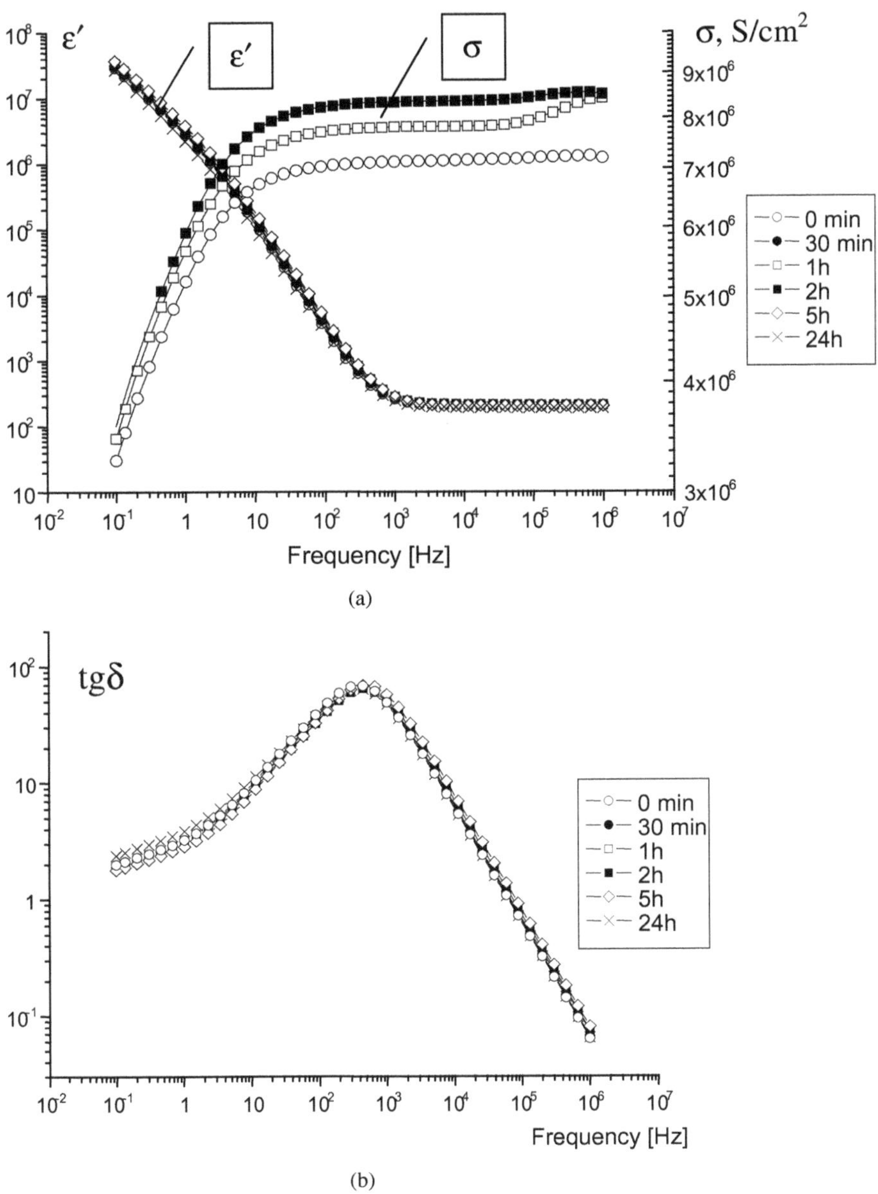

Figure 3.4. Study of water activated for 30 min and stored at a temperature of 5°C. The given numbers correspond to the additional water storage duration after the completion of its activation prior to the start of measurements.

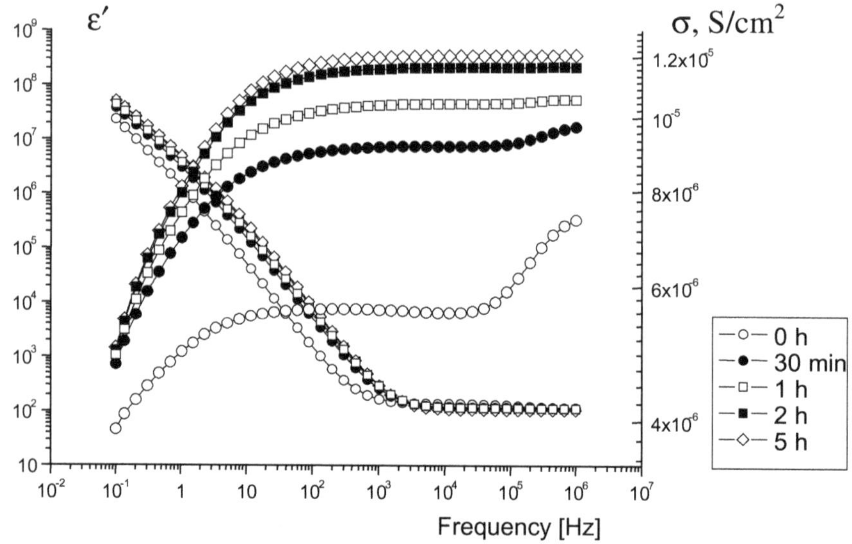

(a)

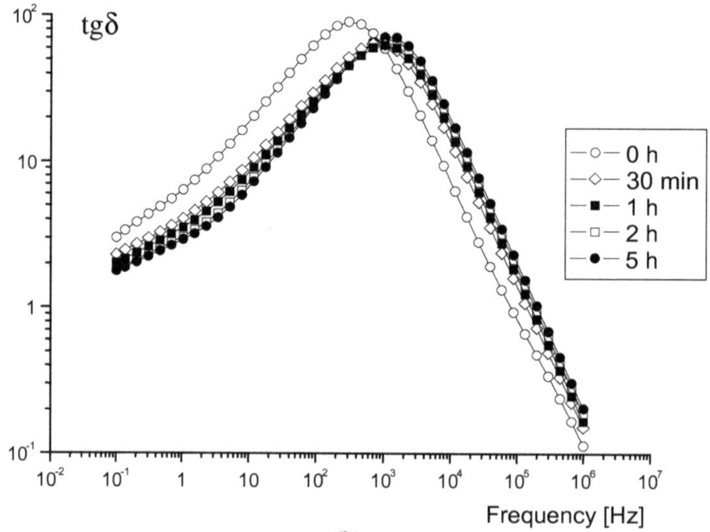

(b)

Figure 3.5. Study of water activated for 30 min and stored at a temperature of 20°C.

some general regularities (a sharp decrease of the dielectric permittivity and the conductivity on the activation of water and their gradual recovery in the process of storage) and, at the same time, reveals the significant differences such as a significantly faster relaxation of the parameters with increase of the storage duration. In particular, for five hours of the storage, the maximum value of the dielectric permittivity and the conductivity grew by 200%.

It should be noted that the actual change in the studied parameters $\varepsilon'(\omega)$ and $\sigma(\omega)$ at the initial time moment (immediately after the activation) can be more in this case. This assertion is related to the following obvious circumstance because the process of change occurs for a sufficiently long time, the formal assertion that water was studies immediately after its activation is not completely correct. In fact, we may only assert that this value characterizes the properties of water to the time point of the termination of measurements.

The still faster relaxation of the electrodynamic parameters of activated water corresponds to the samples stored at a temperature of 40°C. These data are presented in Fig. 3.6. The comparison of the results presented in Figs. 3.5 and 3.6 demonstrates clearly the exceptionally strong influence of the temperature of water on the duration of preservation of its anomalous properties after the activation.

To a still greater degree, this influence is seen in Fig. 3.7. In this figure, we present the dependence of the same parameters of activated water, $\varepsilon'(\omega)$ and $\sigma(\omega)$, on the frequency in the case where water was placed in the measuring cell immediately after activation. In this case, the process of measurement was realized on the simultaneous heating of water to a high temperature of 72°C. Though the activation duration for water in this case was close to the optimum one (30 min), and the measurements were carried out directly after the activation without additional storage, the dependences of the quantities $\varepsilon'(\omega)$ and $\sigma(\omega)$ on the frequency are not practically different from the analogous properties of initial nonactivated water, the data for which are presented in Fig. 3.3.

It follows from this result that "the duration of the memory" of water at such high temperature turns out essentially less than the duration of the measurement, being equal to approximately five minutes.

The other result follows from the analysis of properties of water activated for 60 min. In Fig. 3.8, we present the results of measurements of the properties of water stored after such an activation at a temperature of 20°C. It is seen that the activation in this case causes much less (by 10%) changes in the dielectric permittivity and the conductivity of water than those for 30 min activation.

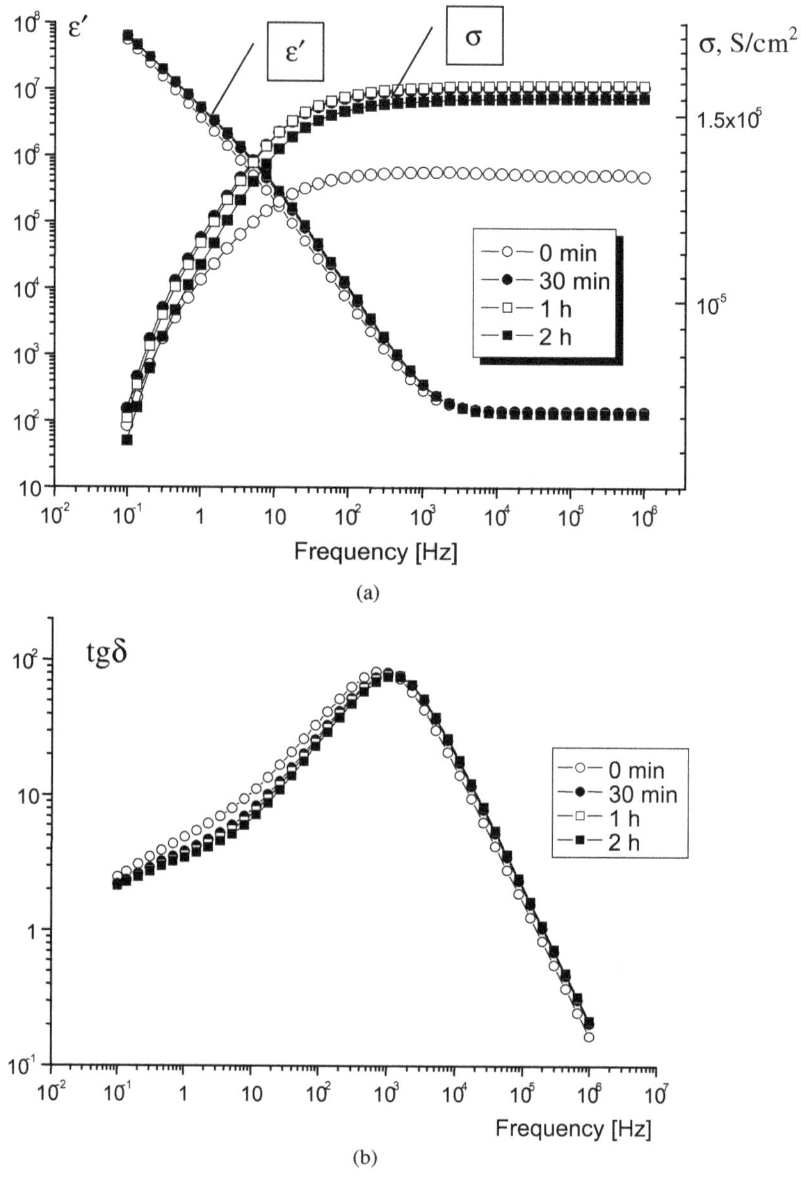

Figure 3.6. Study of water activated for 30 min and stored at a temperature of 40°C.

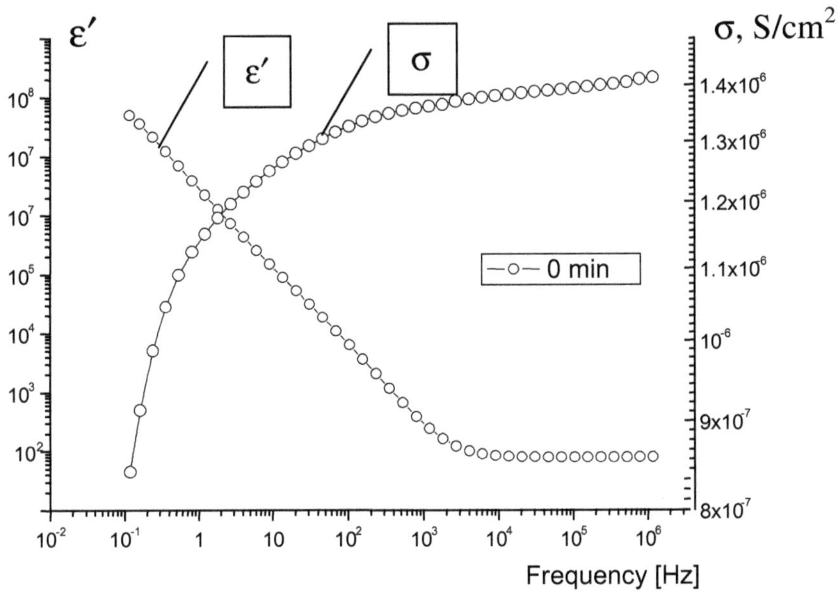

(a)

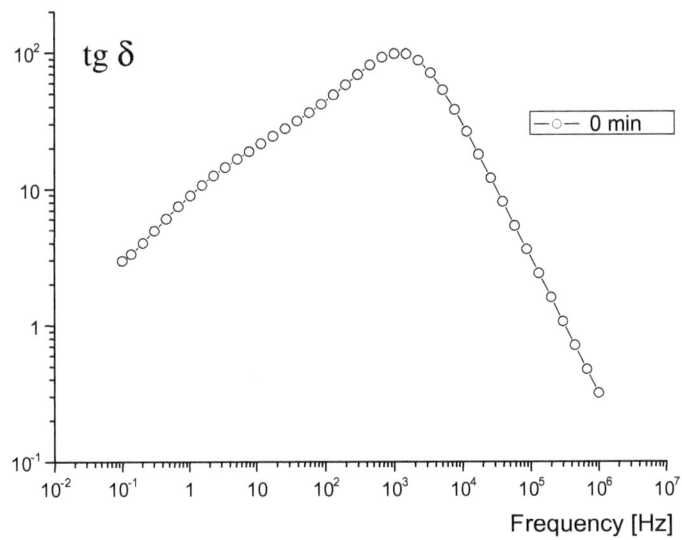

(b)

Figure 3.7. Study of water activated for 30 min. Measurements were carried out at a temperature of 72°C immediately after the completion of activation.

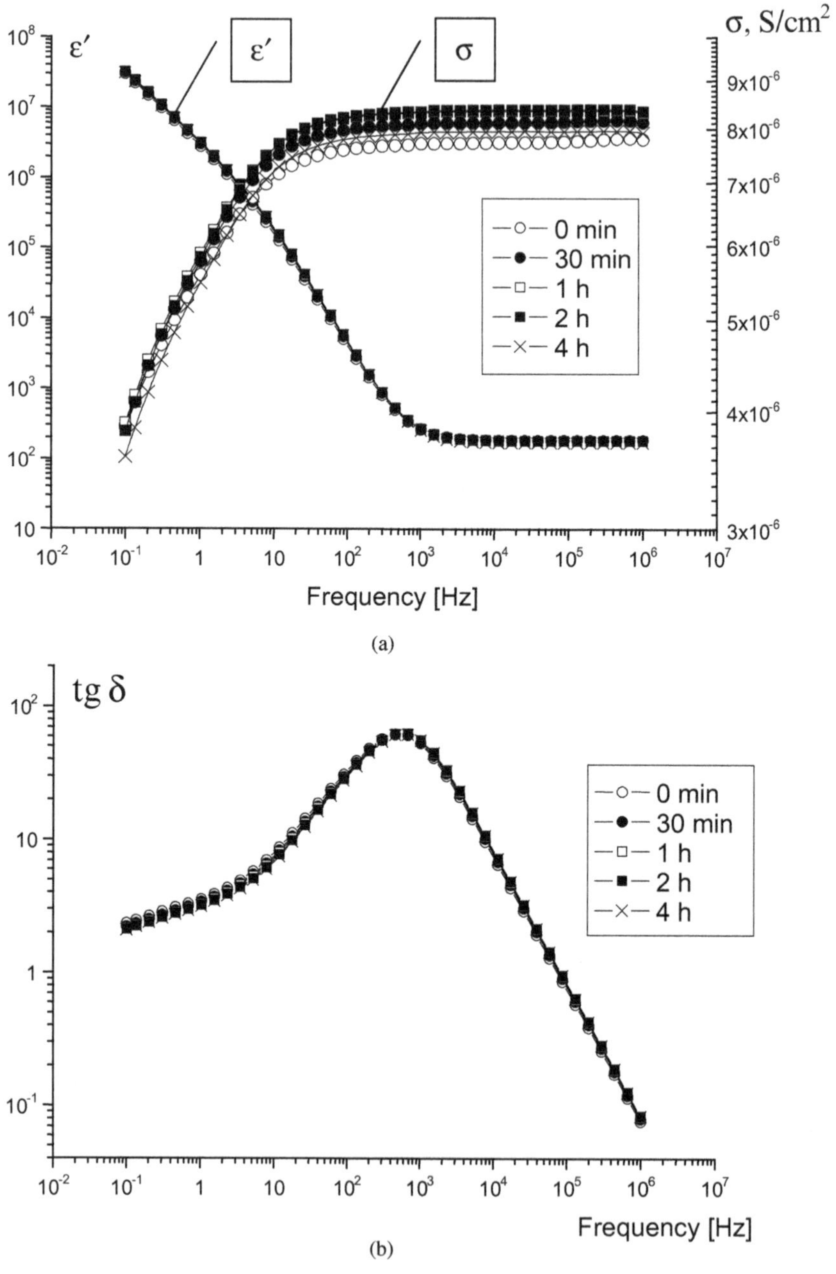

Figure 3.8. Study of water activated for 60 min and stored at a temperature of 20°C.

This result demonstrates very clearly the importance of the use of the optimum duration of the activation of water. One more manifestation of a significant dependence of the properties of water on the duration of activation will be considered below in the analysis of changes of the viscosity of such water.

One of the most important problems is the determination of the duration of "the memory of water" on the basis of the performed experiments.

In Fig. 3.9, we present the dependence of the relaxation change of $\varepsilon'(\omega, t)$ on the storage duration of MRET Activated Water after the activation. These results correspond to the frequency $\omega = 10\,\text{Hz}$ and are obtained from the direct processing of the above-discussed spectra obtained for water stored for the different time intervals at different temperatures after the activation. Water was activated for 30 min in all cases.

It is found that at a very great storage duration, the anomalous electrodynamic properties of activated water gradually disappear. The duration of preservation of these properties depends very significantly on the storage temperature.

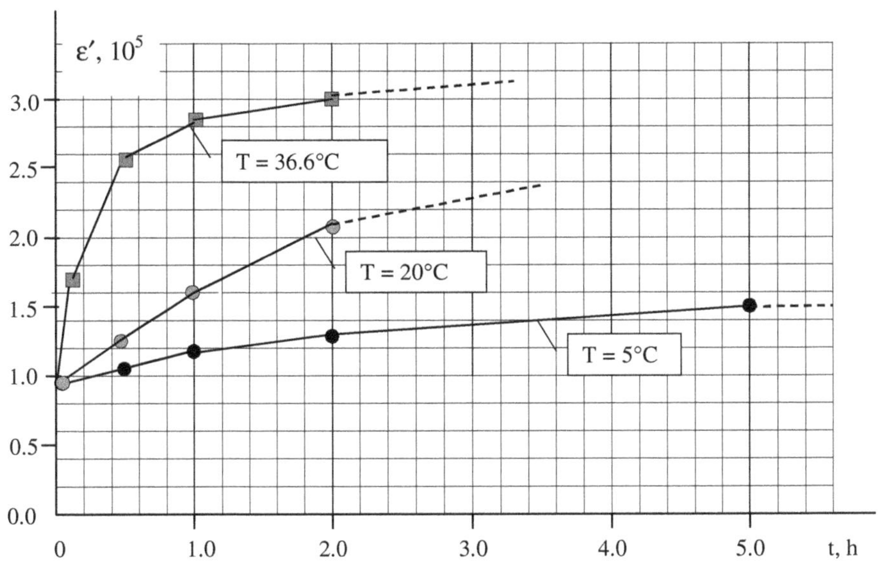

Figure 3.9. Relaxation of the dielectric permittivity of activated water $\varepsilon'(\omega)$ at the frequency $f = 10\,\text{Hz}$ ($\omega = 2\pi f \approx 62\,\text{s}^{-1}$) versus the storage duration after the activation. Three curves correspond to three temperatures of the storage of water ($T = 5°\text{C}$, $20°\text{C}$, and $36.6°\text{C}$).

By extrapolating the obtained dependences of the relaxation change of the parameters of water $\varepsilon'(\omega)$ and $\sigma(\omega)$ by the exponential law

$$\varepsilon'(\omega, t) \approx \varepsilon'(\omega, 0) + \Delta\varepsilon'(\omega)[1 - e^{-t/T_w}], \tag{3.10}$$

$$\sigma(\omega, t) \approx \sigma(\omega, 0) + \Delta\sigma(\omega)[1 - e^{-t/T_w}], \tag{3.11}$$

we can estimate the duration of relaxation T_W.

It is seen from Fig. 3.9 that the duration of relaxation of the mentioned electrodynamic characteristics of activated water becomes very great (at least 5–7 days) at a comparatively low temperature (at 5°C). With increase of the temperature of the storage, this duration rapidly decreases. The duration of relaxation is 10–15 h at a temperature of 20°C and is equal to at most 4–6 h at a temperature of 40°C close to the temperature of a human body. These results qualitatively agree with the calculations performed in Vysotskii's work (2004), whose results are presented in Chap. 1 in Tables 1.1 and 1.2.

The additional studies of ordinary and activated water were carried out in visible, ultraviolet, and infrared regions of the spectrum.

The studies in the visible and ultraviolet ranges in the interval of wavelengths $\lambda = 190$–1100 nm were performed with the use of a spectroscope Helios Alpha (USA).

The mechanism of a sharp increase of the absorption of the UV emission in the region of wavelengths shorter than 190 nm is related to several mechanisms.

The main mechanism in the region of the near-UV emission is conditioned by transitions in the electron subsystem of a H_2O molecule which lead to the dissociation of this molecule and the formation of the ground (nonexcited) state of the system H + OH. Under the action of the emission with shorter wavelengths, the following becomes essential: the processes leading the dissociation of molecules of water with the formation of excited molecules OH^* and radicals H, as well as the radiolysis products H_2 and O.

In Fig. 3.10, we present the results of measurements of the optical density in the region of near-UV emission for ordinary nonactivated water and for water activated for 30 min.

Visually, both plots in Fig. 3.10 coincide. This corresponds to that almost no difference between the spectra of initial (nonactivated) water and activated one was observed in the visible and UV regions of the absorption spectrum in the limits of accuracy of the method (0.5%), except for a small increase of the absorption, at most 1–2%, in the UV region of the spectrum at wavelengths near $\lambda = 190$–210 nm.

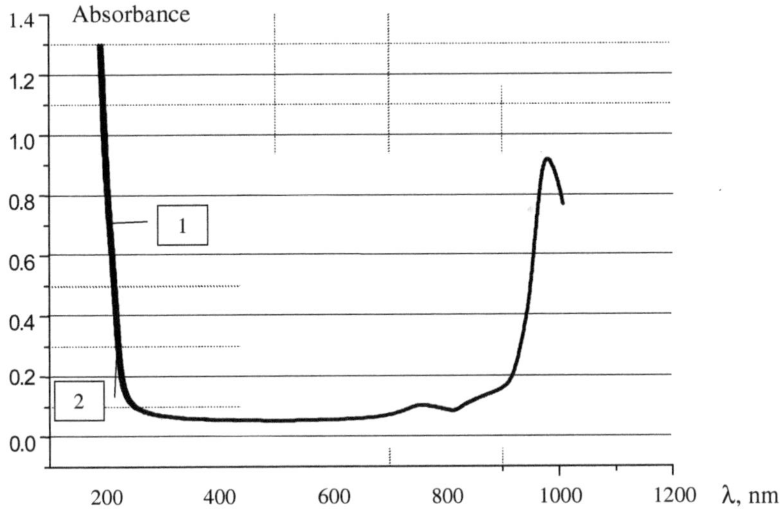

Figure 3.10. Optical density of ordinary water (1) and water activated for 30 min (2) versus the wavelength in the visible and UV ranges. Both plots visually coincide with each other.

This result differs basically from the above-presented very-essential changes in the electrodynamic characteristics of activated water in the region of low frequencies presented in Figs. 3.4–3.8. It shows that the process of activation by means of a MRET activator does not lead to a significant change of the electron subsystem of atoms and involves only structural and configurational changes of the system of molecules.

This result is also essentially different from the data presented in Chap. 1 in Figs. 1.5 and 1.6 which demonstrate a great change in the optical density in the short-wave part of the visible range and in the UV range, but rather under the action of a powerful pulse or low-power continuous SHF emission with a sufficiently great total energy of the acting field for the whole duration of irradiation (in the interval from 10 to 1000 J). This is related to the fact that the total energy of the acting field and its frequency were by many orders less in the case of the studied mechanism of the activation of water on the basis of the MRET technology.

The same fraction of activated water (the duration of activation was equal to 30 min) was studied by the method of infrared Fourier-spectroscopy with the help of a device "Nexus" produced by the firm "Thermo Nicolet". The studies were performed in the range of wavenumbers from 50 to 4000 cm^{-1}.

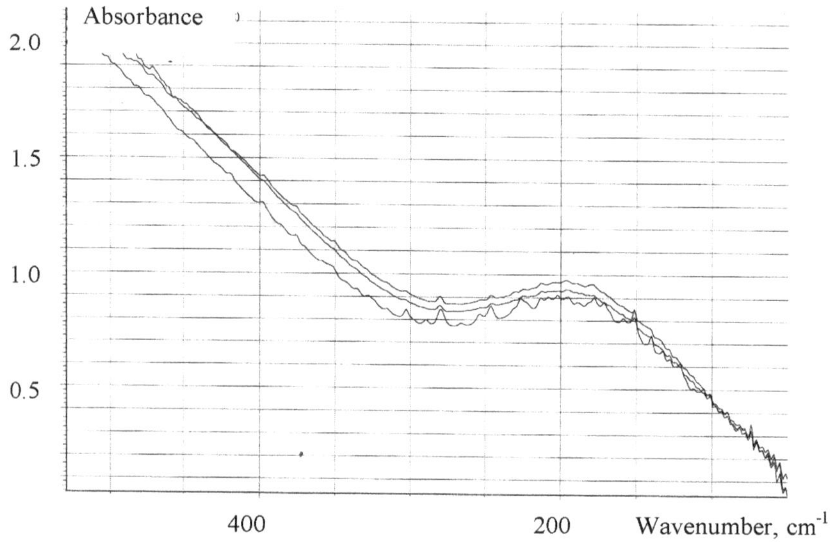

Figure 3.11. IR-spectra of nonactivated water (upper curve) and activated water which was studied in 5 h after the activation (middle curve) and in 1 h (lower curve).

The results of studies in various sections of the IR-range are presented in Figs. 3.11–3.12.

In Fig. 3.11, we give the IR-spectra in the range 50–550 cm^{-1} for the initial water and the water activated for 30 min.

The studies were carried out with three samples of water derived from the identical distilled water:

- initial distilled water;
- activated water stored after the activation prior to the measurement for 30 min at room temperature; and
- analogous activated water which was stored after the activation prior to the measurement for 5 h at room temperature.

Such studies allow us to determine the dependence of the influence of the activation on the IR-spectrum of water, as well as the influence of the storage duration of this water on its properties.

It is seen from this figure that a decrease of the absorption of water (a decrease of the imaginary part of the dielectric permittivity) in the long-wave part of the IR-range occurs in the process of activation. This decrease

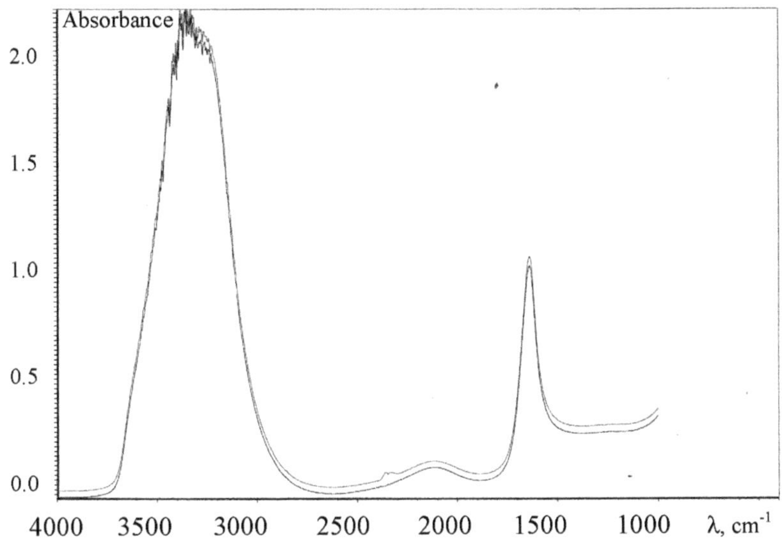

Figure 3.12. IR-spectrum of nonactivated water (upper curve) and water activated for 30 min (lower curve). The spectrum was studied in 1 h after the activation.

exceeds 10% for the range of wavenumbers less than 300 cm^{-1}. With increase of the wavenumber, the absolute value of a change in the coefficient of absorption remains approximately constant, and the relative change decreases, respectively.

With increase of the duration of storage of activated water, the discovered change in the optical density in the IR-range gradually decreases.

The effect of a small decrease in the optical density of activated water is also observed in the region of great wavenumbers 1000–4000 cm^{-1}. This result is presented in Fig. 3.12 for nonactivated water and water activated for 30 min (the spectrum of the latter was studied in 1 h after the activation).

Activated water was also studied by Raman scattering spectroscopy. The Raman scattering spectra for the initial (nonactivated) water and those for activated water were studied in 24 h after the completion of the activation and are presented in Fig. 3.13. The studies were performed with the same fraction of water (activation for 30 min). The spectra determine the Raman scattering characteristics in the region from 1 to 4000 cm^{-1}.

It is seen from the data presented in Fig. 3.13 that the activation of water leads to a small increase of the Raman scattering amplitude in the whole range from 1 to 4000 cm^{-1}.

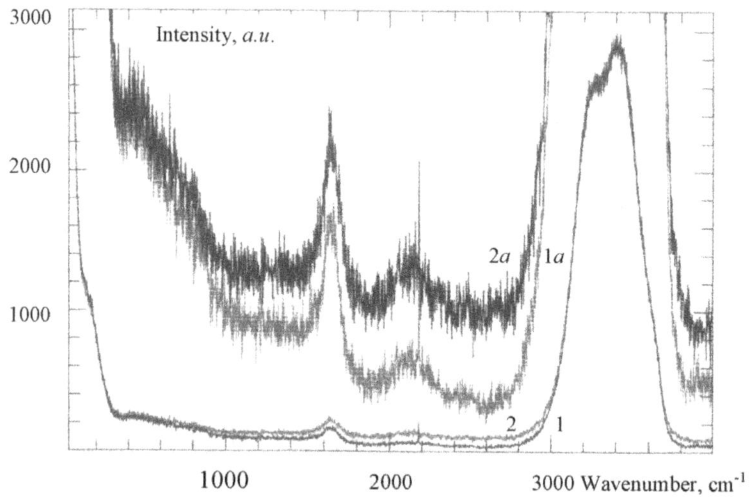

Figure 3.13. Raman scattering spectra of nonactivated (1) and activated (2) water. Fragments of the spectra (1*a*) and (2*a*) correspond to the increase in the scale by 10 times. All spectra were measured in 24 h after the activation.

The totality of the obtained experimental data imply that the specific features of the spectrum of activated water in IR, visible, and UV ranges preserve for a long time after the completion of the activation, though the very changes in the spectrum in these ranges turned out small.

Such effects of changes in the optical properties of activated water which are small in magnitude, but with the long time of their preservation can be satisfactorily substantiated, if we consider that they are related to a change in the properties of a relatively small number of separate molecules of water isolated from the external action and corresponding to the hindered relaxation. Based on the above-considered model of the memory of water, we can refer such changes to the molecules of water being in the volume of clathrate microcavities.

It was indicated earlier that there exists a strong repulsive electrostatic field in such microcavities. For this reason, the internal "walls" of microcavities possess the hydrophobic properties and do not form hydrogen bonds with molecules of water placed in the volume of microcavities. These molecules, due to the absence of a hydrogen bond with molecules of water forming the "walls" of microcavities, have a changed configuration of the electron shell (it corresponds to a free H_2O molecule, rather than that of a bound molecule) and, respectively, somewhat different optical properties.

The process of activation renders a much stronger influence on the dynamics of the mechanical fluctuation motion of stable global structural elements in the water bulk. The formation and the long-term existence of such structural elements depend on the activation of water and confirm the possibility for the memory of water to exist.

With the purpose to discover the presence of such structural elements in the water bulk, we carried out additional study of the optical characteristics of activated water with the use of a laser correlation analyzer AKVA-01. The principle of action of this device is based on the analysis of the characteristics of a mutual temporal coherence function of the laser emission scattered in the volume of the studied water in the direction perpendicular to the initial laser beam. In the study, we used the emission with a wavelength of $\lambda = 0.63\,\mu$m generated by a He–Ne laser. The longitudinal coherence length of such an emission prior to its scattering is very large and, in any case, exceeds hundreds of meters. The width of the spectrum of this emission is very small.

The parameters of the scattered emission depend on the motion of scattering molecules of water and the ensembles of these molecules. Due to the presence of the fluctuation (diffusive) mechanical motion of the scattering objects, the phase of the scattered emission continuously fluctuates, which decreases very sharply the coherence length of the scattered laser field. A change in the coherence length of the emission and the associated change in the spectrum of the emission are those parameters of which the measurement allows us to determine the characteristics of the motion of scattering objects (the coefficient of diffusion) and, hence, the mass and size of these objects. The coherence length can be found with the help of the autocorrelation function.

We now consider briefly the quantitative aspect of the above-discussed processes and substantiate the possibility of their experimental realization.

Consider the characteristics of the emission scattered by the studied sample of water. Let the size of the scattering region be much less than the distance from it to the place of the registration of the scattered light.

If a coherent laser emission falls on a part of the volume of water containing N scattering objects, then the amplitude of the scattered electric field intensity of such an emission is characterized by the quantity

$$E(\vec{k}, \theta, R, t) = E_0 \sum_{i=1}^{N} \{f_i(\vec{k}, \theta)/R\} e^{-i\{\omega_0 t - \vec{k}\vec{r}_i(t)\}}. \qquad (3.12)$$

Here, E_0 is the amplitude of an incident wave, $f_i(\vec{k}, \theta)$ is the scattering amplitude of the emission with the wave vector $\vec{k}$ scattered by an angle θ, R is the distance from the region center in the water bulk, where the objects scattering the light are positioned, to detectors which registered the scattered light.

The electric field intensity of the scattered laser emission [Eq. (3.12)] depends on the totality of coordinates $\vec{r}_i$ characterizing the position of various scattering objects in the studied volume of water. Since these coordinates are random variables depending on time, the very amplitude of the field [Eq. (3.12)] is also a random variable. In order to determine the regularities of the process of scattering, it is necessary to determine those quantities which are not random and describe unambiguously the scattering.

The deterministic (nonrandom) characteristics of the scattered emission can be found if the autocorrelation function of field is known:

$$K_E(\vec{k}, \theta, R, \tau) = \langle\{E(\vec{k}, \theta, R, t+\tau) - \langle E(\vec{k}, \theta, R, t+\tau)\rangle\}\{E^*(\vec{k}, \theta, R, t) - \langle E^*(\vec{k}, \theta, R, t)\rangle\}\rangle = (NE_0^2/R^2)\langle|f(\vec{k}, \theta)|^2\rangle|e^{-(i\omega_0+k^2D)\tau}. \quad (3.13)$$

Here, D is the diffusion coefficient of scattering objects which is determined from the equation

$$\langle|\vec{r}(t+\tau) - \vec{r}(t)|^2\rangle = D\tau. \quad (3.14)$$

The spectrum of the scattered emission is calculated with the help of the Wiener–Khinchin formula

$$S_E(\vec{k}, \theta, R, \omega) = \frac{1}{2\pi}\int_{-\infty}^{\infty} K_E(\vec{k}, \theta, R, \tau)e^{-i\omega\tau}d\tau = \langle|f(\vec{k}, \theta)|^2\rangle\frac{NE_0^2}{\pi R^2}\frac{k^2D}{(\omega-\omega_0)^2+(k^2D)^2}. \quad (3.15)$$

It is seen from formula (3.15) that the spectrum of the scattered emission depends on the coefficient of diffusion D defining the mean characteristics of the motion of scattering objects. This coefficient, in turn, depends on the size and mass of these objects. Thus, the study of the correlation spectrum [Eq. (3.15)] allows us to determine the mean characteristics of the motion of scattering objects in the water bulk and, on this basis, to perform the qualitative analysis of the spatial structure of water. In addition, such studies allow one to determine the dependence of these characteristics on the activation of water.

The experiments were carried out on the basis of four different samples of the initial water:

1. Natural Mineral Water "AVIAN" (produced in France).
2. Fresh Water Life "JINBO/Seok Su" (produced in the Republic of Korea).
3. Natural Mineral Water "HAITAI/Gangwondo/Pyeong Chang" (produced in the Republic of Korea).
4. Natural Mineral Water "JEJU/Sa Da Soo" (produced in Iceland).

As a quantitative characteristic, we used the integral index of the molecular dynamics of water W which is inversely proportional to the coefficient of diffusion D of the scattering objects, includes the normalizing factor (in order to make W dimensionless), and is directly related to the parameters of the autocorrelation function (3.13). An increase of the index W corresponds to a decrease of the coefficient of diffusion and characterizes uniquely the increase of the mass and size of a stiffly bound molecular complex which is present in the water bulk and scatters the laser emission.

We applied the following procedure and the sequence of studies, whose total duration was equal to 71 h.

Firstly, we carried out three measurements of the index W for 3 h for each of the studied types of water. The purpose of these measurements was related to the determination of the stability of a measuring unit. The results of this "testing period" correspond to the first three points on all the plots presented in Fig. 3.14.

The very small variance of the index W confirms a high degree of stability, a small value of the apparatus error, and the error related to the procedure of measurements. It is obvious that the differences of the indexes of molecular dynamics W for different types of water prior to the activation are related to a great extent to the differences of their salt composition.

This is conditioned by the fact that the ions dissolved in water influence the mass and size of molecular complexes in water and, hence, the coefficient of diffusion of these complexes and the spectrum of the scattered laser emission. We may assume that such an influence can be realized in at least two ways. On the one hand, the dissolved ions break the chemical homogeneity of water and therefore affect the formation of molecular complexes formed from molecules of water. On the other hand, the union of dissolved ions with these complexes changes the density and the mass of the latter. It is obvious that both mechanisms lead to the essential influence of dissolved

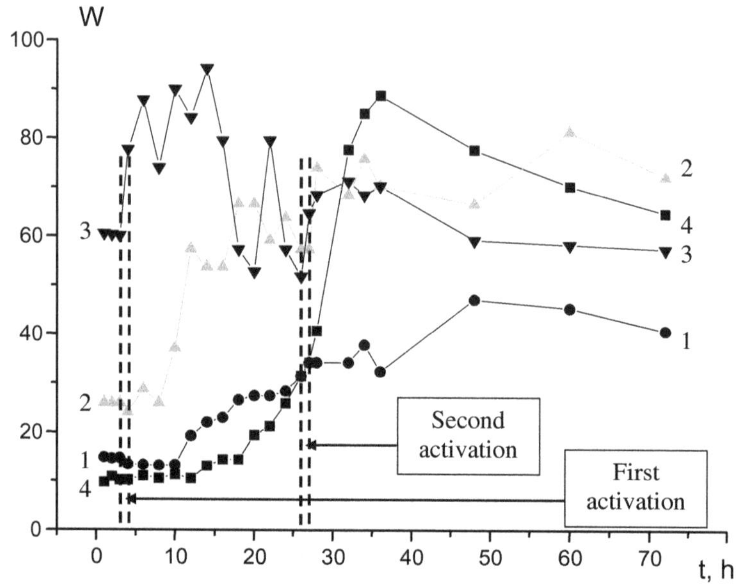

Figure 3.14. Integral index *W* of the molecular dynamics of activated water of different types versus time. The numbers stand for the types of water given in the text.

salts on the spectrum of the scattered emission, which was observed in the study of various samples of water.

We then performed the activation of water of each type for one hour. Immediately after the completion of the activation, we carried out the next study. From Fig. 3.14, it is seen that the index *W* was changed comparatively slightly (except for water 3) for the time of this first activation.

Thereafter for 24 h (with an interval of two hours), we measured the index *W* for all types of water. It is seen that, for this time interval, there occurred a systematic increase of the index *W* for all types of water except for water 3. In this case, a change in *W* was accompanied by very strong oscillations for the water samples 2 and 3. At the same time, the index *W* for all the remaining types of water was changed as a monotonously increasing function of time. In 24 h after the first activation, we performed the second activation of water with the same duration of one hour. The results of this activation also turned out ambiguous. After the activation, we observed a very sharp increase of the index *W* for water sample 4. At the same time, this index was practically constant for the remaining samples of water. The measurements after the second activation were performed for 42 h (firstly, we carried out three measurements with an interval of two hour and then the other three measurements with an interval of 12 h).

After the completion of the whole cycle of measurements, the values of the index W for all samples of water (except for water specimen 3) turned out on a level somewhat exceeding the initial value W_0 (prior to the first activation). The characteristics of water specimen 3 after the completion of all actions and all measurements turned out to be equal to the initial values of the index W_0 which were determined prior to the beginning of the cycle of measurements and were anomalously great.

Based on the results of the analysis of the performed cycle of measurements, we can made some conclusions.

First of all, it should be noted that, upon the activation of water, there occur both a very significant increase of the degree of correlation of the scattered emission and a decrease of the coefficient of diffusion of scattering complexes in water, which testifies unambiguously to the formation of very large and stable clusters in water after the activation.

The second important conclusion consists in that the processes of formation and change of these clusters do not cease after the termination of the activation but continue for many hours and days. Such an effect can occur in the case where the activation of water induces such a change of the properties of separate water molecules and strongly-bound molecular groups, at which their further joining in great clusters becomes energy-gained.

The third conclusion is related to the unambiguous confirmation of the assumption about the presence of the stable water memory, the duration of existence of which can be equal to many hours and days, by the correlation analysis of the scattered laser emission. This memory can be interpreted as the existence of stable supermolecular clusters in the volume of water.

One more conclusion consists in that the process of formation of stable clusters in the volume of water depends significantly on the salt composition of water. In this case, different types of mineral water are characterized by different dependences of the parameter W on the time interval after the activation.

3.3. Procedure and Results of the Measurement of the Viscosity of Activated Water

Viscosity is one of the most important characteristics of water. The particular meaning of the viscosity of water in any biological system is related to the transport functions of water and to the dependence of the motion and the evolution of macromolecules, cells, viruses, and other microobjects on the properties of water, in which they are placed in living organisms.

As will be shown in the following chapters, a sharp change in the viscosity of activated water can be one of the basic reasons for the anomalous influence of activated water on biological systems.

In our experiments, the viscosity of water was measured with the help of a rheometer RS 150 L of the firm Haàke. The studies were executed on the basis of a measuring cell. It consisted of two immovable coaxial cylinders 1 and 2. In the region between them, there is a rotor (a rotating coaxial cylinder 3 coaxial with cylinders 1 and 2) (Fig. 3.15). Water under study is placed in the free region between the surfaces of immovable cylinders 1 and 2 and the rotor. With the help of this system of coaxial cylinders, we performed the measurements at low shear stresses.

The essence of the process of measurement is as follows. In the process of measurements, the moment of forces M is applied to the rotor which should turn. The moment of forces M is proportional to the tangential shear stress τ

$$\tau = aM, \tag{3.16}$$

and the coefficient of proportionality a depends only on the surface area and the form of a measuring cell.

We will start from the obvious fact that the relative strain of a sample γ (the strain of the layer of water between the cylinders) determines directly

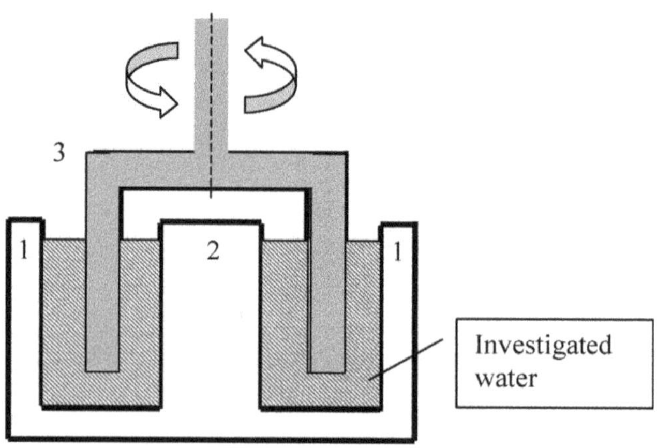

Figure 3.15. Measuring system of coaxial cylinders.

a rotation angle of the rotor

$$\varphi = b\gamma, \tag{3.17}$$

to which a small tangential stress is applied. Here, b is the coefficient, also depending on the area and the form of a measuring cell.

In the region of small stresses, the viscosity coefficient η connects the rate of the relative shear strain of a sample $d\gamma/dt$ and a value of the viscous tangential stress τ with the help of the relation

$$\tau = \eta\frac{d\gamma}{dt} = \frac{\eta}{b}\frac{d\varphi}{dt}. \tag{3.18}$$

The coefficient of shear viscosity is numerically equal to the mechanical momentum transferred in unit time across unit area.

In view of the above-given formula connecting τ and M, we obtained the following final expression for the viscosity coefficient:

$$\eta = abM\left(\frac{d\varphi}{dt}\right)^{-1}. \tag{3.19}$$

This formula implies that the viscosity coefficient for a specific measuring cell depends on the moment of forces applied to the rotor and on the instantaneous angular velocity of the rotor, $d\varphi/dt$, induced by the action of this moment of forces.

This device allows us to measure a deviation of the rotor by a certain angle φ and the time, for which this deviation occurs. On the basis of these values and the values of the applied moment of forces, we can determine the rheological characteristics of solutions.

The measurements were performed in the mode of a piecewise constant shear stress.

In this case, the temporal variation of the shear stress τ applied to a sample is shown in Fig. 3.16.

The action of this shear stress leads to a turn of the rotor. By measuring the angular velocity of the rotation of the rotor at a given value of the applied stress, we can determine the viscosity η by assuming the constancy of a strain over the sample volume.

The characteristic temporal dependence of the viscosity coefficient of a studied liquid sample corresponding to the applied piecewise constant stress is presented in Fig. 3.17.

To study the influence of the activation of water on its viscosity, we executed several series of measurements of the dynamical viscosity η as a

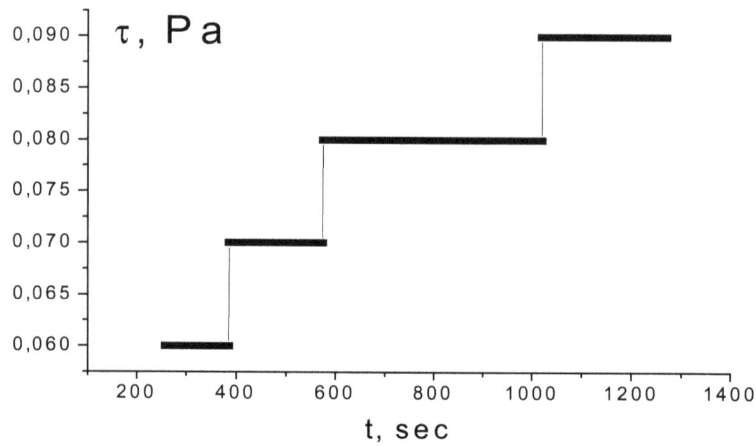

Figure 3.16. Applied shear stress τ versus time.

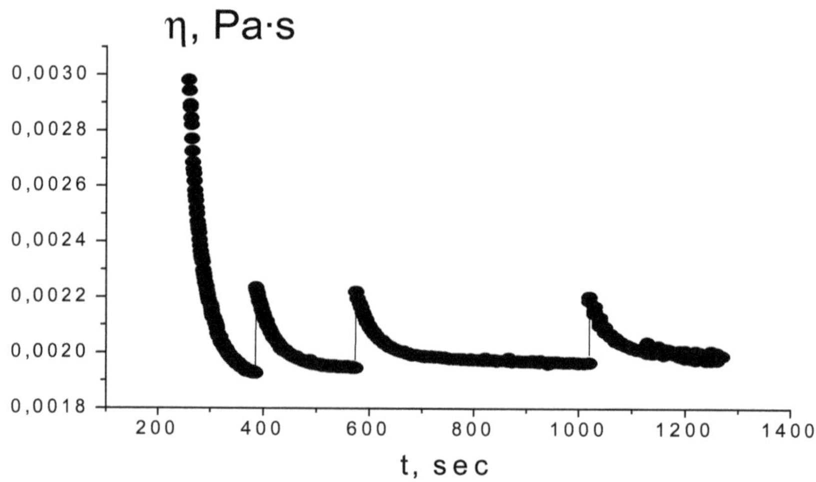

Figure 3.17. Characteristic curve of the viscosity η of a studied water sample under the action of a piecewise constant stress shown in Fig. 3.16 versus time.

function of the applied stress τ with different fractions of water (the initial nonactivated distillated water and an analogous water activated for different time intervals) at two temperatures (20°C and 36.6°C). The resulting dependences of the viscosity coefficient on the shear stress at a temperature of

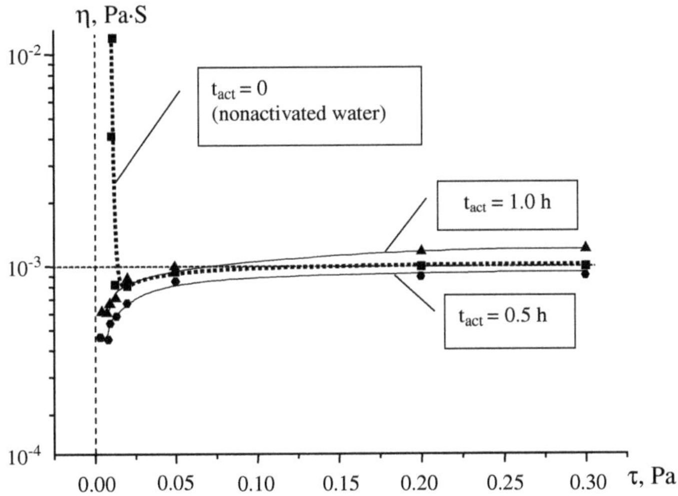

Figure 3.18. Viscosity η of "ordinary" and activated water in the region of small values of the shear stress τ at a temperature of 20°C. The averaged results of two series of measurements for each fraction of water are presented.

20°C for three fractions of water (nonactivated and processed for 30 min and 60 min) are shown in Figs. 3.18 and 3.19 in more details (on a linear scale).

It is seen from the analysis of Fig. 3.18 that the viscosity of all fractions of water at a temperature of 20°C is close to the "standard" value $\eta_0 \approx 10^{-3}$ Pa·s at the comparatively large values of shear stresses τ applied to the surface of the cylinder contacting with the studied water (at $\tau > 0.02$ Pa). With decrease in τ, the viscosity coefficient of "ordinary" water decreases slightly (from the initial η_0 at $\tau > 0.3$ Pa to $\eta \approx 0.7 \times 10^{-3}$Pa·s at $\tau \approx 0.015$ Pa). This decrease occurs by the same law for all fractions of water. In this interval of τ, the difference between the coefficients of viscosity in activated and ordinary water samples is not very large (though the viscosity of ordinary water turned out somewhat greater than that of water activated for 30 min, but less than that of water activated for 60 min).

The most significant difference took place at a very small shear stress, $\tau < 0.015$ Pa. In this region after the attainment of the minimum at $\tau \approx 0.015$ Pa, the viscosity of ordinary (nonactivated) water begins to sharply increase with decrease of the shear stress. From Figs. 3.18 and 3.19, it is seen that the viscosity increases by more than 100 times from the minimum value $\eta \approx 0.7 \times 10^{-3}$ Pa·s under the shear stress $\tau \approx 0.015$ Pa to

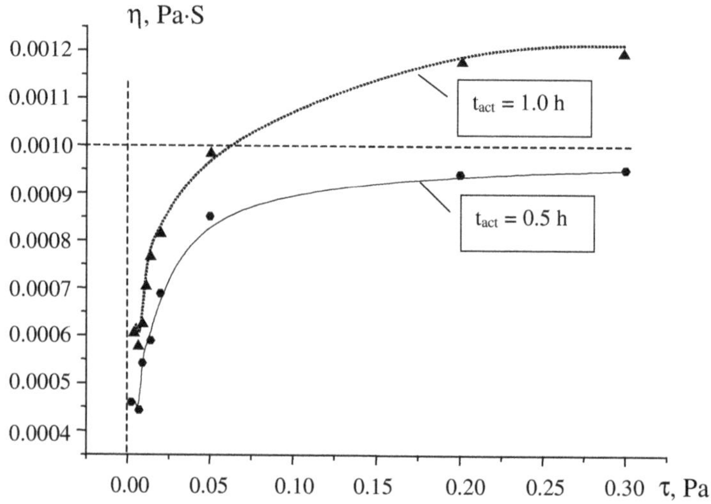

Figure 3.19. Viscosity of activated water η in the region of small shear stresses τ at a temperature of 20°C.

$\eta \approx 0.1$ Pa·s at $\tau \approx 0.004$ Pa. Such a tendency is logically sufficiently justified and is related to the fact that a very small value of the shear stress (or, respectively, a small pressure) corresponds to a very small velocity of the relative motion.

Under such a condition, the molecular bonds between water molecules and the surface which is moving relatively to water turn out to be maximally strong, and the viscosity coefficient and the force of viscous friction, respectively, are maximum.

It is necessary to note that this effect (the presence of a local minimum of the viscosity coefficient of water under a small shear stress) is well known. It is observed if the temperature of water is lower than 25°C. The general regularities of a change of the viscosity of activated water qualitatively correspond to those of ordinary water, but not for the quantitative characteristics. It is seen from Fig. 3.19 that, with decrease in the shear stress, the viscosity of activated water continues to drop from $\eta_{60} \approx 10^{-3}$ Pa·s and $\eta_{30} \approx 0.9 \times 10^{-3}$ Pa·s at $\tau \approx 0.1$ Pa (for water with the duration of activation of 60 min and 30 min) to $\eta_{60} \approx 6.5 \times 10^{-4}$ Pa·s and $\eta_{30} \approx 5.5 \times 10^{-4}$ Pa·s at $\tau \approx 0.01$ Pa. For water activated for 30 min, the minimum viscosity $\eta_{30(\mathrm{min})} \approx 4.3 \times 10^{-4}$ Pa·s is reached at $\tau_{30(\mathrm{opt})} \approx 0.0045$ Pa. It is seen that the viscosity of water in the region of very small shear stresses

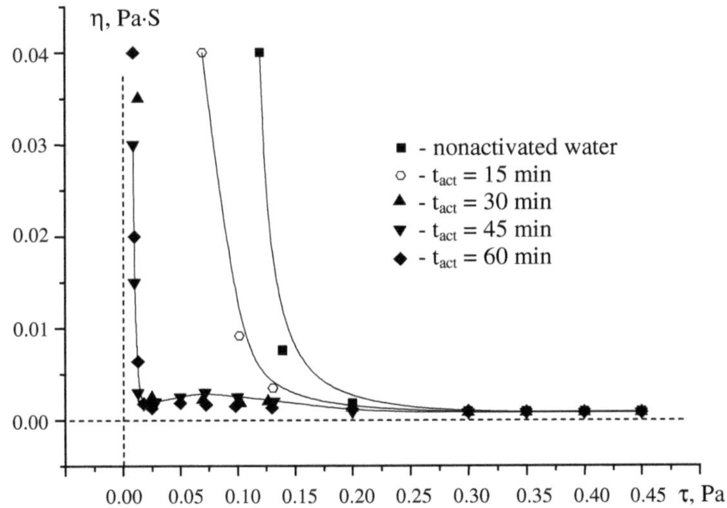

Figure 3.20. Viscosity η of nonactivated water and four fractions of activated water in the region of small shear stresses τ at a temperature of 36.6°C.

(at $\tau_{\text{opt}} \approx 0.004$–$0.005$ Pa) decreases upon the activation (as compared to that of nonactivated water) by 200–250 times. In this case, water activated for 30 min has the least viscosity.

Of a great interest is the dependence of the viscosity coefficient on the applied mechanical stress for water being at a temperature of 36.6°C. In this case, we carried out the more detailed studies with regard for the influence of the water activation duration.

In Fig. 3.20, we present the results of studies of the viscosity coefficient for initial distillated nonactivated water and for the samples of water which were obtained for the duration of activation of 15 min, 30 min, 45 min, and 60 min and were at a temperature of 36.6°C in the process of measurement. With regard for the fact that the anomalous properties of activated water relax rapidly with increase of the temperature, the measurements were carried out immediately after the activation of water.

It is seen from the obtained data that an increase of the water activation duration leads to a very significant decrease of the viscosity coefficient in the region of small mechanical stresses at this temperature (relative to the initial nonactivated water).

It is necessary to note that the absolute values of the viscosities of non-activated and activated water samples in the region of small mechanical stresses increase significantly with the temperature. For example, for the viscosity coefficient of nonactivated water equal to its minimum value $\eta \approx 0.7 \times 10^{-3}$ Pa·s at 20°C for $\tau \approx 0.015$ Pa, we determined at a temperature of 36.6°C for the same $\tau \approx 0.015$ Pa that $\eta \approx 5 \times 10^{-3}$ Pa·s, which corresponds to the increase by 7 times. The close changes of the viscosity with increase of the temperature are also characteristic of activated water. This is clearly seen from the comparison of Figs. 3.19 and 3.20.

It is seen from Fig. 3.20 that a change (increase) of the duration of activation in the limits of 1 h leads to a monotonous displacement of the curve of the viscosity coefficient as a function of the applied mechanical stress to the side of less viscosities. The position of the minimum of the viscosity coefficient is also shifted with increase of the duration of activation from $\tau \approx 0.1$ Pa for the duration of activation of 30 min to $\tau \approx 0.05$–0.02 Pa for the duration of activation of 45–60 min.

Very important is the circumstance that the viscosity of activated water in the region of very small shear stresses at the vitally important temperature 36.6°C, e.g. 20°C, is by several orders less than that of ordinary (nonactivated) water.

The measurements of the viscosity of activated water at much less values of the shear stress on the basis of the used procedure is associated with very great errors. It is obvious that, in any case, the viscosity of activated water for very small shear stresses turns out to be much less than the viscosity of nonactivated water.

The revealed exceptionally sharp decrease of the viscosity of activated water under a small shear stress leads naturally to the similar sharp increase of the fluidity of water with decrease in the shear stress to $\tau \approx \tau_{\text{opt}}$. It is obvious that this is related to a change of the state of water and to a change of the character of the interaction of such water with the surface adjacent to water. Such "superfluid" water has extremely small friction with walls and can penetrate through very small pores, capillaries, and channels. In such water, the process of division of cells is changed. With regard for the fact that the small pressure of a liquid is a characteristic sign of living systems, such "superfluidity" of activated water allows us to explain many effects of the anomalous influence rendered by activated water on biological objects. Such effects will be considered comprehensively in Chaps. 4–6, where the results of studies of the influence of water on plants, microorganisms, and animals are given. The possible mechanisms of the direct action of activated water on the development of these living objects will be analyzed in Chap. 7, while analyzing the specific features of the biological action of activated water. We will show that a change in the viscosity of activated water plays one of the decisive roles in these processes.

3.4. Influence of the Activation of Water on Hydrogen Index pH

It is well known that a value of the hydrogen index pH of the intracellular and intercellular liquid water–salt medium is one of the essential factors of the influence on the course of biochemical processes and the development of living organisms. With the purpose to study the influence of the activation of water on pH, we studied the dependence of pH on the duration of activation of water (water was activated for 15 min, 30 min, 45 min, and 60 min), on the temperature (at 4°C and 20°C), and the duration of storage of this water after the completion of the activation.

In experiments, we use water obtained from distillation directly prior to the studies. Then, the 100-ml samples of activated water were stored in standard graduated flasks of 250 ml closed by rubber plugs. For the first 3.5 h after the activation, all samples including the control ones were placed at room temperature (20°C). Subsequently, according to the conditions of the experiment, half of flasks were placed into a cooler and stored there at 4°C, and the other half remained at 20°C. The control samples with analogous but nonactivated water were present in both versions.

The measurements of pH were carried out with the help of a universal ionometer EV-74. In order to enhance the accuracy and reliability, all measurements were carried out on three identical samples of water which corresponded to the given duration of activation. To this end, water was distributed immediately; after the total activation in three different flasks which were stored and studied under the same conditions. Thus, each measurement of water activated for a definite time interval was repeated three times. Then, we determined the mean value. The final error of measurements in each series was at most $\Delta(\mathrm{pH}) \approx 0.1$.

The detailed studies of all the samples of water obtained for the different durations of activation were performed with a short time interval (in each 30 min) for the first 3.5 h after the completion of its activation at water temperature of 20°C. The subsequent measurements were performed regularly for 16 days with an interval of one to two days for different fractions of water stored at temperatures of 4°C and 20°C. The samples of water taken from a cooler, where they were stored at a temperature of 4°C, were placed prior to the measurement in a thermostat, where their temperature rapidly increased to 20°C. Then we measured the hydrogen index.

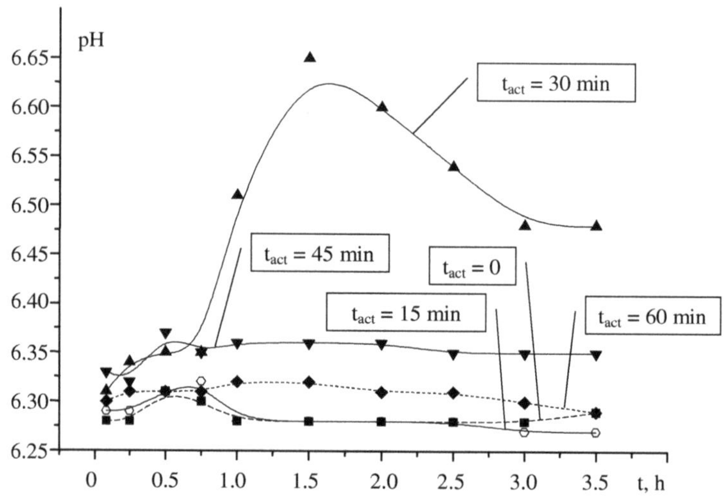

Figure 3.21. Hydrogen index pH of different fractions of activated water versus the duration of its storage at a temperature of 20°C in the first few hours after the activation.

In addition, the separate measurements were carried out by using the samples of nonactivated water and activated water stored after the activation at a temperature of 4°C for 120 days.

The results of measurements are presented in Figs. 3.21 and 3.22.

It was established that a quite insignificant alkalization of water occurred during the direct action of a MRET activator on water. In particular, when the hydrogen index for the initial distillated nonactivated water $(\mathrm{pH})_0 = 6.28$, the value of pH immediately after the completion of the activation for 15 min, 30 min, 45 min, and 60 min was equal to $(\mathrm{pH})_{15} = 6.29$, $(\mathrm{pH})_{30} = 6.31$, $(\mathrm{pH})_{45} = 6.33$, and $(\mathrm{pH})_{60} = 6.30$. Further studies showed that the process of alkalization of activated water became gradually more intense for the first 1.5–2 h of its storage after the activation. These data are presented in Fig. 3.21.

The effect of the increase in pH is especially clearly manifested for water activated for 30 min. In this case, the value of pH increased from pH = 6.31 to pH = 6.65 during the first 1.5 h after the completion of the activation and then decreased gradually for 2 h down to pH = 6.45. A small increase of pH was also observed for water activated for 45 min. For the other fractions of activated water, the changes for the first few hours of the storage were insignificant.

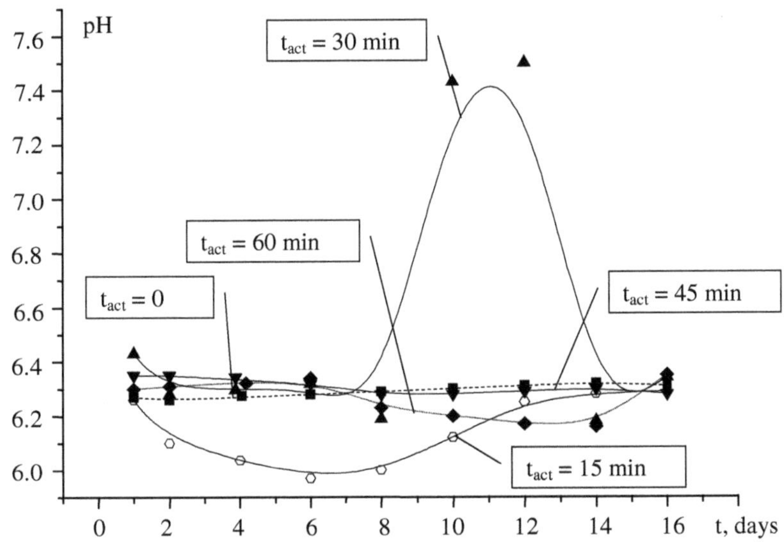

Figure 3.22. Influence of the water activation duration on pH depending on the time of its storage at a temperature of 4°C (directly before the measurements, the water samples were heated to 20°C).

The study of the samples of activated water and control at a longer time interval of the storage at two temperatures (4°C and 20°C) showed that the anomaly in the behavior of pH observed directly after the activation of water during 30 min was not unitary. In the process of storage, we registered additional spontaneous increase of pH to a value exceeding the first maximum. The greatest changes in pH were discovered in the same fraction of water in 9–11 days of the storage at various temperatures. In particular, in 9 days of the storage, the value of pH grew to pH = 7.4–7.5 [as compared with pH = 6.30–6.36 in control samples of nonactivated water which were stored at the same temperatures 4°C (Fig. 3.22) and 20°C (Fig. 3.23)]. A half-width of this maximum (the duration of its existence) corresponded to $\Delta t \approx 4$ days at a water storage temperature of 4°C and $\Delta t \approx 6$ days at 20°C. Thereafter, a spontaneous decrease of pH occurred which returned it to its initial value, pH $\approx$ 6.3–6.4, only in 14 days. This maximum was observed on all samples of water stored and measured separately which corresponded to the duration of activation of 30 min and the storage temperatures of 4°C and 20°C. There are weighty reasons to suppose that similar oscillations can also exist for a greater duration of the storage of water.

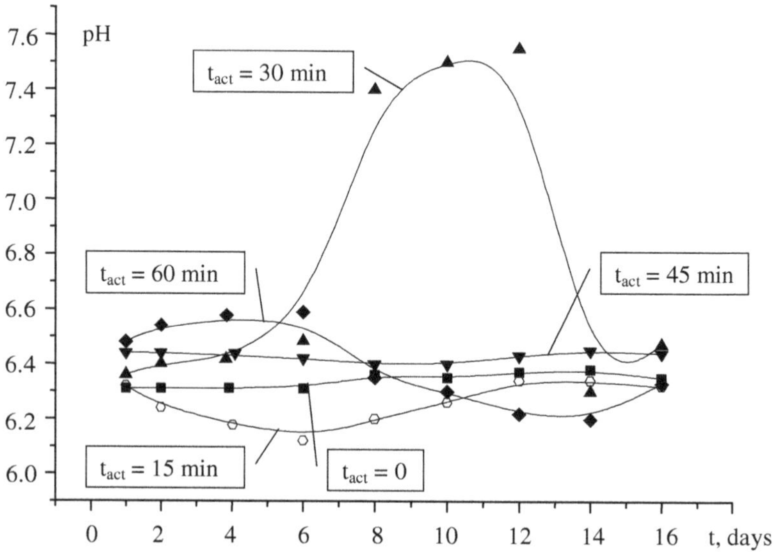

Figure 3.23. Influence of the water activation duration on pH depending on the time of its storage at a temperature of 20°C.

Among other anomalous phenomena, we note the effect of a decrease of pH to values 5.91–6.12 after two days of the storage of water treated for 15 min and stored at 4°C. This deviation to the side of an increased acidity disappeared (pH was restored to the control level) only in 14 days. An analogous deviation (though it was somewhat less in scale) was observed for water stored at a temperature of 20°C. These data are presented in Figs. 3.22 and 3.23.

We note that the fluctuations of pH for control (nonactivated) water stored under the same conditions were practically invariable (with regard for the errors of measurements).

It is worth noting the essential fact that the displacement of pH to the alkaline side was fixed also at a long-term (for 120 days) storage of the samples of activated and control water. All these samples were obtained at the same time on the basis of identical distillated water.

In particular, the value of pH for the samples of water activated for 30 min and 60 min and stored thereafter for 120 days in a cooler at a temperature of 4°C was equal to 6.48–6.52, whereas the value of pH for control samples stored for the same time interval both at 4°C and 20°C was identical and equal to 6.37.

In conclusion, we note that the presence of great ordered spontaneous changes in pH for the great durations of storage of water activated in a certain manner testifies directly to the existence of the "memory" of water. The presence of such great ordered changes in the properties of activated water is related, most likely, to certain phase transitions. Very curious is the fact that the time moment of the realization of these phase transitions, as well as the amplitude of changes in pH, do not depend (or weakly depend) on the temperature of the storage of activated water and is approximately equal to 10–11 days. At the same time, the half-width of the region of this spontaneous change depends rather sharply on the temperature.

CHAPTER 4
Influence of MRET Activated Water on the Growth of Higher Plants

4.1. General Principles and Methods of the Study of the Influence of Activated Water on Plants

As was shown in Chapter 3, activated water has unique physicomolecular properties.

Upon activation, important dynamical characteristics such as the dielectric permittivity and conductivity of water are changed very significantly, which is of great importance for the processes of ion transport. The same properties have determining significance for the interaction of charged components of biological molecules (including molecules of DNA and RNA).

Sharp changes of both the optical density and the refractive index of activated water in the UF-range can essentially vary the specific features of the interaction of UF-radiation with the surface of any biological object.

Activated water (with optimum duration of activation) is characterized by an essential decrease of the viscosity, which can result in the distinctive "super fluidity" of such water at small motion velocities of water and small mechanical loads. It is obvious that such features influence the metabolism and other processes. It will be shown below that these features can influence importantly the process of cellular division and, finally, the growth and development of biological systems.

Upon activation of water, there occurs a notable change of the hydrogen index pH, which can change many characteristics and, in particular, influence both the system of recognition and the immune system (Pinchuk, Vysotskii, 2001; Vysotskii, Pinchuk and Kornilova, 2002).

Changes of all the above-listed parameters are preserved for a great period of time. At low temperature, activated water can keep its anomalous properties for many days and even months.

The above-considered stable changes of the physicomolecular characteristics of water are partially justified in the model of the clathrate-based memory of water, according to which the information "written down" in the volume of water as a changed clathrate structure can be kept and used for a very long time. The performed studies of physicomolecular characteristics have shown that the duration of water "memory" calculated on the basis of the clathrate model is in well agreement with the results of experiments.

It is quite evident that the totality of marked anomalies of activated water should influence the characteristics of biological objects, whose life cycle is closely related to such water.

Chapters 4–7 are devoted to the presentation of the methods and results of experimental studies of the influence of different fractions of activated water (water subjected to the activation for different time intervals) on various biological objects. The studies were carried out in the natural order connected to the complication of the internal organization of objects (firstly, we investigated plants; then, microorganisms; and finally, animals). Such an order of presentation seems to be natural and logically justified. It allows us to find the common regularities in the operation of activated water.

In each of the experiments with activated water under identical conditions, an additional study of the influence of analogous water not subjected to the activation process (control experiment) was carried out.

The current Chap. 4 is devoted to the study of the influence of activated water on plants of different types.

Before considering specific results, let us discuss briefly the general facts concerning a specific role of water in the development and the vital activity of plants.

Water makes up about 90% of all the contents of plant cells. Alongside macro- and micro-elements, it is a necessary part of any soil mixture for the cultivation of plants under natural conditions, as well as the cultural medium for the cultivation of plants under sterile conditions. In the vessels of xylem and phloem and in the lacticifers of higher plants, there is a kind of vegetative juice, which is analogous in function to blood plasma of higher animals and even human. The composition of this juice includes a big complex of organic substances and inorganic elements. The composition and concentration of these substances and elements differ strongly not only in different plants, but also in different organs of the same plant. This is the main essential difference from blood plasma, which is characterized by high stability of its composition within the scope of a whole organism.

The main characteristic that unites all kinds of cellular juice is the large content of water which reaches 98% in relative concentration (Villee, 1967).

With the lack of water, all growing processes are violated, which can cause the death of plants. We may say without exaggeration that water is the main component of any plant, and its presence is the main condition for the plant's existence as a living being. This implies at once that water is the main element of a biological system, whose influence allows one to realize the strongest influence on this system. Thus, the bringing of water with any particular characteristics into soil (irrigation) or into a cultural medium will definitely influence the basic growing parameters of plants. Such water can render a positive or negative influence on all developmental phases of a plant, including the peculiarities of the germination of seeds (speed of germination, percentage of germinating capacity), the growth of the overground part of a plant e.g. a sprout (the growth of its height and weight), and the growth of a root system. It can also influence the size of leaves (the area of a leave surface), and can cause the appearance of atypical (by their form, size, and color) leaves.

In addition to the results of studying the influence of activated water on the growth of plants under natural conditions, we also discuss here its influence on the growth of sterile cultures under conditions of a special cultural medium.

The usage of sterile cultures for the study of the influence of activated water is caused by certain circumstances, among which the most important is the increase of reliability and authenticity of the obtained results. The matter is that a sterile culture does not practically contain any bacteria. Therefore, the influence of extraneous factors (for example, infection or contamination) is practically excluded. This circumstance allows us to judge the direct effect rendered by the additional factors connected, in particular, the usage of activated water.

The results of studies of the physicomolecular properties of activated water stated above allow us to predict the influence of water activation on the development of plants. In particular, the discovered change (a reduction) of the viscosity of activated water should influence essentially the movement of cellular juice in the vessels of xylem and phloem. Change of the conductivity and the dielectric permittivity should render a strong influence on the movement and the characteristics of ions in water.

In addition, water with special characteristics can accelerate or inhibit the processes of cellular division, which can be observed visually in sterile cultures.

The specific reasons and the possible physicomolecular mechanisms of the influence of activated water on plants will be considered in details in Chap. 8 which is devoted to the general analysis of such mechanisms.

It is possible to note one more important aspect of the studied problem.

The analysis of the influence of activated water on the growth and the characteristics of biomass allows us to solve the inverse task, namely to develop the method of identification of the properties of activated water on the basis of the analysis of the influence of this water on biological objects. It is evident that the usage of such biological gauges on the basis of plants, for example, can be of great importance on the wide usage of the stimulating and prophylactic action of activated water on animals and man.

Thus, studying some key parameters, which characterize the growth of plants, allows us to make certain conclusions about the influence of some particular properties of water on plants and, eventually, to find the optimum ways of using such water.

Economic aspects of this problem related to the opportunity to use particular properties of activated water to increase the efficiency of agricultural production have doubtless importance.

In the experiments discussed below, the influence of water activated for 30 min and 60 min on the growth of plants in soil and under conditions of a sterile cultural medium was investigated. Water was activated directly before its usage or it was kept after the activation in a cooler at a temperature of 4°C no longer than one day.

Control plants were irrigated with analogous, nonactivated water.

To prepare sterile culture, a sterile concentrate of the medium was prepared beforehand, which was then diluted with distilled water which was later activated. To preserve the sterility of experiments, special measures were undertaken. In particular, in the case of a liquid cultural medium, it was activated after the sterilization directly in sterile test tubes.

Control and experimental plants were cultivated under identical conditions of illumination and at the same temperature. The studies of the influence of activated water on plants were carried out under the general scientific guidance of Dr. N. A. Matveeva at the Institute of Cellular Biology and Gene Engineering of the National Academy of Sciences of Ukraine.

4.2. Influence of MRET Activated Water on the Germination of Seeds of Vegetable Crops

For studying the influence of different fractions of activated water on the germination of seeds and the formation of leaves, seeds of the following vegetable crops were used: radish "Red giant" and "Krasa rannyaya", peas "Alpha", string beans "Valentino", cabbage of grade "Dymerskaya" and pumpkin of grade "Zhdana". These seeds were sown into compositionally similar soil and were periodically irrigated with a certain fraction of water. The observation of the appearance of shoots gives the following results.

The study of the germination of above-mentioned vegetable seeds irrigated with two different fractions of activated water and ordinary distilled water has shown that, practically for all tested plants (except for string beans), irrigation with water activated for 60 min promoted a much faster germination of seeds at the initial stage.

The example of such stimulating influence of activated water for the first 10 days on the germination of seeds of radish "Red giant" is presented in Figs. 4.1–4.3. Thirty seeds were sown in a Petri dish. The first shoots of this plant have appeared in all variants in four days after sowing. The sequence and characteristics of the seed germination are presented in Table 4.1.

From the photos given in Figs. 4.1–4.3 and the data presented in Table 4.1, it follows that the activation of water renders essential stimulating influence and promotes a much earlier germination of radish seeds. The strongest effect of stimulation corresponds to the water activated for 1 h.

One more example of the stimulating influence of activated water for the first 10 days on the germination of seeds of other radish grade is presented in Figs. 4.4 and 4.5. Thirty seeds of radish of grade "Krasa rannyaya" were sown in a Petri dish. The first shoots of this plant have appeared in all variants of the experiment in 4 days after sowing into soil. In this case, the number of shoots essentially depended on the type of activated water, with which the soil with sown seeds was irrigated.

The sequence of the germination of seeds is presented in Table 4.2.

From the presented data, it follows that the activation of water renders a stimulating influence and promotes a much earlier germination of radish seeds. The strongest effect corresponds to water activated for 1 h. Irrigation with such water increased the early germination of seeds by 80%. This result is of importance and can be directly used, for example, for the accelerated germination of radish.

Figure 4.1. Shoots of radish "Red giant" in 4 days after sowing into soil. Immediately after the sowing of plants, the soil was irrigated with ordinary (control) or activated water with the duration of activation of 1 h and 0.5 h.

Figure 4.2. Shoots of radish "Red giant" in 5 days after sowing into soil (soil was irrigated with control or activated water).

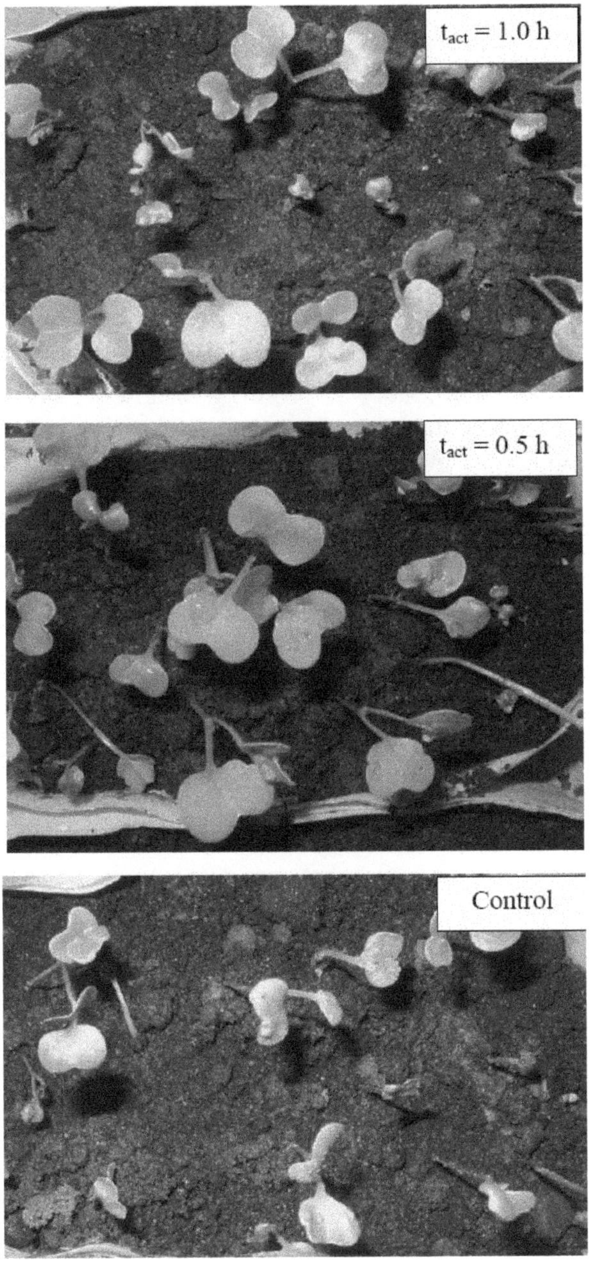

Figure 4.3. Shoots of radish "Red giant" in 9 days after sowing into soil (soil was irrigated with control or activated water).

Table 4.1. Germination of seeds of radish "Red giant" irrigated with activated and control water.

Fraction of water for irrigation	Number of sprouts in 4 days after sowing	Number of sprouts in 5 days after sowing	Number of sprouts in 6 days after sowing	Number of sprouts in 9 days after sowing
$t_{act} = 1.0\,h$	15	23	25	28
$t_{act} = 0.5\,h$	12	20	22	25
Control	5	14	16	24

The final data determining the presence or absence of the stimulating effect on the growth of the shoots of various vegetable crops irrigated with water activated for 1 h are presented in Table 4.3.

It is clear that the usage of water activated for 1 h renders a very large stimulating effect on the majority of vegetable crops at the beginning of the cultivation period.

In 10–20 days after the beginning of the cultivation, the effect of the stimulation was noticeably reduced, and the final difference between the numbers of sprouts using activated and control (nonactivated) water was insignificant.

Somewhat different results were obtained using other water fraction. The generalized data characterizing the stimulating effect of water activated for 0.5 h are presented in Table 4.4.

It is clear that water activated for 0.5 h increased the number of sprouts in comparison with the control for radish and peas. For other crops, no positive effect was observed.

In 10–20 days of the cultivation, the effect of activated water ($t_{act} = 0.5\,h$) on the development of plants was analogous to that of water activated for 1 h.

4.3. Influence of MRET Activated Water on the Growth of Stalk and Leaves of Vegetable Crops

The stage of the appearance of shoots in the studied vegetable crops is followed by the stage of the formation and growth of a stalk and leaves. To determine the efficiency of the influence of activated water on these processes, a periodic registration of the basic plant characteristics and the measurement of the height of their overground part were carried out. This part of studies was executed in the following way.

Figure 4.4. Shoots of radish "Krasa rannyaya" in 4 days after sowing into soil (soil was irrigated with control or activated water).

Figure 4.5. Shoots of radish "Krasa rannyaya" in 9 days after sowing into soil (soil was irrigated with control or activated water).

Table 4.2. Germination of radish seeds "Krasa rannyaya" irrigated with activated and control water.

Fraction of water for irrigation	Number of sprouts in 4 days after sowing	Number of sprouts in 5 days after sowing	Number of sprouts in 6 days after sowing	Number of sprouts in 9 days after sowing
$t_{act} = 1.0\,h$	18	22	22	28
$t_{act} = 0.5\,h$	10	20	24	24
Control	10	14	22	22

Table 4.3. Influence of water activated for 1 h on the number of shoots of different vegetable crops.

Plant	Relative change of the number of shoots irrigated with water with $t_{act} = 1.0\,h$ relative to the control at the beginning of the seed germination	Relative change of the number of shoots irrigated with water with $t_{act} = 1.0\,h$ relative to the control at the end of the seed germination
Cabbage of grade "Dymerskaya"	+62%	−3.8%
Pumpkin "Zhdana"	+200%	−20%
Radish "Red giant"	+200%	+16%
Radish "Krasa rannyaya"	+80%	+27%
Peas "Alpha"	+200%	+18%
String beans "Valentino"	0%	+10%

After the completion of the certain growth phase, plants were taken out from soil, and then stalks with leaves were separated from roots. After that, the weighting of the overground parts of plants was made. The procedure to determine the weight of overground parts is presented in Fig. 4.6.

Then, the average surface area of leaves of one plant was determined. For this purpose, fragments with an area of 1 cm^2 were firstly cut from several typical leaves of plants. Based on these fragments, the average weight of a surface of 1 cm^2 of the leaf was measured. Then, the average weight of each plant (not including the stalk weight) was divided by this size, which results in the value of the average area of the surface of leaves of one plant.

Table 4.4. Influence of water activated for 0.5 h on the number of shoots of different vegetable crops.

Plant	Relative change of the number of shoots irrigated with water with $t_{act} = 0.5$ h relative to the control at the beginning of the seed germination	Relative change of the number of shoots on the irrigation with water with $t_{act} = 0.5$ h relative to the control at the end of the seed germination
Cabbage of grade "Dymerskaya"	−12%	−23%
Pumpkin "Zhdana"	0%	0%
Radish "Red giant"	+140%	+4%
Radish "Krasa rannyaya"	0%	+9%
Peas "Alpha"	+100%	−9%
String beans "Valentino"	−100%	−10%

The data on the change of the characteristics of the overground part of various plants are presented below.

4.3.1. *Radish "Red giant"*

The parameters determining a change of the absolute and relative characteristics of the overground part of a radish plant for the period from 9 to 25 days after the appearance of shoots are presented in Tables 4.5 and 4.6. The general appearance of the plants of radish "Red giant" in 9 days after the sowing is presented in Fig. 4.7. The general form of the plants of radish immediately after the extraction from soil in 25 days is presented in Fig. 4.8.

For 25 days of the observation of the growth of sprouts of radish "Red giant", a gradual and accumulating change of the intensities of the growth of plants was registered for those irrigated with water activated for 1 h.

At the end of the observation period, the plants which were irrigated with such activated water exceeded substantially the control plants in all parameters (as for the increment of the average height of the overground part of a plant by 21.9%, the average weight of the overground part of a plant by 57.1%, and the average area of the surface of leaves of a plant by 37.6%).

At the same time, the differences between the plants irrigated with water activated for 0.5 h and control plants turned out to be insignificant, and they correspond to statistical errors.

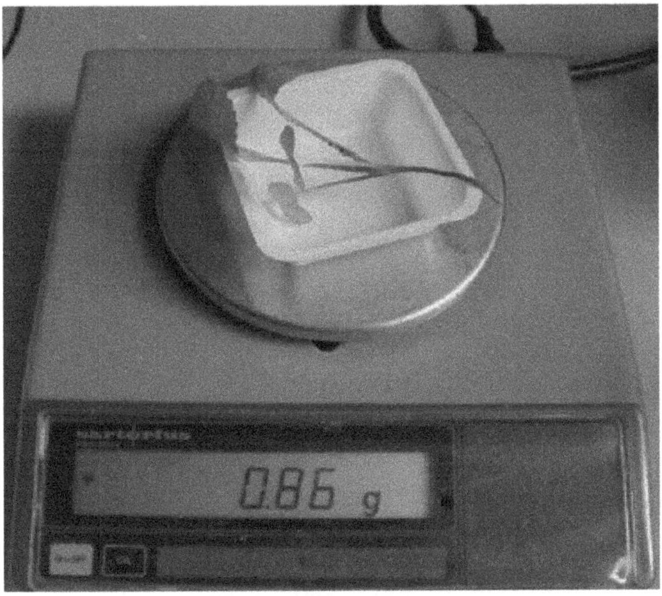

Figure 4.6. Weighing of plants for the determination of the weight of the overground part.

Table 4.5. Characteristics of the overground part of radish "Red giant" in 9–25 days after the appearance of shoots.

Fraction of water	Average height of the overground part of a plant in 9 days after the appearance of shoots	Average height of the overground part of a plant in 15 days after the appearance of shoots	Average height of the overground part of a plant in 25 days after the appearance of shoots	Average weight of the overground part of a plant in 25 days after the appearance of shoots	Average area of leave surface of a plant in 25 days after the appearance of shoots
$t_{act} = 1.0$ h	2.2 cm	6.7 cm	9.08 cm	0.44 g	7.65 cm^2
$t_{act} = 0.5$ h	2.4 cm	6.2 cm	7.97 cm	0.29 g	5.24 cm^2
Control	2.6 cm	6.0 cm	7.45 cm	0.28 g	5.56 cm^2

Table 4.6. Characteristics of radish "Red giant" irrigated with activated water compared with control water in 25 days after the appearance of shoots.

Fraction of water	Average height of the overground part of a plant in 25 days after the appearance of shoots	Average weight of the overground part of a plant in 25 days after the appearance of shoots	Average area of the surface of leaves of a plant in 25 days after the appearance of shoots
$t_{act} = 1.0\,h$	121.9%	157.1%	137.6%
$t_{act} = 0.5\,h$	106.9%	103.5%	94.2%
Control	100%	100%	100%

Figure 4.7. Plants of radish "Red giant" in 9 days after the sowing.

4.3.2. *Radish "Krasa rannyaya"*

The changes of the absolute and relative characteristics of the overground part of the plants of radish "Krasa rannyaya" for the period from 9 to 25 days after the appearance of shoots are presented in Tables 4.7–4.9.

For 25 days of the observation of the growth of sprouts of radish "Krasa rannyaya", we registered the distinctions in the growth intensity of plants irrigated with different fractions of activated water. At the end of the observation, the height of plants in all variants differed insignificantly. However, the plants which were irrigated with activated water exceeded the control plants substantially in the following parameters: 11–22% increment of the average weight of the overground part of a plant and 9.7–23.4% in average area of the surface of leaves of a plant. The higher parameters correspond to water activated for 1 h.

Figure 4.8. Plants of radish "Red giant" in 25 days after the sowing.

Table 4.7. Average height of the overground part of radish "Krasa rannyaya" in 9–25 days after the appearance of shoots.

Fraction of water	Average height of the overground part of a plant in 9 days after the appearance of shoots	Average height of the overground part of a plant in 15 days after the appearance of shoots	Average height of the overground part of a plant in 25 days after the appearance of shoots
$t_{act} = 1.0\,h$	2.4 cm	6.8 cm	10.2 cm
$t_{act} = 0.5\,h$	2.5 cm	7.2 cm	10.26 cm
Control	2.1 cm	6.9 cm	10.71 cm

Table 4.8. Characteristics of the overground part of radish "Krasa rannyaya" irrigated with activated and control water in 25 days after the appearance of shoots.

Fraction of water	Average height of the overground part of a plant in 25 days after the appearance of shoots	Average weight of the overground part of a plant in 25 days after the appearance of shoots	Average area of the surface of leaves of one plant in 25 days after the appearance of shoots
$t_{act} = 1.0\,h$	10.20 cm	0.55 g	14.53 cm^2
$t_{act} = 0.5\,h$	10.26 cm	0.50 g	12.92 cm^2
Control	10.71 cm	0.45 g	11.77 cm^2

Table 4.9. Characteristics of radish "Krasa rannyaya" irrigated with activated water compared with control water in 25 days after the appearance of shoots.

Fraction of water	Average height of the overground part of a plant in 25 days after the appearance of shoots, control is 100%	Average weight of the overground part of a plant in 25 days after the appearance of shoots	Average area of the surface of leaves of one plant in 25 days after the appearance of shoots
$t_{act} = 1.0\,h$	95.29%	122%	123.4%
$t_{act} = 0.5\,h$	95.8%	111%	109.7%
Control	100%	100%	100%

4.3.3. *Peas "Alpha"*

The parameters determining the changes of absolute and relative characteristics of the overground part of peas in the period from 14 to 34 days after the appearance of shoots are presented in Tables 4.10–4.12. The general appearance of the plants of peas appear in 14 and 29 days after its sowing into soil is presented in Figs. 4.9 and 4.10.

For 34 days of the observation of the growth of sprouts of peas "Alpha", the gradual change in the intensity of growth of plants was registered. In the first 25 days after the germination, the heights of plants irrigated with activated water exceeded those of control plants. In this case, the plants irrigated with water activated for 1 h and for 0.5 h were higher than the control ones, by 21.3% and 17.1% respectively. At the end of the observation period, no distinctions in the heights of plants, weights of the overground part, and average areas of leaves were observed for experimental and control plants.

4.3.4. *String beans "Valentino"*

In Tables 4.13–4.15, we give the parameters determining the growth of the overground part of string bean for the period from 14 to 34 days after the appearance of shoots. The general appearance of the plants of string bean in 20 days after sowing in soil is presented in Fig. 4.11.

Table 4.10. Average height of the overground part of peas "Alpha" in 14–29 days after the appearance of shoots.

Fraction of water	Average height of the overground part of a plant in 14 days after the appearance of shoots	Average height of the overground part of a plant in 20 days after the appearance of shoots	Average height of the overground part of a plant in 29 days after the appearance of shoots
$t_{act} = 1.0$ h	7.7 cm	14.2 cm	22.5 cm
$t_{act} = 0.5$ h	5.8 cm	13.7 cm	23.0 cm
Control	4.5 cm	11.7 cm	23.0 cm

Table 4.11. Characteristics of the overground part of peas "Alpha" irrigated with activated and control water in 34 days after the appearance of shoots.

Fraction of water	Average height of the overground part of a plant in 34 days after the appearance of shoots	Average weight of the overground part of a plant in 34 days after the appearance of shoots	Average area of the surface of leaves of one plant in 34 days after the appearance of shoots
$t_{act} = 1.0$ h	36.9 cm	2.22 g	111 cm^2
$t_{act} = 0.5$ h	37.2 cm	2.19 g	122 cm^2
Control	37.2 cm	2.4 g	122 cm^2

Table 4.12. Characteristics of peas "Alpha" irrigated with activated water compared with control water in 34 days after the appearance of shoots.

Fraction of water	Average height of the overground part of a plant in 34 days after the appearance of shoots	Average weight of the overground part of a plant in 34 days after the appearance of shoots	Average area of the surface of leaves of one plant in 34 days after the appearance of shoots
$t_{act} = 1.0$ h	99.19%	95.5%	90.98%
$t_{act} = 0.5$ h	100%	91.25%	100%
Control	100%	100%	100%

Figure 4.9. Plants of peas "Alpha" in 14 days after the sowing.

Figure 4.10. Plants of peas "Alpha" in 29 days after the sowing.

Table 4.13. Average height of the overground part of string bean " Valentino" in 14–29 days after the appearance of shoots.

Fraction of water	Average height of the overground part of a plant in 14 days after the appearance of shoots	Average height of the overground part of a plant in 20 days the appearance of shoots	Average height of the overground part of a plant in 29 days after the appearance of shoots
$t_{act} = 1.0$ h	8.0 cm	21.3 cm	24.9 cm
$t_{act} = 0.5$ h	7.5 cm	18.4 cm	27.7 cm
Control	6.8 cm	18.6 cm	20.7 cm

Table 4.14. Characteristics of the overground part of string bean "Valentino" irrigated with activated and control water in 34 days after the appearance of shoots.

Fraction of water	Average height of the overground part of a plant in 25 days after the appearance of shoots	Average weight of the overground part of a plant in 25 days after the appearance of shoots	Average area of the surface of leaves of one plant in 25 days after the appearance of shoots
$t_{act} = 1.0\,h$	30.0 cm	4.14 g	123 cm^2
$t_{act} = 0.5\,h$	31.37 cm	4.60 g	158.5 cm^2
Control	27.09 cm	3.67 g	114.5 cm^2

Table 4.15. Characteristics of string bean "Valentino" irrigated with activated water compared with water in 34 days after the appearance of shoots.

Fraction of water	Average height of the overground part of a plant in 34 days after the appearance of shoots, control is 100%	Average weight of the overground part of a plant in 34 days after the appearance of shoots	Average area of the surface of leaves of one plant in 34 days after the appearance of shoots
$t_{act} = 1.0\,h$	107.5%	112.8%	107.42%
$t_{act} = 0.5\,h$	115.79%	125.3%	138.42%
Control	100%	100%	100%

Figure 4.11. Plants of string bean "Valentino" in 20 days after the sowing.

For 34 days of the observation of the growth of sprouts of string bean "Valentino", we fixed some distinctions in the intensity of growth of plants irrigated with activated and control water. For the whole period of the observation (34 days), the heights of plants irrigated with activated water exceeded those of control plants. In this case, the plants irrigated with water activated for 1 h and for 0.5 h in 34 days were higher than the control ones by 7.5% and 15.79%, heavier by 12.8% and 25.3%, and had the bigger area of leaves by 7.42% and 38.42%, respectively.

Thus, the irrigation of string beans with both fractions of activated water resulted in the substantial improvement of the growth of plants.

4.3.5. *Cabbage "Dymerskaya"*

The results of measurements of the parameters of cabbage plants irrigated with activated and ordinary water are presented in Tables 4.16–4.18. The appearance of these cabbage plants in 25 days after the sowing is presented in Fig. 4.12.

For 25 days of the observation of the growth of sprouts of cabbage "Dymerskaya", the gradual increase of distinctions in the intensities of growth of plants irrigated with water activated for 1 h and with ordinary water was registered. As a result, at the end of the period of the observation, the plants irrigated with activated water exceeded the control ones by the increment of the average height of the overground part (by 11.8%) and by the average weight of the overground part (6.2%). The differences in the average areas of the surface of leaves were within the limits of statistical error. At the same time, the distinctions between the plants irrigated with water activated for 0.5 h and the control plants appeared much smaller and did not exceed 4.8–5.2%.

Table 4.16. Average height of the overground part of cabbage "Dymerskaya" in 9–25 days after the appearance of shoots.

Fraction of water	Average height of the overground part of a plant in 9 days after the appearance of shoots	Average height of the overground part of a plant in 15 days after the appearance of shoots	Average of the overground part of a plant in 25 days after the appearance of shoots
$t_{act} = 1.0$ h	2.0 cm	5.79 cm	11.15 cm
$t_{act} = 0.5$ h	2.0 cm	5.26 cm	10.45 cm
Control	1.8 cm	4.78 cm	9.71 cm

Table 4.17. Characteristics of the overground part of cabbage "Dymerskaya" irrigated with activated and control water in 25 days after the appearance of shoots.

Fraction of water	Average height of the overground part of a plant in 25 days after the appearance of shoots	Average weight of the overground part of a plant in 25 days after the appearance of shoots	Average area of the surface of leaves of one plant in 25 days after the appearance of shoots
$t_{act} = 1.0$ h	11.15 cm	0.51 g	13.82 cm^2
$t_{act} = 0.5$ h	10.45 cm	0.48 g	12.92 cm^2
Control	9.71 cm	0.48 g	13.66 cm^2

Table 4.18. Characteristics of cabbage "Dymerskaya" irrigated with activated water compared with control water in 25 days after the appearance of shoots.

Fraction of water	Average height of a plant in 25 days after the appearance of shoots	Average weight of the overground part of a plant in 25 days after the appearance of shoots	Average area of the surface of leaves of one plant in 25 days after the appearance of shoots
$t_{act} = 1.0$ h	111.8%	106.2%	101.2%
$t_{act} = 0.5$ h	104.8%	100%	94.6%
Control	100%	100%	100%

4.3.6. *Pumpkin "Zhdana"*

The examples considered above show quite significant positive (stimulating) influence of activated water on the growth and the development of plants. For the sake of justice, it should be noted that we observed the example of the negative influence as well. It turned out that activated water rendered inhibiting influence on the growth of pumpkin. The quantitative characteristics determining this effect are presented below.

In Tables 4.19–4.21, we present the characteristics of the growth of the overground part of pumpkin "Zhdana" for the period from 9 to 27 days after the appearance of shoots irrigated with different fractions of water.

Figure 4.12. Plants of cabbage "Dymerskaya" in 25 days after sowing into soil. During the growth, the plants were irrigated with ordinary and activated water.

The general appearance of plants of pumpkin in 9 and 27 days after sowing into soil are shown in Figs. 4.13 and 4.14. From these photos, it is evident that the application of activated water inhibits the growth of pumpkin plants.

> The observation of the growth of sprouts of pumpkin "Zhdana" for 27 days showed the absence of a positive effect by the two fractions of activated water. For all parameters (except for the average area of leaves of the plants irrigated with water activated for 1 h), the experimental plants grew worse than the control plants. The average height of the plants irrigated with activated water was less than that of the control ones by 12–20%, and the average weight of the overground part of a plant was less by 11–32%.
>
> Thus, the results of experiments have shown that both investigated fractions of activated water render a negative influence on the initial stage of the growth of the overground part of pumpkin.

Table 4.19. Average height of the overground part of a pumpkin "Zhdana" in 9–23 days after the appearance of shoots.

Fraction of water	Average height of the overground part in 9 days after the appearance of shoots	Average height of the overground part in 18 days after the appearance of shoots	Average height of the overground part in 23 days after the appearance of shoots
$t_{act} = 1.0$ h	10.5 cm	16.25 cm	16.3 cm
$t_{act} = 0.5$ h	4.6 cm	12.5 cm	13.25 cm
Control	11.0 cm	17.8 cm	18.4 cm

Table 4.20. Characteristics of the overground part of pumpkin "Zhdana" irrigated with activated and control water in 27 days after the appearance of shoots.

Fraction of water	Average height of the overground part in 27 days after the appearance of shoots	Average weight of the overground part in 27 days after the appearance of shoots	Average area of the surface of leaves in 27 days after the appearance of shoots
$t_{act} = 1.0$ h	19.0 cm	5.59 g	125.8 cm^2
$t_{act} = 0.5$ h	17.6 cm	4.23 g	102.91 cm^2
Control	21.6 cm	6.26 g	124.16 cm^2

Table 4.21. Characteristics of pumpkin "Zhdana" irrigated with activated water compared with control water in 27 days after the appearance of shoots.

Fraction of water	Average height of the overground part in 27 days after the appearance of shoots, control is 100%	Average weight of the overground part in 27 days after the appearance of shoots	Average area of the surface of leaves in 27 days after the appearance of shoots
$t_{act} = 1.0$ h	87.96%	89.29%	101.32%
$t_{act} = 0.5$ h	80.36%	67.57%	82.88%
Control	100%	100%	100%

Figure 4.13. Plants of pumpkin "Zhdana" in 9 days after sowing into soil.

Figure 4.14. Plants of pumpkin "Zhdana" in 27 days after sowing into soil. During the growth period, plants were irrigated with ordinary and activated water.

4.4. Influence of MRET Activated Water on the Growth of Plants in a Sterile Cultural Medium

Earlier, we considered the specific features of the influence of activated water on higher plants growing in soil. In the following sections, we present the results of studies of the influence of activated water on the cellular structures forming callus tissues and growing in a cultural nutrient medium. It is obvious that such a circle of objects and systems does not exhaust all possible variants of the growth of plants. In the previous case, there was an additional and insufficiently controllable factor of influence, the soil. In the present case, even if the studies were carried out in a sterile cultural medium, they were limited to the cellular system, rather than to a whole higher plant.

In order to study the influence of activated water on higher plants, sterile higher plants such as *Solanum tuberosum* of grade “Lugovskoi” and *Solanum rickii* capable of growth in sterile cultural media (in sterile test tubes) were used as the object of study. The Murashige–Skoog standard sterile cultural medium (Murashige and Skoog, 1962) was prepared for the growth of plants. For the preparation of an agar-based medium, we took one part of ordinary distilled water and four parts of the corresponding activated water. The following way of activation was used:

Firstly, a concentrated solution of the cultural medium dissolved in a small amount of ordinary distilled water was prepared. After sterilization and cooling, this solution was diluted with the corresponding fraction of activated water in the ratio of 1:4. In such a way, the cultural medium was enriched with a specific fraction of activated water by 80%.

For the use of the liquid cultural medium, it was activated for 0.5 h or 1 h after the sterilization process directly in sterile test tubes under conditions of a sterile box.

Then we carried out the basic studies of the features of the growth of a culture.

In test tubes with the sterile medium, the parts of plants (shoots) having one bud were sown. Plants were cultivated in the presence of light at a temperature of 20°C with the following light/dark period within each day: 16 h of light — 8 h of darkness. In three weeks after the sowing, the evoluted plants were taken from test tubes, and the measurement of their major parameters — the height of plants, weight of the overground part, and weight of leaves — was carried out, and the surface area of leaves was evaluated.

In addition, after the completion of each experiment, the following coefficients of action of activated water which characterize the average parameters of evoluted plants were determined:

(1) Coefficient of inhibitory action of activated water on the growth of plants in the sterile culture $K_1 = N_{1a} : N_{1c}$.
    Here, N_{1a} is the amount of segments survived from the cultivation in the medium with activated water, and N_{1c} is the amount of segments survived from the cultivation in the control medium.
(2) Coefficient of stalk formation $K_2 = N_{2a} : N_{2c}$.
    Here, N_{2a} is the amount of segments, of which sprouts in the medium with activated water were formed, and N_{2c} is the amount of segments, of which sprouts in the control medium were formed.
(3) Coefficient of the length of formed sprouts $K_3 = L_a : L_c$.
    Here, L_a is the average length of the sprouts formed on one segment cultivated with activated water, and L_c is the average length of a sprout formed on one segment in the control.
(4) Coefficient of a change of the coloring of leaves $K_4 = N_{4a} : N_{4c}$.
    Here, N_{4a} is the number of plants with atypical coloring of leaves in the medium with activated water, and N_{4c} is the number of plants with atypical coloring of leaves in the control medium.
(5) Coefficient of root formation $K_5 = N_{5a} : N_{5c}$.
    Here, N_{5a} is the amount of segments, on which roots were formed in 3 weeks of the cultivation in the medium with activated water, and N_{5c} is the amount of segments on which roots in the control medium were formed.

For studying the influence of activated water, the growth of plant *Solanum tuberosum* (one plant in a separate test tube for each fraction of activated water and for the control) was investigated. To get reliable data and the necessary statistics, we carried out each experiment simultaneously (in parallel) as a series of 11 recurrences. In general, 33 plants were studied. For a plant *Solanum rickii*, the study was carried out according to an analogous scheme (one plant in a separate test tube for each fraction of activated water and for the control) with the use of a series of 20 recurrences. In total, 60 plants were studied.

The data of experiments, the metrological parameters of evoluted plants *Solanum tuberosum* of grade "Lugovskoi", and the general view of evoluted plants are presented in Figs. 4.15 and 4.16 and in Tables 4.22 and 4.23.

Figure 4.15. General view of sterile higher plants *Solanum tuberosum* in test tubes after the cultivation during three weeks in the sterile cultural medium on the basis of ordinary and activated water.

Figure 4.16. Sterile higher plants *Solanum tuberosum* taken from test tubes after the cultivation during three weeks in the sterile cultural medium on the basis of ordinary and activated water.

Table 4.22. Influence of activated water on the growth of sterile higher plants *Solanum tuberosum*.

Fraction of water	Average height of a plant	Average weight of the overground part of a plant	Average weight of leaves of a plant	Average area of leaves of a plant
$t_{act} = 1.0\,h$	7.8 cm	0.072 g	0.013 g	$0.93\,cm^2$
$t_{act} = 0.5\,h$	6.8 cm	0.068 g	0.014 g	$1.00\,cm^2$
Control	4.4 cm	0.058 g	0.013 g	$0.93\,cm^2$

Table 4.23. Coefficients of action of activated water on the growth of sterile higher plants *Solanum tuberosum*.

Fraction of water	K_1	K_2	K_3	K_4	K_5
$t_{act} = 1.0\,h$	1	1	1.77	0	1
$t_{act} = 0.5\,h$	1	1	1.54	0	1

Respectively, the data for the plant *Solanum rickii* are presented in Figs. 4.17 and 4.18 and in Tables 4.24 and 4.25.

The analysis of the results of the experiments on the cultivation of plants *Solanum rickii* and *Solanum tuberosum* of grade "Lugovskoi" in the sterile medium with activated water allows us to make the following conclusions:

Activated water is not toxic for plants of the sterile culture; the survivability of the sowed material in all variants made 100% and did not differ from that in a cultural medium prepared with "ordinary" water.

Activated water does not interfere with the normal growth of the overground part of plants and the root system and does not result in the appearance of atypical coloring of leaves.

Activated water in the dilution of 1:4 rendered no essential stimulating influence on the growth of plants *Solanum rickii* in the sterile culture.

For plants *Solanum tuberosum*, a small increase in the height of plants and the weight of their overground part was observed for the growth in the medium with activated water in comparison with those of the control.

4.5. Feature and Paradoxes of the Influence of Activated Water on Shaping and Growth of Callus Tissue

The above-considered features of the influence of different fractions of activated water on the growth of higher plants which were cultivated *in vivo*

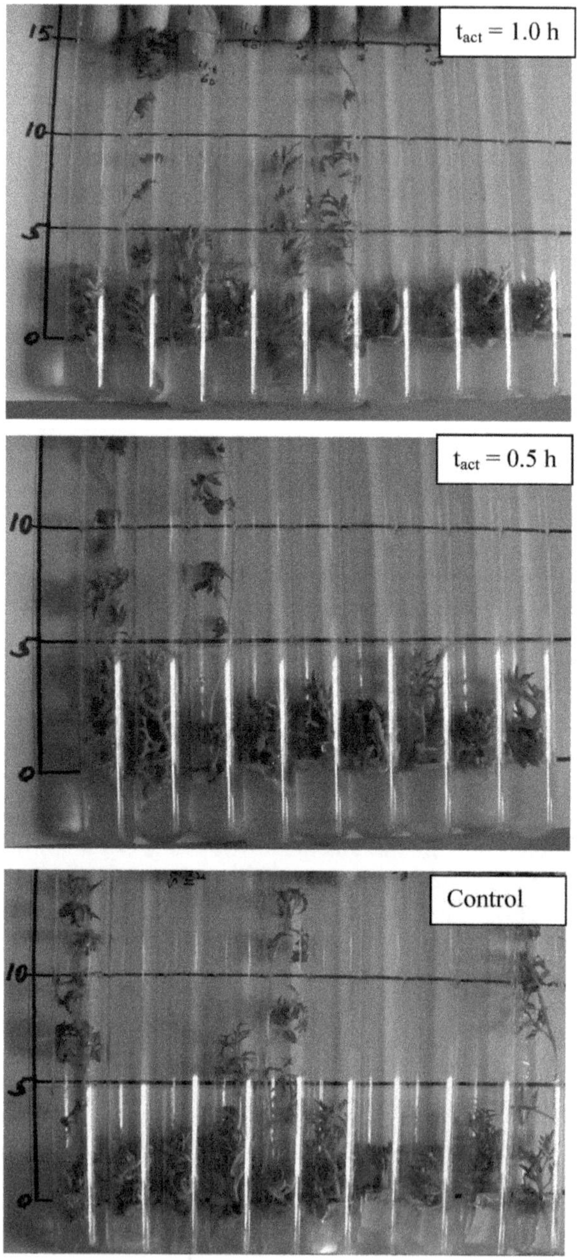

Figure 4.17. General view of sterile higher plants *Solanum rickii* in test tubes after the cultivation during three weeks in the sterile cultural medium on the basis of ordinary and activated water.

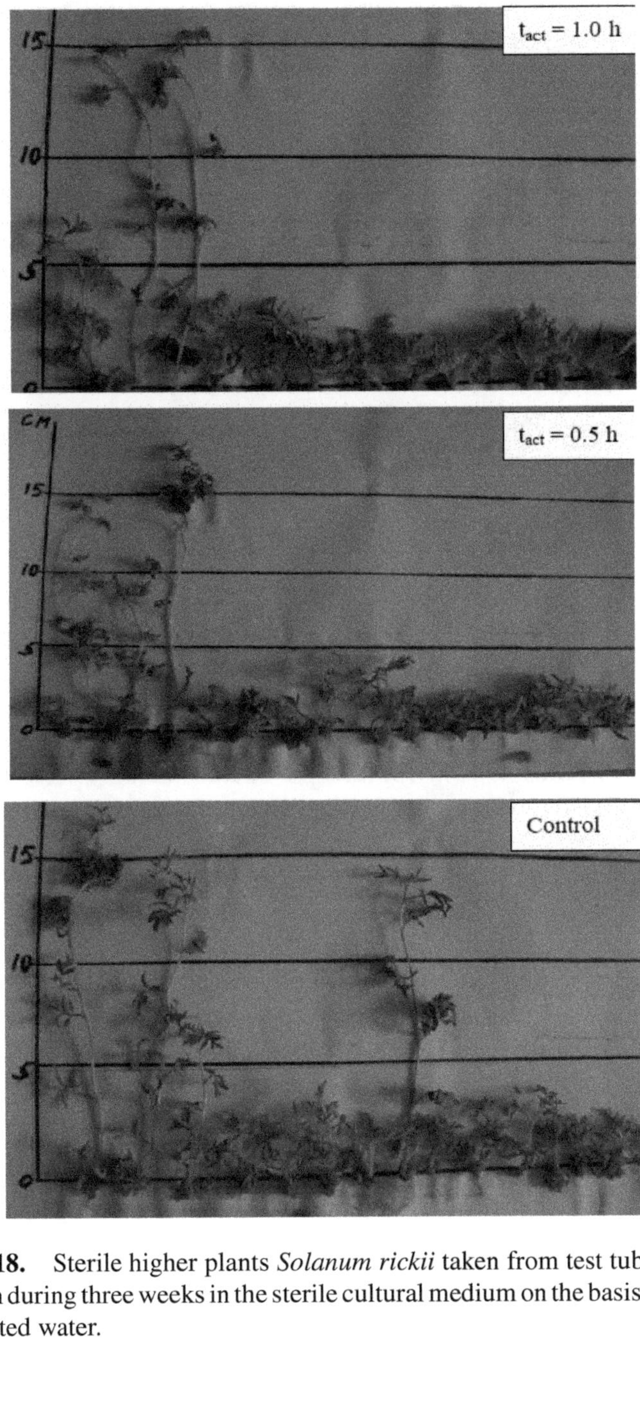

Figure 4.18. Sterile higher plants *Solanum rickii* taken from test tubes after the cultivation during three weeks in the sterile cultural medium on the basis of ordinary and activated water.

Table 4.24. Influence of activated water on the growth of sterile higher plants *Solanum rickii*.

Fraction of water	Average height of a plant	Average weight of the overground part of a plant	Average weight of leaves of a plant	Average area of a leaves of a plant
$t_{act} = 1.0$ h	4.5 cm	0.254 g	0.15 g	15 cm^2
$t_{act} = 0.5$ h	4.7 cm	0.321 g	0.18 g	18 cm^2
Control	5.2 cm	0.279 g	0.15 g	15 cm^2

Table 4.25. Coefficients of action of activated water on the growth of sterile higher plants *Solanum rickii*.

Fraction of water	K_1	K_2	K_3	K_4	K_5
$t_{act} = 1.0$ h	1	1	0.85	0	1
$t_{act} = 0.5$ h	1	1	0.90	0	1

(in soil), as well as *in vitro* (in an agar-based medium), have proved the essential influence of the activation of water on all the stages of development of higher plants. However, such studies leave many questions open, in particular, those concerning the very mechanism of the influence of activated water on the growth of higher plants. Especially, the question about the structural elements of a plant (beginning from DNA and a genome up to structural elements such as roots, stalks, or leaves), on which activated water renders the maximal influence, remains obscure.

For the partial answer to such topical question, we carried out the studies of the influence of activated water on the development of irregularly growing nondifferentiated vegetative cells *in vitro*.

Such objects are referred to as *callus tissues*. They are characterized by a nonspecific growth of nondifferentiated cells. As a result, an unsystematically growing biological tissue, in which there are no specific and functional attributes, is formed. Such properties of callus tissues are related to the fact that the growth and the cell fission in initial vegetative fragments are very sharply accelerated by the influence of special chemical preparations.

In such a mode of the accelerated development, cells do not reproduce the functional organs of a plant.

The special importance of the study of callus tissues consists in that they, in a certain sense, are indirect analogs of some heavy diseases characteristic of animals and men (for example, psoriasis or cancer).

In order to study the influence of two fractions of activated water (the duration of activation was 1 h and 0.5 h), we prepared the agar-based sterile Murashige–Skoog medium (Murashige and Skoog, 1962) with the addition of 2 mg/L of 2,4-dichlorophenoxyacetic acid and 0.5 mg/L of kinetin. These biologically active substances are the initiators of accelerated cellular divisions resulting in the formation of quickly and irregularly growing nondifferentiated cells from the ordinary differentiated cells of plants.

The experiments on the formation of callus tissue were carried out on segments of a stalk of the plant *Solanum rickii* studied above in the sterile cultural medium.

The way of activation was the same as that in the case of the growth of plants in the sterile cultural medium considered in the previous section.

Firstly, we prepared a concentrated solution of the cultural medium on the basis of a small amount of ordinary distilled water. After the sterilization and the cooling, this solution was diluted with the corresponding fraction of activated water up to the necessary ratio of activated and ordinary water. The further studies showed that the degree of such a dilution is very important and appreciably determines the efficiency of this "water–water mixture".

The example of the medium sterilization will be considered in details below in Sec. 5.1.

Totally, several series of experiments were carried out.

In the first series of experiments, for the preparation of the studied medium, we took one part of ordinary distilled water and four parts of water activated for 1 h or 0.5 h.

In the control experiment, only initial nonactivated distilled water was used for the preparation of the sterile cultural medium.

In each Petri dish containing the cultural medium and the necessary biologically active substances, we sowed 20 segments of a stalk of *Solanum rickii* for each variant. The data on the average weight of initial segments are presented in Table 4.26. The photos with the general view and the magnified parts of Petri dishes with stalk segments before the beginning of the first experiment on the cultivation of callus tissue are presented in Figs. 4.19 and 4.20. From the data presented in Table 4.26 and the photos, it is clear that the initial stalk segments were approximately identical.

Table 4.26. Increment of callus tissue in activated water.

Initial average weight of one segment, mg	Fraction of water (the duration of activation)	Final average weight of one segment, mg	Average increment of the weight of one segment, mg	Coefficient of inhibition of the growth of callus tissue
20.2 ± 0.1	1 h	20.5 ± 0.1	0.3 ± 0.2	$350 \cdots 1800$
20.8 ± 0.1	0.5 h	21.2 ± 0.1	0.4 ± 0.2	$300 \cdots 900$
19.6 ± 0.1	the control	193.5 ± 0.1	173.9 ± 0.2	1

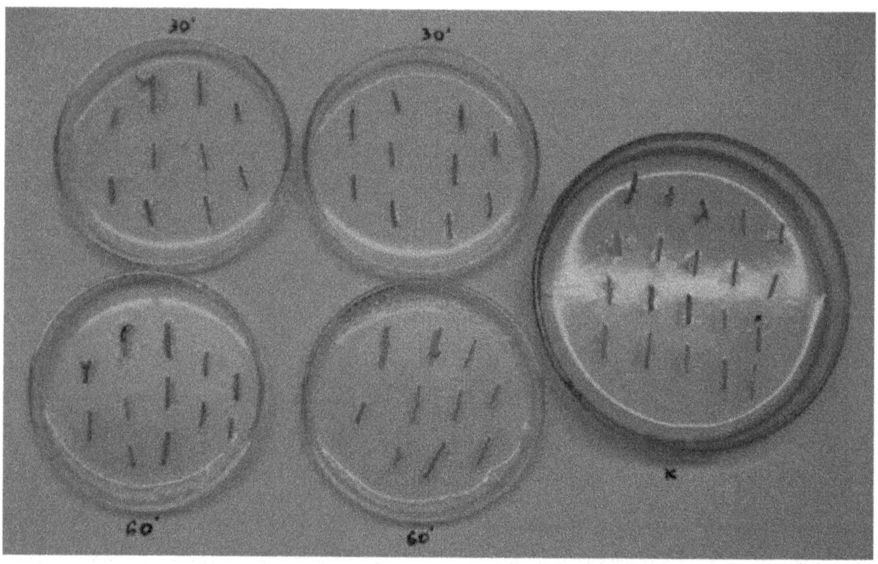

$t_{act} = 1.0$ h $\quad$ $t_{act} = 0.5$ h $\quad$ Control

Figure 4.19. General view of Petri dishes with stalk segments before the beginning of the first series of experiments on the cultivation of callus tissues. The medium composition in different Petri dishes: control is the cultural medium on the basis of initial nonactivated water; $t_{act} = 1.0$ h means the cultural medium containing 20% of nonactivated water and 80% of water activated for 1 h; $t_{act} = 0.5$ h means the cultural medium containing 20% of nonactivated water and 80% of water activated for 0.5 h.

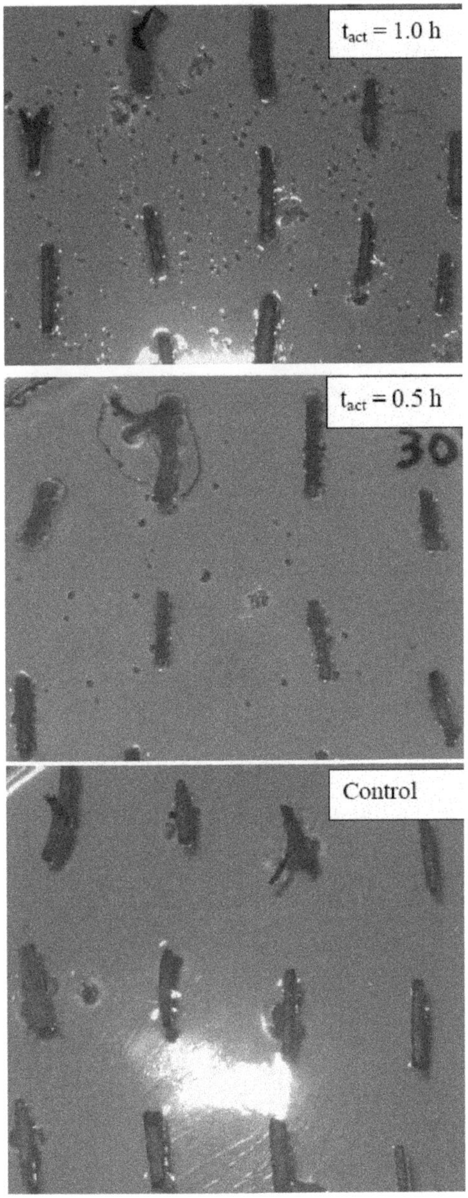

Figure 4.20. Magnified parts of a working field of Petri dishes containing the initial segments of a stalk before the beginning of the first series of experiments on the cultivation of callus tissue. The cultural medium composition in different Petri dishes corresponds to the data presented in Fig. 4.19.

First of all, it is worth to note that the survivability of stalk segments in all Petri dishes was identical and equal to 100% in all the experiments without exception. This result has confirmed that activated water is not toxic and the stalk segments of *Solanum rickii* are normally growing.

The results of experiments on the cultivation of callus tissue for 14 days at a temperature of 20°C are presented in Table 4.26 and Fig. 4.21.

The last column in Table 4.26 characterizes the coefficient of growth inhibition K of callus tissue determined as the ratio of the relative increment of the weight of one segment in the control experiment $(\Delta M/M)_{\text{control}}$ to the analogous increment of the weight of one segment $(\Delta M/M)_{\text{act}}$ for a specific fraction of activated water.

It follows from these data that, at a given degree of liquid dilution in the composition of the Murashige–Skoog cultural medium (20% of nonactivated water and 80% of water subjected to the preliminary activation), both fractions of activated water render the extremely strong inhibiting influence on the growth of callus tissue.

The great value of the coefficient of inhibition of the development of callus tissue ($K = 300 \cdots 1000$) testifies to the practically full suppression of the process of cellular division of nondifferentiated cells and the termination of the growth of callus tissue in the volume of the activated medium.

To confirm this effect, the second (repeated) series of experiments was carried out under completely identical conditions, with the same medium, at the same temperature, and with the same full duration of experiments. The results of these studies are presented in Table 4.27 and in Fig. 4.22.

It is clear that the results of the second (repeated) series of experiments performed under identical conditions completely confirm (within statistical error) the conclusion about both the extremely strong inhibition of the process of cellular division of nondifferentiated cells and the practically full termination of the growth of callus tissue on both fractions of the activated medium.

The third series of experiments was devoted to the study of the influence of a degree of mixing of ordinary and activated water on the inhibition of the growth of callus tissue. In these experiments, for the preparation of the studied medium, we took one part of ordinary distilled water and 2.5 parts of activated water (activated, respectively, for 1 h or 30 min). In the control experiment, only initial nonactivated distilled water was used. The duration of this series of measurements was also equal to two weeks. The results of these measurements are presented in Table 4.28 and in Fig. 4.23.

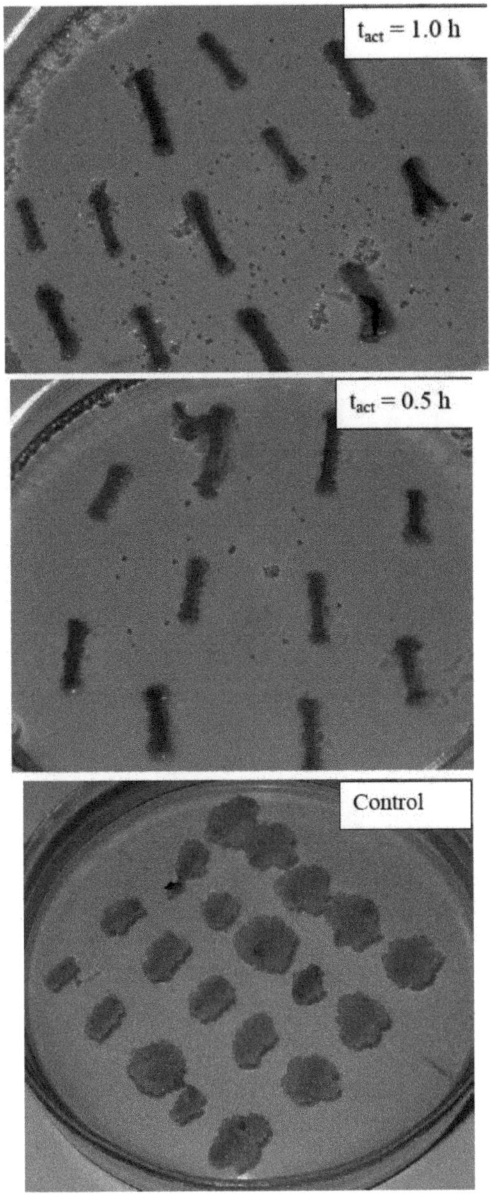

Figure 4.21. Magnified parts of a working field of Petri dishes containing segments of a stalk in two weeks after the beginning of the first series of experiments on the cultivation of callus tissue. The composition and the arrangement order of Petri dishes are completely identical to those presented in Fig. 4.20.

Table 4.27. Increment of callus tissue in activated water (the repeated series of experiments).

Fraction of water (the duration of activation)	Initial average weight of one segment, mg	Final average weight of one segment, mg	Average increment of the weight of one segment, mg	Coefficient of inhibition of the growth of callus tissues
1 h	22.2 ± 0.1	22.7 ± 0.1	0.4 ± 0.2	$470 \cdots 1000$
0.5 h	29.8 ± 0.1	30.1 ± 0.1	0.3 ± 0.2	$550 \cdots 2500$
Control	24.6 ± 0.1	257.9 ± 0.1	233.3 ± 0.2	1

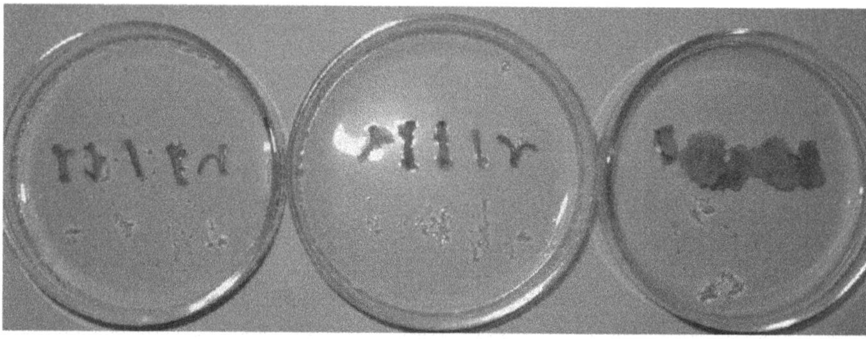

$t_{act} = 1.0$ h $t_{act} = 0.5$ h Control

Figure 4.22. General view of Petri dishes containing segments of a stalk in two weeks after the beginning of the second (repeated) series of experiments on the cultivation of callus tissue. The cultural medium composition and the ratio of nonactivated and activated water are completely identical to the first series (see the caption of Fig. 4.19).

In conclusion, we make several additional remarks concerning the features and results of our studies.

Before realizing the third series of measurements, we would expect *a priori* that a rather small reduction of the relative concentration of activated water from the initial value of 80% (in the first and second series of experiments) up to 71% (in the third series) should result in the same small (within the limits of 10%) predicted reduction of the inhibition effect of the growth of callus tissue for both fractions of activated water. However, such

Table 4.28. Increment of callus tissue in a diluted mixture of activated and ordinary water (the third series of experiments).

Fraction of water (the duration of activation)	Initial average weight of one segment, mg	Final average weight of one segment, mg	Average increment of one segment, mg	Coefficient of inhibition of the growth of callus tissue
1 h	22.9 ± 0.1	131.8 ± 0.1	108.9 ± 0.2	1.39
0.5 h	20.8 ± 0.1	26.0 ± 0.1	5.2 ± 0.2	$25.5 \cdots 27.5$
Control	22.6 ± 0.1	171.8 ± 0.1	149.2 ± 0.2	1

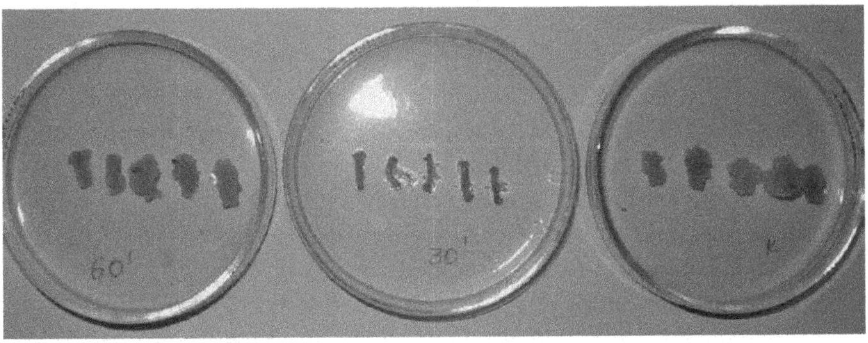

$t_{act} = 1.0$ h $\quad$ $t_{act} = 0.5$ h $\quad$ Control

Figure 4.23. General view of Petri dishes containing segments of a stalk in two weeks after the beginning of the third series of experiments on the cultivation of callus tissue in the medium prepared with a diluted mixture of activated and ordinary water.

simplified prediction turned out erroneous, and the results turned out to be essentially different.

It follows from the obtained results that the dilution of activated water with ordinary water does not lead to a monotonous reduction of the effect of inhibition identically for both fractions. The effect turned out much more complex and interesting.

From the obtained data, it is clear that the cultural medium prepared on the basis of the mixture of ordinary water and activated water with the duration of activation of 0.5 h in the ratio of 1:2.5 is characterized (as was expected) by the essential, though considerably weaker, effect of inhibition. For this medium, the effect of inhibition remains very big ($K \approx 25$), though its value has decreased 20 times in comparison with that for the medium with the mixture of water in the ratio of 1:4.

Contrary to that, for the cultural medium prepared on the basis of the mixture of ordinary water and activated water with the duration of activation of 1.0 h in the ratio of 1:2.5, the effect of inhibition of the growth of callus tissue and the influence on the cellular division is manifested slightly and is characterized by a rather small coefficient of inhibition, $K \approx 1.4$.

It is possible to state some assumptions concerning this effect. Taking into account that activated water by itself has no toxic properties, we may assume that a small impurity of ordinary nonactivated water essentially alters the character of influence of activated water on the cell surface.

If we start from the concept of the influence of activated water on the total surface energy of dividing cells, which will be advanced below in Sec. 5.3 and considered more accurately in Chap. 7, we can assume that the growth of callus tissue is inhibited in the medium with a great concentration of activated water for the same reason, because of which there is no division of cells of microbiological culture *Escherichia coli* on meat-peptone agar under aerobic conditions.

Within the framework of such a concept, it is necessary to assume that there is a certain threshold for the surface energy, the overcoming of which makes the cell division disadvantageous. In this case, little changes of the relative concentration of activated water result in similar little changes of the surface energy which can appear lower than the threshold in this case, and the process of division becomes allowed. This prediction can have very important consequences and can promote the development of a real mechanism of the essential influence on biological objects without the presence of side effects, which are undoubtedly manifested on the use of various chemical preparations or ionizing radiation for the same purpose.

It is obvious that the determination of the surface energy threshold is one of those perspective tasks of fundamental biology, to which the authors are going to address in the future.

A special importance of the discovered effect of strong inhibition of irregularly growing nondifferentiated vegetable cells is related to the fact that cells of the animal origin have similar properties. There are weighty grounds to assume that activated water can be used for the treatment of psoriasis, which is revealed as the considerably accelerated division of cells of epidermis, and for the treatment of oncological diseases.

In Chap. 6, we will present the experimental results on the study of the influence of activated water on the prophylaxis and the treatment of two lines of oncological diseases *in vitro* and *in vivo* (mice after the inoculation of tumors) which unequivocally confirm this assumption. Certain fractions of activated water render strong antitumoral action and promote the reduction of a size of tumors and the increase of the lifetime of infected mice. These results allow us to assert that the same effect can be used for the treatment of human oncological diseases.

CHAPTER 5

Effects of MRET Activated Water on Microbial Culture and Natural Microbial Associations

5.1. The Problem and Methods of Studying the Influence of Activated Water on Microbial Cultures and Microbial Associations

5.1.1. *General statement of the problem and initial biological test-objects*

In Chap. 5, we consider the results of studying the influence of different sorts of activated water on microbial cultures and microbial associations.

Such a selection of objects for the study is quite natural. It is related to the general logic of scientific researches which have to combine both in depth investigation of one object and the more general investigation of an aggregate of objects in natural interconnection. This concerns, in particular, microorganisms which live inside animals and men. For them, such a connection is typical when they are in the state of natural symbiosis, supplementing and helping one another. Such symbiotic communities of microorganisms form the syntrophic associations which include a great number of their species.

Based on this conception, we used both a pure bacterial culture and the mixed syntrophic associations of microorganisms, which have composition similar to that of the associations in human, in the study of the action of activated water on microorganisms. The results of studies of the pure cultures are presented in Secs. 5.2–5.4, and the results concerning the syntrophic associations are given in Sec. 5.5.

In the investigation of pure microbiological cultures, the pathogenic strain *Escherichia coli K 12* was used as the test culture.

As a source of syntrophic microbial associations, we used a granulated preparation which consisted of microorganisms taken from the active sludge of an aerotank and the fermented sediment of a methane-tank. The preparation contained the maximal variety of representatives of the physiological groups of microorganisms, and correspondingly represented the variety of metabolic pathways.

Escherichia coli is a stable symbiont of the digestive tracts of warm-blooded animals, as well as the most prevalent test-culture in microbiology. Because of this, the data on the action of a medium activation to culture *Escherichia coli* are sufficiently representative.

The testing of syntrophic associations reflects the influence of activated water on the natural and artificial microbial cenoses and supercenoses.

The experiments on the study of the effect of activated water on pure cultures and microbiological associations were executed under the guidance of Dr. A. B. Tashirev at D. K. Zabolotny Institute of Microbiology and Virology of the National Academy of Sciences of Ukraine.

5.1.2. *Methods of microbiological studies and equipment*

It is well known that any living system is a self-stabilizable object, for which the composition of chemical element content is genetically determined in DNA and is almost invariable during the whole life. Basing on the classification by relative concentration, the chemical elements are usually parted into vitally necessary macro- and micro-elements.

The vitally necessary elements are: oxygen O (the typical concentration in a living culture is about 24%), hydrogen H (64%), carbon C (9%), and nitrogen N (0.13%).

The elements necessary for the growth of biological cultures include: Na (7×10^{-3}%), K (4.5×10^{-2}%), Ca (7.5×10^{-2}%), Mg (2×10^{-2}%), Fe (8×10^{-4}%), P (1.3×10^{-2}%) and Si (3.5×10^{-2}%).

The specificity of the growth and the life of biological objects also requires the presence of a number of trace microelements such as Cl (7×10^{-3}%), Al (6×10^{-3}%), B (6×10^{-4}%), Ti (1×10^{-4}%), Zn (3×10^{-5}%), Li (1×10^{-4}%), Cu (1×10^{-5}%), Sr (1×10^{-5}%), Ba (5×10^{-6}%), F (3×10^{-5}%), Br (6×10^{-6}%), Rb (4×10^{-6}%), Sn (1×10^{-6}%), Ni (5×10^{-6}%), Mo (1×10^{-6}%), and Co (1×10^{-6}%).

For some microbiological cultures, the microelements S, Mn, J, and Hg are essential as well.

The percentages given are the mean values and can differ by some orders for separate cultures. At the same time, the requirement for the element composition to be stable is one of the necessary conditions for normal vital activity.

It is known that culture *Escherichia coli* and the majority of microorganisms that are included into syntrophic associations fed heterotrophically. For this reason, we used the combined meat-peptone medium with glucose which contained organic and mineral substances at high concentrations (the sources of nitrogen, phosphorus, and essential microelements) in experiments.

To accelerate the growth of microorganisms and to increase metabolic activity, we introduced a 40% concentrated solution of glucose into the nutrient medium [meat-peptone broth (MPB) or meat-peptone agar MPA)] to the final concentration of 2%.

Glucose was introduced several times into the liquid medium due to a decrease of the metabolic activity of microorganisms (these characteristics will be discussed during the analysis of specific experiments). We added glucose one time into the agarized medium before its transfer into Petri dishes.

MPB was used to cultivate *Escherichia coli* and to determine the metabolic parameters of synthrophic associations.

In the determination of the influence of activated water on the morphological parameters of colonies and cells of *Escherichia coli* and their stability to antibiotics, we used MPA. Under sterile conditions, we inroduced sodium resazurinate, which is simultaneously the indicator of both the metabolic activity and the redox-potential (a redox-indicator), into all the variants of nutrient media.

To grow the microbiological culture *Escherichia coli* and synthrophic microbial associations under anaerobic conditions, we used 50-ml colorless glass flasks with a thread on the neck.

During the growth of the microbiological culture *Escherichia coli,* the 30-ml portion of MPB was introduced into each flask. The flasks were closed with plugs made of flexible rubber and were screwed with metal hoods with thread. The sowing material (2 ml of metabolically-active 24-h culture) was injected into the flasks with a syringe through rubber plugs.

For the cultivation of associations, the 5-g portion of a dry microbial granular preparation and a 30-ml portion of MPB were introduced into each flask, and then these flasks were closed with rubber plugs and screwed with metal hoods.

To grow *Escherichia coli* culture under aerobic conditions, we used 20-ml glass test-tubes with cotton-gauze plugs. The sowing mterial (1 ml of the active 24-h culture) was introduced into tubes with a pipette under sterile conditions with a burner flame. Such a method ensured the required level of sterility. Other protective measures needed to ensure the necessary sterility will be considered in what follows.

In order to grow *Escherichia coli* culture on an agarized medium, we used glass Petri dishes. Each dish was filled by 15 ml of melted MPA. To control the sterility, these dishes were held for 24 h in a thermostat at 30 dishes. Each dish was filled with 15 ml of melted MPA. To control the sterility, these dishes were then held for 24 h in a thermostat at 30°C.

During the study of the morphology of colonies and cells of *Escherichia coli*, we prepared tenfold dilutions of the inoculum (1 ml of active 24-h culture of *Escherichia coli*). The corresponding aliquots were then spread on the agarized medium with a Drigalsky spatula to obtain the isolated colonies. The inoculation into dishes were performed from 3–5 dilutions: we uniformly spread 0.1 ml of the cell suspension over the surface of the agarized medium using a glass spatula. In the experiments on the study of the influence of an activated medium on the stability of *Escherichia coli* to antibiotics, 0.1 ml of the concentrated suspension of microorganisms was spread over the medium surface with a glass spatula in order to obtain a uniform bacterial film, the so-called "total lawn".

In the examination of the effect of activated water on the action of toxic elements, heavy metals were used.

It is known that heavy metals such as chromium, mercury, and copper are already toxic for microorganisms at their concentration of 1–5 mg/L in a medium. Their toxicity is caused by the high oxidation-reduction potential (redox-potential) and by such fact that they can replace the metals (cobalt, molybdenum, calcium, magnesium, etc.) in the active centers of enzymes of the structural and energetic metabolism. In turn, this leads to the inactivation of enzymes and the inhibition of the growth of microorganisms. The mechanisms of the detoxication of heavy metals become apparent in the enzymatic reduction of the high-potential toxic forms of metals to low- or non-toxic compounds. Moreover, the detoxication occurs as a result of the excretion of redox-metabolites (proteins, polysaccharides, etc.) into the medium by microorganisms. Therefore, the reduction rate of metals is the important criterion that characterizes the metabolic activity of microorganisms and their ability to maintain homeostasis under the influence of stress factors.

To study the effect of activated water on the ability of *Escherichia coli* and syntrophic associations to reduce heavy metals, we used a Cr(VI) compound — potassium chromate K_2CrO_4. In aqueous solution, potassium chromate dissociates into K^+ and CrO_4^{2-}.

The CrO_4^{2-} anion is a high-potential compound that creates the redox-potential $Eh' = +550\,mV$ in a medium. By interacting with CrO_4^{2-}, the metabolically active microorganisms reduce it to insoluble (and, hence, nontoxic) chromium(III) hydroxide:

$$CrO_4^{2-} + (n-1)H_2O + 5H^+ + 3e = Cr(OH)_3 \cdot nH_2O.$$

The quantitative determination of Cr(VI) was made by the colorimetric method with diphenylcarbazide (DPK). In an acid medium with Cr(VI) compounds, DPK creates the complex compounds colored in red or crimson. The coloring power of the analytic solution is proportional to the concentration of Cr(VI). From a flask or test-tube, 2 ml of the cultural liquid were taken away, then these samples were put to a chemically pure test-tube, and 3 drops of concentrated sulphuric acid and 0.5 ml of a 1% alcoholic solution of DPK were added.

As a control measurement, we carried out the analytic determination of chromate in distilled water and in a sterile nutrient medium that contained 50 mg/L of Cr(VI). In the experiment, a sterile solution of Cr(VI) was introduced in the flasks containing the metabolically active culture up to the final concentration of 20 or 50 mg/L, by using a syringe.

5.1.3. *Method of activation of nutrient media and means of registration of the processes of vital activity of microbiological cultures*

In all the experiments, we used three variants of nutrient media: the control one (a medium without activation) and the media activated during 0.5 h and 1.0 h.

To provide the necessary level of sterility, special measures were undertaken. In this case, it is necessary to take into consideration the following circumstance:

Our investigation of the physical properties of activated water showed that the anomalous characteristics of such water disappear when heated up to a temperature of 60°C. This circumstance leads to the unambiguous conclusion about the extreme inefficiency of the technological cycle, when the medium is firstly activated and then is sterilized. More proper is such a method where the liquid medium and all technological equipment destined for the contact with the medium (laboratory glassware, in particular) are sterilized first and only then activated. Such a technique imposes certain limitations on the preparation of experiments.

We started from the fact that the medium is to be activated under sterile conditions. For this purpose, we sterilized the external surface of the activation block directly before each activation cycle, by treating it sequentially with a 3% solution of hydrogen peroxide followed with 96% ethanol.

For the activation of MPB (in order to grow syntrophic microbial associations and culture *Escherichia coli* under anaerobic conditions), we used 1-liter sterile glass jars with a sterile activation block in its neck (Fig. 5.1). After the activation, the medium was poured into flasks next to the flame of a gas burner.

To grow *Escherichia coli* culture under aerobic conditions, MPB was poured into sterile test-tubes. The open test-tubes, in the burner flame, were

Figure 5.1. Activation of the medium for the growth of microbial syntrophic associations and *E. coli* under anaerobic conditions (before the introduction into flasks).

placed, with sterile tweezers in a jug precisely at right angle to the bottom surface for 100% efficient irradiation of the medium in test-tubes during activation. The general view of their mutual position in a jug used for the activation is shown in Figs. 5.2 and 5.3.

After the activation, the test-tubes were closed up with cotton-gauze plugs and then filled with the suspension of a metabolically active culture.

The dishes with the agarized medium, which were ready for inoculation, were activated in the following way. A dish was opened and placed on a sterile glass plate. Then, it was covered with a sterile plastic hood with the built-in activation block, as shown in Fig. 5.4.

The activation block was placed above a dish at a distance of about 4–5 cm from the agarized medium surface. The introduction of microorganisms in all variants of the experiment was carried out after the activation

Figure 5.2. Activation of meat-peptone broth with glucose for the cultivation of *E. coli* under aerobic conditions in test-tubes (side view).

Figure 5.3. Activation of meat-peptone broth with glucose for the cultivation of *E. coli* under aerobic conditions in the test-tubes (view from above).

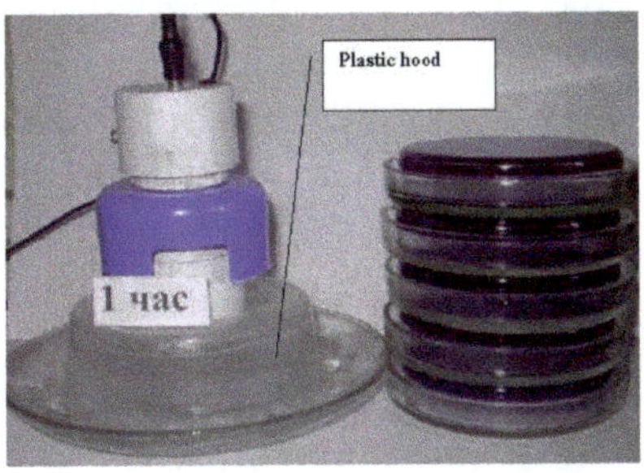

Figure 5.4. Activation of the agarized medium for the cultivation of *E. coli* in Petri dishes. An open dish with the medium is covered with a sterile plastic hood.

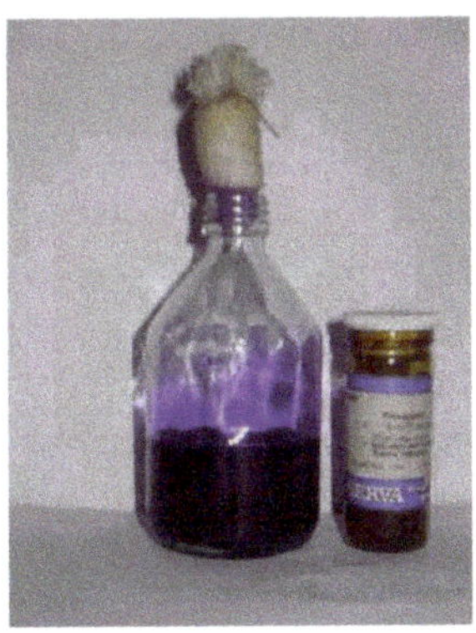

Figure 5.5. Redox-indicator, sodium resazurinate, has a blue-violet color. In the studies, sodium resazurinate made by "SERVA" was used.

of the medium and the following 24-h exposure at 30°C. Such an exposure is necessary to control the sterility during the activation.

The determination of the influence of activated water on the cultural-morphological and physiological properties of microorganisms was made by the following parameters:

- reductase activity,
- volume of a synthesized gas,
- optical density of the cultural liquid (an increase of the concentration of cells),
- cultural-morphological properties of microorganisms (the morphology of cells, the shape and color of colonies, etc.).

In each variant of the experiment, we carried out five identical studies to ehnance the reliability of experimental data.

The *reductase activity* is an indicator of the integral metabolic activity of microorganisms. The reductase activity is an evidence for the ability of microorganisms to change the oxidation-reduction potential (redox-potential, in millivolts) of a cultural medium. It is known that each physiological group of microorganisms, or a separate species in some cases, has its distinctive redox-potential, at which the culture growth begins. The preparatory metabolism of microorganisms (in particular, the excretion of reduction equivalents) is directed to the establishment of the physico-chemical conditions optimal for this culture in a medium. The phase of logarithmic growth (the most active) is characterized by the high reductase activity, which leads to a fast decrease of the redox-potential. In addition, the reductase activity correlates with the integral metabolic activity of microorganisms, i.e. with the activity of multienzymatic systems of the energetic constructional metabolism.

To evaluate the reductase activity, we used the indicator of oxidation-reduction potential, sodium resazurinate, which is widely applied in biology (Fig. 5.5).

Sodium resazurinate (named as resazurin in what follows) is nontoxic for microorganisms and allows one to unambiguously visually determine the basic redox-indices of a cultural medium. The indicator has three colors for different redox forms:

(i) *resazurin* — violet, redox-potential $Eh' > -50\,\mathrm{mV}$ (high-potential, aerobic conditions in most cases),
(ii) *resorufin* — red, $Eh' \leq -50\,\mathrm{mV}$ (facultative anaerobic conditions),

(iii) *leukoresazurin* — colorless, $Eh' \leq -100$ MB (low-potential, obligatory anaerobic conditions).

The reductase activity index was determined visually by a change of the medium coloration using the color scale (to within 5%). The technique applied in this work (we made N = 5 simultaneous parallel studies) allowed us to increase the accuracy of measurements of the reductase activity by $\sqrt{N} \approx 2.24$ times, which corresponds to a final accuracy of 2.24%.

In test-tubes and flasks, the reductase activity was determined by the coloration intensity of resazurin in the cultural liquid. In the agarized medium with resazurin, the reductase activity of *E. coli* was determined by the size of the zone, where resazurin becomes colorless (diameter, in mm). In the study of the stability to antibiotics, the size of the growth inhibition zone (GIZ) was determined by the size of a spot around the disk with antibiotic that was colored in violet or red with resazurin.

The *gas synthesis intensity* during the growth of microorganisms under anaerobic conditions allows one to determine the integral activity of the redox-enzymatic complexes of microorganisms, as well as the rates of consumption of a substrate and accumulation of final gaseous metabolites. On the whole, the volume of synthesized gas is proportional to the metabolite activity of microorganisms. During the cultivation of *Escherichia coli* and syntrophic associations under anaerobic conditions, we measured the evolved gas volume in definite time intervals (they are indicated in the tables and diagrams). For this purpose, a rubber plug hermetically closing a flask was pierced with a syringe needle (the 20-ml volume). The pressure of a superfluous gas moved the syringe piston. The evolved gas volume was determined by the graduation marks on a syringe cylinder. If the gas volume was greater than 20 ml, the second measurement was made. The metabolic activity of microorganisms was determined by the gas evolution dynamics during the cultivation and by the total gas yield during the whole experiment.

The *optical density* is the index of yield (harvest) of the biomass and the increase of the concentration of the cells of microorganisms in the cultural liquid. The growth of microorganisms (the division of cells) is accompanied by a turbidity of the cultural liquid, which leads to the increase of its optical density (D). Therefore, the optical density is the representative index of the growth of *E. coli* in a liquid nutrient medium.

The optical density of the cultural liquid was determined in the following way: the test-tubes and flasks with the cultural liquid were placed right up to a white screen with four black lines with different thicknesses (the thickness of the lines grows from No 1 to No 4).

By the bacteria-induced turbidity standards, it is determined that

(i) the thickness of the cultural liquid layer in the test-tubes, when line No 1, 2, 3, and 4 becomes sequentially invisible, corresponds, respectively, to 0.3, 0.6, 0.9, and 1.2 optical density units.
(ii) the thickness of the cultural liquid layer in the flasks, when line No 1, 2, 3, and 4 becomes sequentially invisible, corresponds, respectively, to 0.2, 0.4, 0.6, and 0.8 optical density units.

The *cultural-morphological properties* (size, shape and color of colonies, size and shape of cells) of *Escherichia coli* culture grown in the activated medium were investigated on the base of special preparations suitable for microscope photorecording. The methods of preparation and usage of such preparations will be considered in detail below in Sec. 5.3.

5.2. Effect of the Activation of the Aqueous Medium on Metabolic Parameters of the Microbiological Culture *Escherichia coli*

5.2.1. *Effect of the duration of activation on the metabolic parameters of culture Escherichia coli grown under aerobic conditions*

The metabolic activity of any biological object is the base of its viability. In turn, the reductase activity is the indicator of integral parameters of the metabolic activity. The reductase activity provides evidence of the ability of microorganisms to change the oxidation-reduction potential of a cultural medium. In particular, it reflects the degree of readiness of the enzymatic system to "serve" the metabolic processes.

The reductase activity is the parameter that describes, on the whole, such integral general index of the functioning of a microbial population (or a "macroorganism") as the rate of metabolic processes in living organisms. In particular, it characterizes the substrate consumption rate (the assimilation rate of nutrient substances) and the oxygen consumption rate for aerobic macro- and micro-organisms.

To determine the influence of the state of water on the reductase activity, we used the oxidation-reduction potential indicator, resazurin. The technique of such an analysis was considered in Sec. 5.1.

In our studies, the index of reductase activity was determined visually by the medium coloration change with the use of the standard color scale. Such a method allows us to determine the correspondence of colors and the belonging to one or another redox-form.

In the test-tubes and flasks, the reductase activity was determined by the intensity of the coloration of resazurin in a cultural liquid.

In the experiments, the following parameters were measured:

K_{ster} — coefficient of sterility that was determined by the coloration intensity of resazurin in the control medium (the discoloration of resazurin, in %);

D_c, $D_{0.5}$, and $D_{1.0}$ — average values of the optical density of a medium in the control experiment and in the media that were activated for 0.5 h and 1.0 h;

V_c, $V_{0.5}$, and $V_{1.0}$ — average values of the gas volume in the control experiment and in the media that were activated for 0.5 h and 1.0 h;

R_c, $R_{0.5}$, and $R_{1.0}$ — average values of the reductase activity index (the discoloration of resazurin, in %) in the control experiment and in the media that were activated for 0.5 h and 1.0 h; and

$K_{0.5} = R_c/R_{0.5}$, $K_{1.0} = R_c/R_{0.5}$ — inhibition coefficients for the reductase activity in the medium activated for 0.5 h and 1.0 h.

In addition to the characteristics considered above, it is expedient to introduce the efficiency coefficient of reductase activity inhibition $K_{0.5;1.0}$.

This coefficient is determined by the proportion

$$K_{0.5;1.0} = \frac{K_{0.5}}{K_{1.0}}$$

and shows how many times the medium activation for 0.5 h intensifies the effect of reductase activity inhibition of *E. coli* culture in comparison with that of the activation for 1.0 h.

In Figs. 5.6–5.13, we present the photos that demonstrate the sequence of temporal changes of the reductase activity for the microbiological *E. coli* (for 15 h) culture grown under aerobic conditions in media that were prepared with different fractions of activated water. The measurement method was considered above (in Sec. 5.1.3).

We recall that the changes of resazurin colors correspond to three color redox-forms:

(i) *resazurin* — violet, redox-potential $Eh' > -50\,\text{mV}$,
(ii) *resorufin* — red, $Eh' \leq -50\,\text{mV}$,
(iii) *leukoresazurin* — colorless, $Eh' \leq -100\,\text{mV}$.

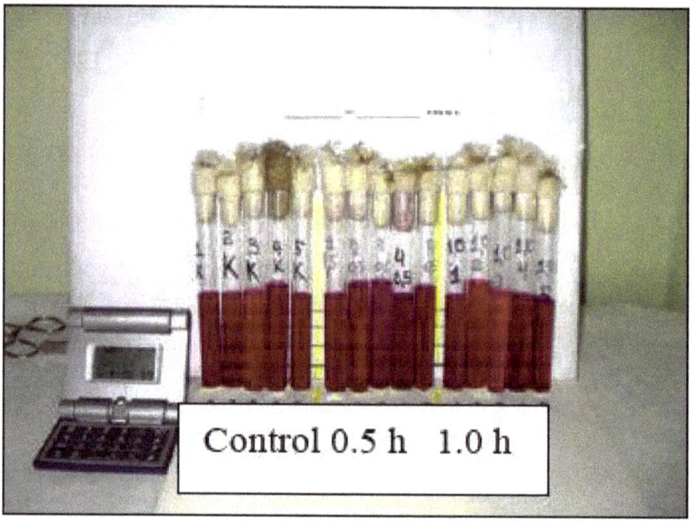

Figure 5.6. 1st hour of the experiment. $R_c = 0.2$, $R_{0.5} = 0.1$, $R_{1.0} = 0.4$, $D_c = 0.05$, $D_{0.5} = 0.05$, $D_{1.0} = 0.05$, $K_{0.5} = 2.0$, $K_{1.0} = 0.5$, $K_{0.5;1.0} = 4.0$.

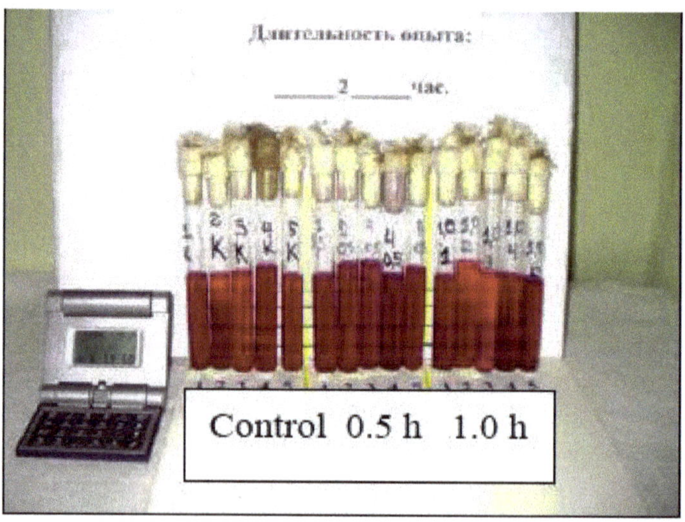

Figure 5.7. 2nd hour of the experiment. $R_c = 10$, $R_{0.5} = 5$, $R_{1.0} = 12$, $D_c = 0.05$, $D_{0.5} = 0.05$, $D_{1.0} = 0.05$, $K_{0.5} = 2.0$, $K_{1.0} = 0.83$, $K_{0.5;1.0} = 2.4$.

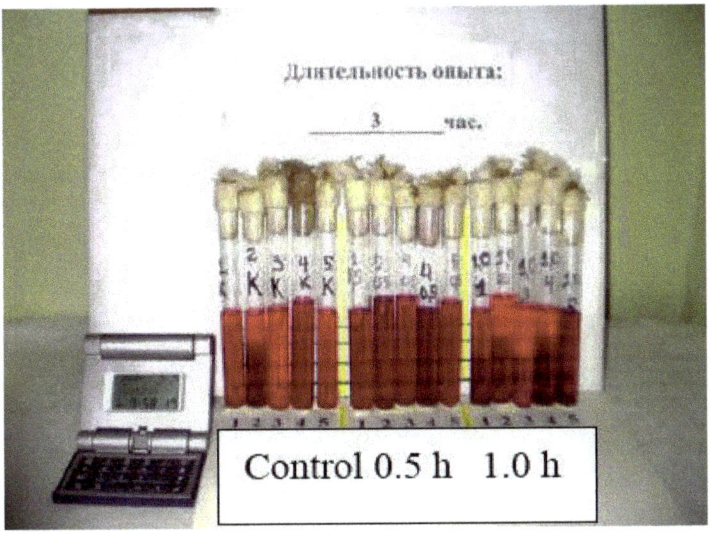

Figure 5.8. 3rd hour of the experiment. $R_c = 25$, $R_{0.5} = 9$, $R_{1.0} = 19$, $D_c = 0.05$, $D_{0.5} = 0.05$, $D_{1.0} = 0.05$, $K_{0.5} = 2.8$, $K_{1.0} = 1.3$, $K_{0.5;1.0} = 2.2$.

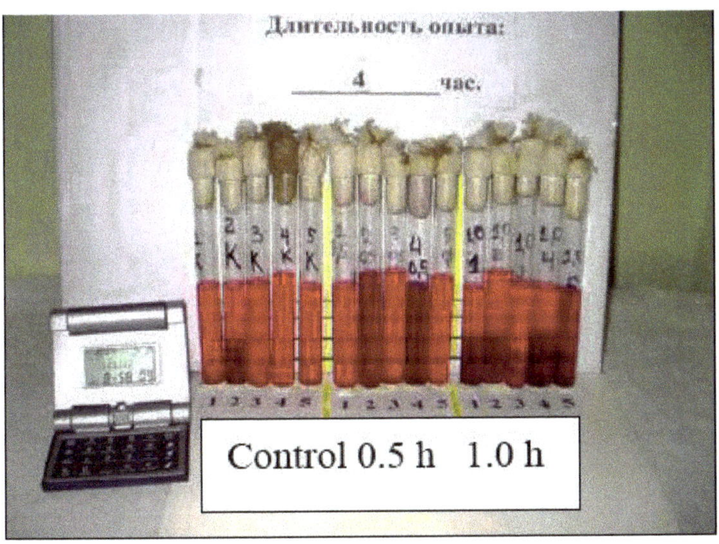

Figure 5.9. 4th hour of the experiment. $R_c = 47$, $R_{0.5} = 34$, $R_{1.0} = 32$, $D_c = 0.05$, $D_{0.5} = 0.05$, $D_{1.0} = 0.05$, $K_{0.5} = 1.4$, $K_{1.0} = 1.5$, $K_{0.5;1.0} = 0.9$.

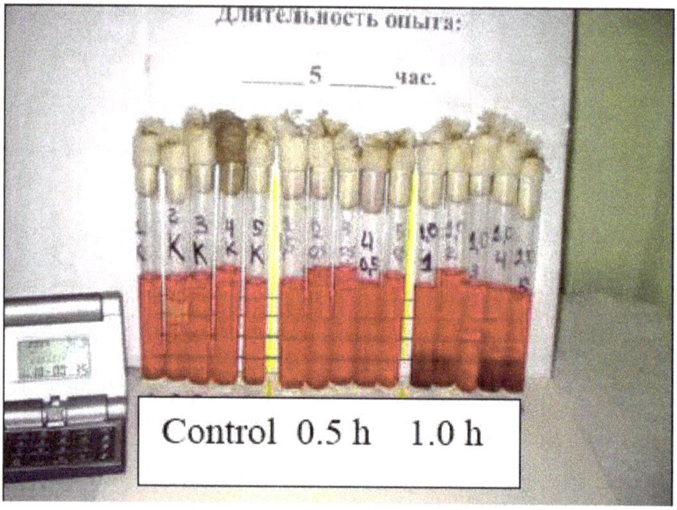

Figure 5.10. 5th hour of the experiment. $R_c = 52$, $R_{0.5} = 50$, $R_{1.0} = 46$, $D_c = 0.1$, $D_{0.5} = 0.1$, $D_{1.0} = 0.1$, $K_{0.5} = 1.04$, $K_{1.0} = 1.13$, $K_{0.5;1.0} = 0.92$.

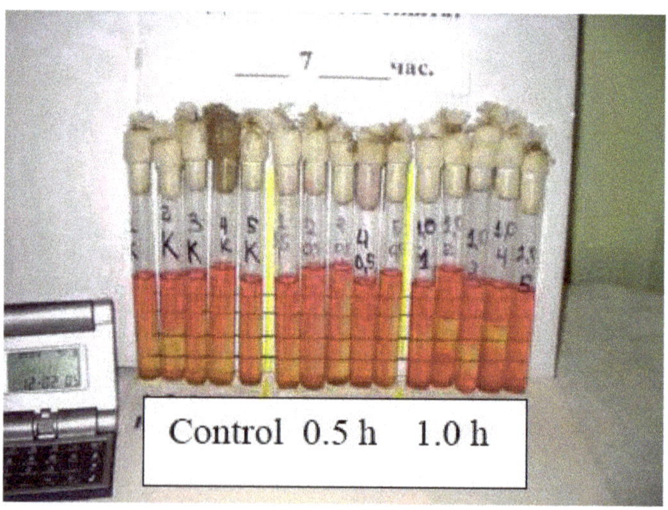

Figure 5.11. 7th hour of the experiment. $R_c = 65$, $R_{0.5} = 65$, $R_{1.0} = 65$, $D_c = 0.6$, $D_{0.5} = 0.6$, $D_{1.0} = 0.6$, $K_{0.5} = 1.0$, $K_{1.0} = 1.03$, $K_{0.5;1.0} = 0.97$. In the test-tubes along the column height of the medium, there appear the local zones of high redox-activity, where resazurin is discoloured completely.

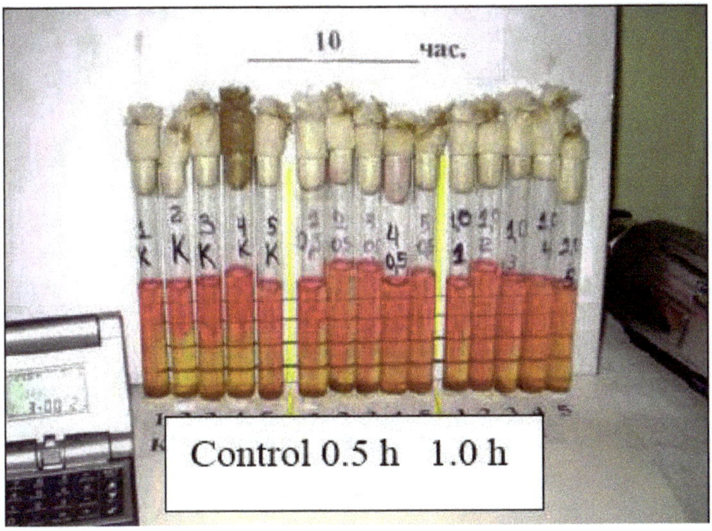

Figure 5.12. 10th hour of the experiment. $R_c = 75$, $R_{0.5} = 70$, $R_{1.0} = 70$, $D_c = 0.6$, $D_{0.5} = 0.6$, $D_{1.0} = 0.6$, $K_{0.5} = 1.04$, $K_{1.0} = 1.04$, $K_{0.5;1.0} = 1.0$.

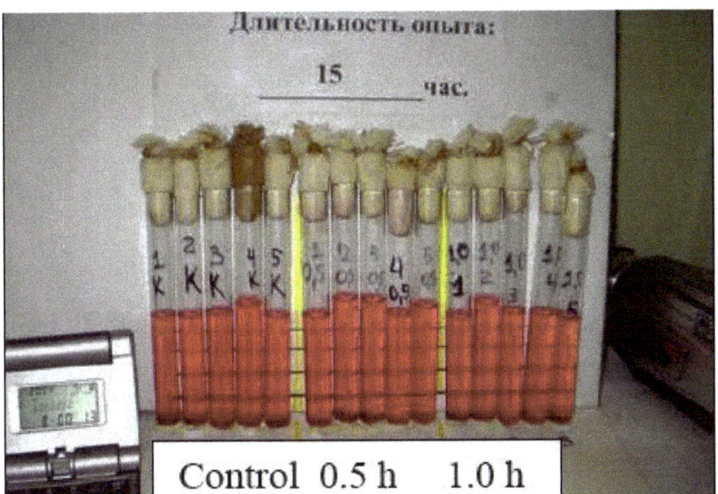

Figure 5.13. 15th hour of the experiment. $R_c = 50$, $R_{0.5} = 50$, $R_{1.0} = 50$, $D_c = 0.1$, $D_{0.5} = 0.1$, $D_{1.0} = 0.1$, $K_{0.5} = 1.0$, $K_{1.0} = 1.0$, $K_{0.5;1.0} = 1.0$.

Below each photo, we present the corresponding information (the time interval after the start of the experiment and the main parameters that describe the state of the medium). If necessary, we gave the notes and comments concerning the specific time after the experiment started. The water-activation duration in this experiment is shown directly on the photos.

The discussion of the results of measurements and the final conclusions are given below.

In Fig. 5.14, we show the above-obtained dependence of the reductase activity on the time from the start of experiment for different media (meat-peptone broth that contains the corresponding water fraction) both for non-activated water (control) and for water activated for 0.5 h and 1 h.

These summary data correspond to the information contained in the figure captions below each photo in Figs. 5.6–5.13.

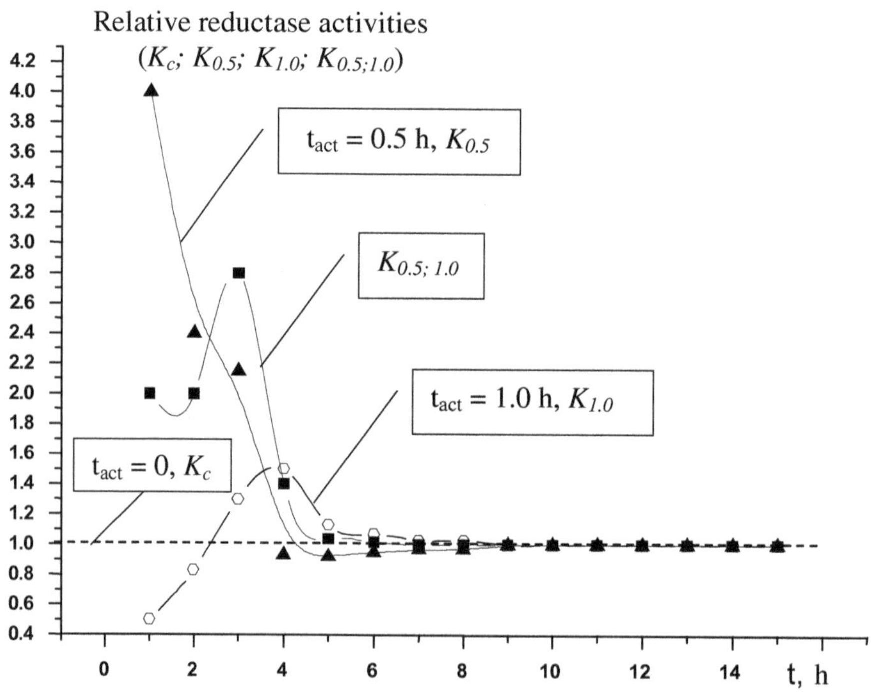

Figure 5.14. Effect of medium-activation duration on the inhibition coefficients of the reductase activity ($K_{0.5}$, $K_{1.0}$, and $K_{0.5;1.0}$) of the microbiological culture *Escherichia coli* grown under aerobic conditions.

Conclusions

1. The activation of the medium leads to both the inhibition and intensification of the reductase activity of the microbiological culture *Escherichia coli* in the initial stage of cultivation under aerobic conditions (for 4–5 h of the cultivation).
2. For the medium activated for 0.5 h, the maximal manifestation of the reductase activity inhibition effect corresponds to the beginning of the cultivation.
3. For the medium activated for 1.0 h, the effect of reductase activity intensification occurs in the initial stage of the cultivation. In 2.5 h, this effect is replaced by the effect of reductase activity inhibition which reaches its maximum in 4 h after the start of the cultivation.
4. Starting from the 5th h of the cultivation, the inhibition ceased in both medium fractions (such a result can be related to the homeostasis of the culture).
5. The medium activation did not influence the biomass increase. The optical density of the cultural liquid was the same in all variants of the experiment.

5.2.2. *Metabolic parameters of Escherichia coli on its growth in the activated water–containing nutrient medium under anaerobic conditions*

Below, we show the sequence of Figs. 5.15–5.36 and other data that describe a change of the reductase activity, volume of an evolved gas, and other metabolic parameters of *Escherichia coli* culture on its growth under anaerobic conditions. To grow *Escherichia coli* under anaerobic conditions, we used 50-ml colorless glass flasks with a thread on their necks. For the cultivation, the flasks were filled with 30 ml of MPB. To create anaerobic conditions, the flasks were hermetically closed with elastic rubber plugs and screwed with metal hoods with a thread. The inoculum (2 ml of the metabolically active 24-h culture) was introduced into the flasks using a syringe, by piercing through a rubber plug.

The whole series of experiments lasted 20 h and included the investigations performed in the presence of different fractions of water. During the studies (in 17 h after the start of the experiment), the strong oxidizing agent containing highly toxic metal such as chromium was carried into the medium. Under each photo, we give the data on main values of the metabolic parameters at the present time moment. If necessary, we present the comments and explanations concerning the specificity of phenomena demonstrated in the photos.

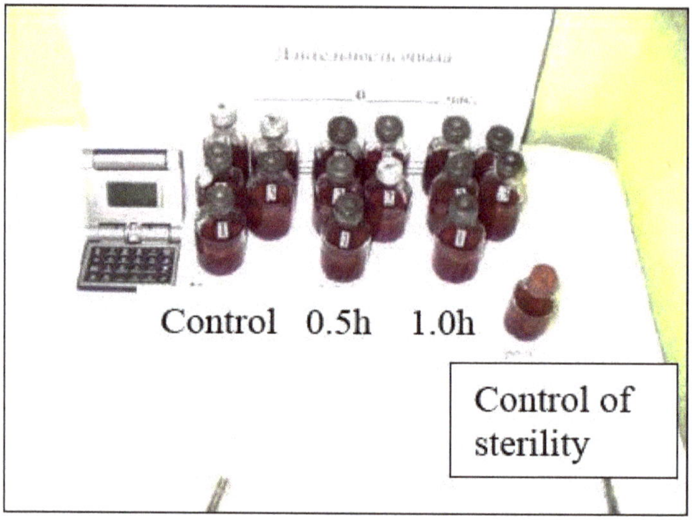

Figure 5.15. Start of the experiment.

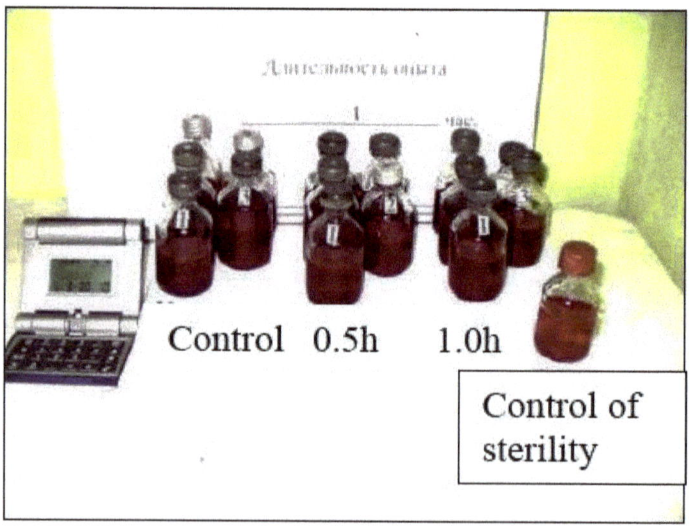

Figure 5.16. 1st hour of the experiment. $R_c = 0$, $R_{0.5} = 0$, $R_{1.0} = 0$, $D_c = 0$, $D_{0.5} = 0$, $D_{1.0} = 0$, $V_c = 0$, $V_{0.5} = 0$, $V_{1.0} = 0$.

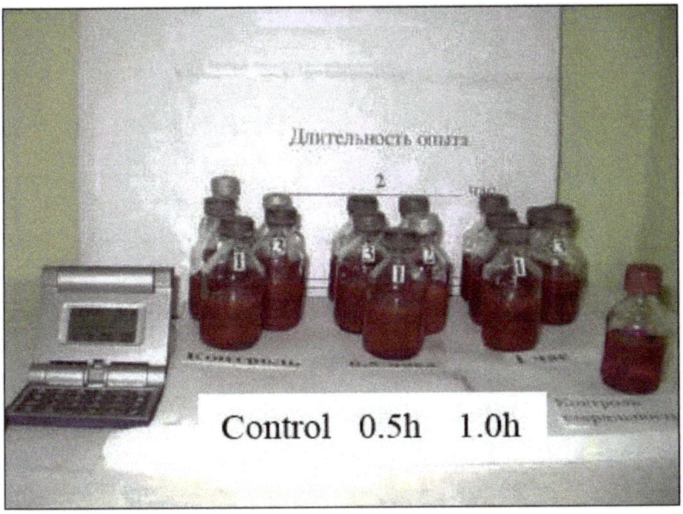

Figure 5.17. 2nd hour of the experiment. $R_c = 35$, $R_{0.5} = 35$, $R_{1.0} = 35$, $D_c = 0$, $D_{0.5} = 0$, $D_{1.0} = 0$, $K_{0.5} = 1$, $K_{1.0} = 1$, $V_c = 0$, $V_{0.5} = 0$, $V_{1.0} = 0$.

Figure 5.18. 3rd hour of the experiment. $R_c = 50$, $R_{0.5} = 50$, $R_{1.0} = 50$, $D_c = 0$, $D_{0.5} = 0$, $D_{1.0} = 0$, $K_{0.5} = 1$, $K_{1.0} = 1$, $V_c = 0$, $V_{0.5} = 0$, $V_{1.0} = 0$.

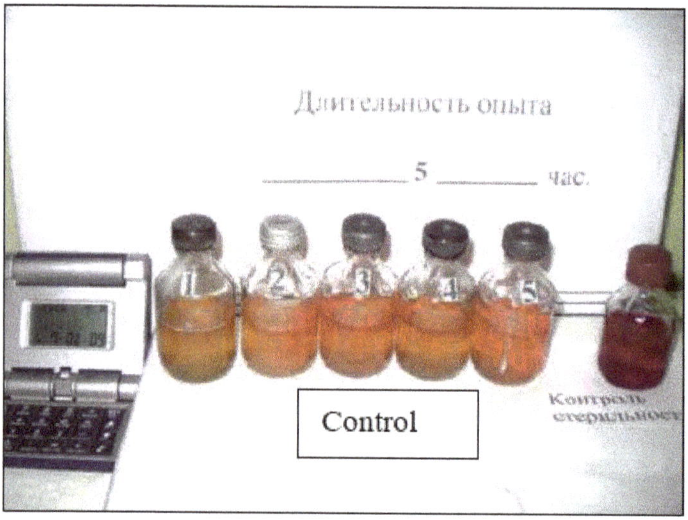

Figure 5.19. 5th hour of the experiment. Control. $R_c = 80$, $D_c = 0$, $V_c = 0$.

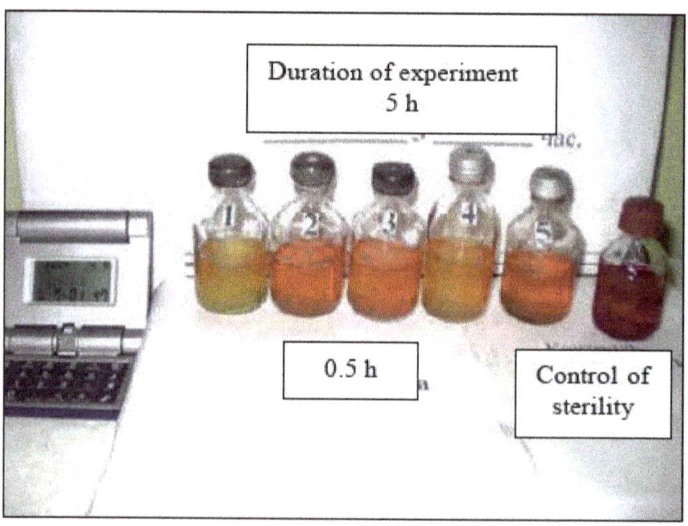

Figure 5.20. 5th hour of the experiment. The activation for 0.5 h. $R_{0.5} = 80$, $D_{0.5} = 0$, $K_{0.5} = 1$, $V_{0.5} = 0$.

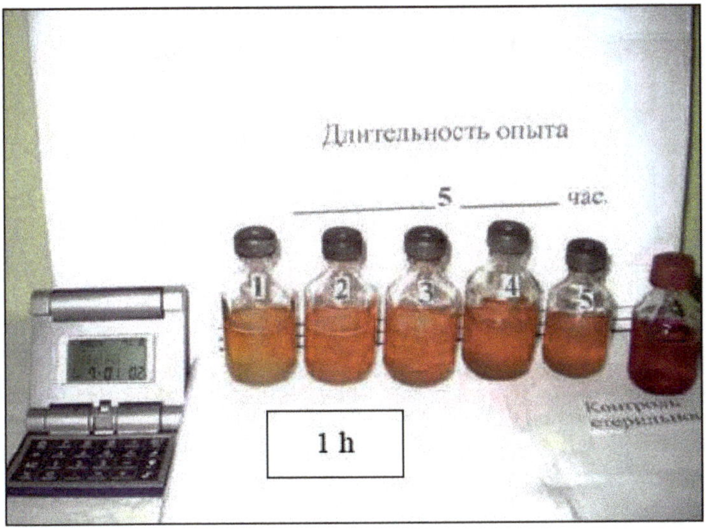

Figure 5.21. 5th hour of the experiment. The activation for 1.0 h. $R_{1.0} = 80$, $D_{1.0} = 0$, $K_{1.0} = 1$, $V_{1.0} = 0$.

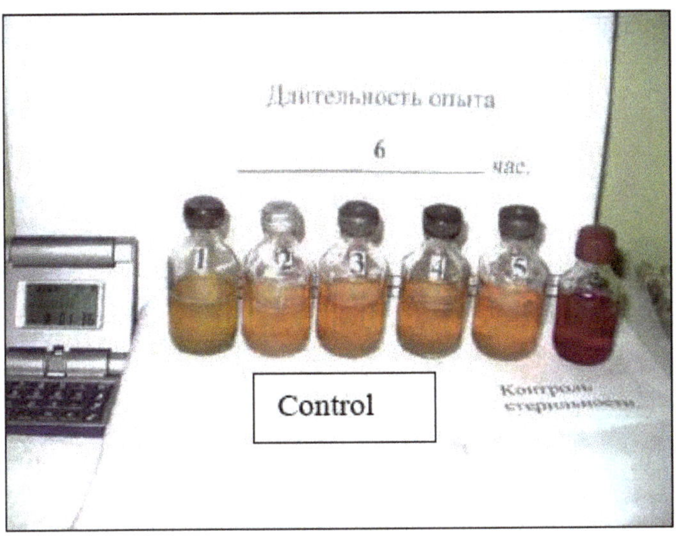

Figure 5.22. 6th hour of the experiment. Control. For all the variants of the experiment (control, the activation for 1.0 h and 0.5 h), glucose is introduced up to a final concentration of 1.8%. $R_c = 80$, $D_c = 0$, $V_c = 0$.

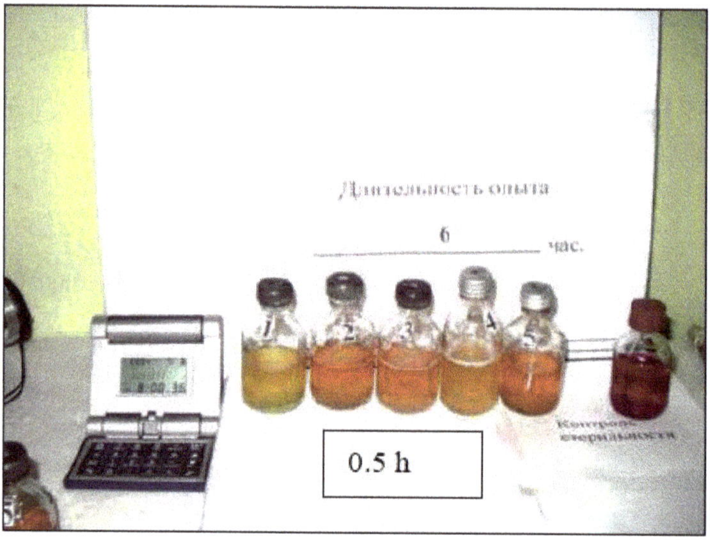

Figure 5.23. 6th hour of the experiment. The activation for 0.5 h. $R_{0.5} = 80$, $D_{0.5} = 0$, $K_{0.5} = 1$, $V_{0.5} = 0$.

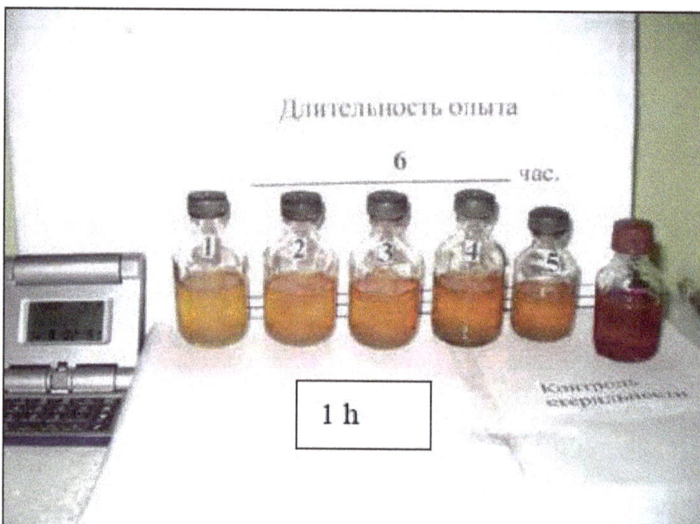

Figure 5.24. 6th hour of the experiment. The activation for 1.0 h. $R_{1.0} = 80$, $D_{1.0} = 0$, $K_{1.0} = 1$, $V_{1.0} = 0$.

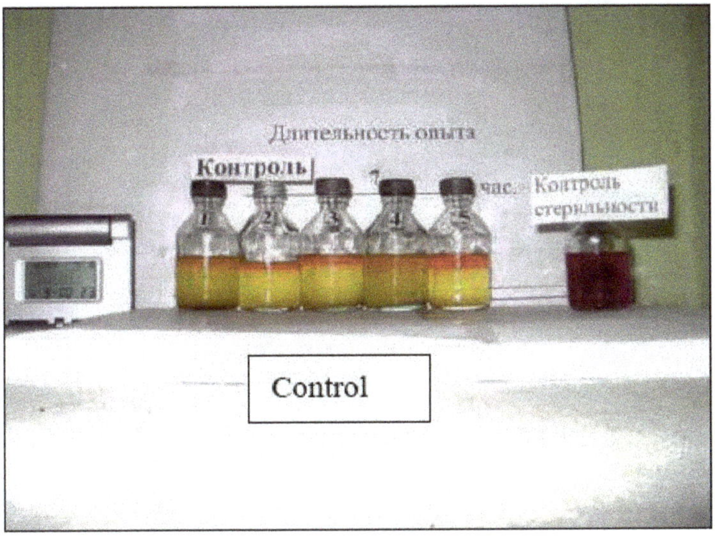

Figure 5.25. 7th hour of the experiment. Control. $R_c = 88$, $D_c = 0.28$, $V_c = 0$.

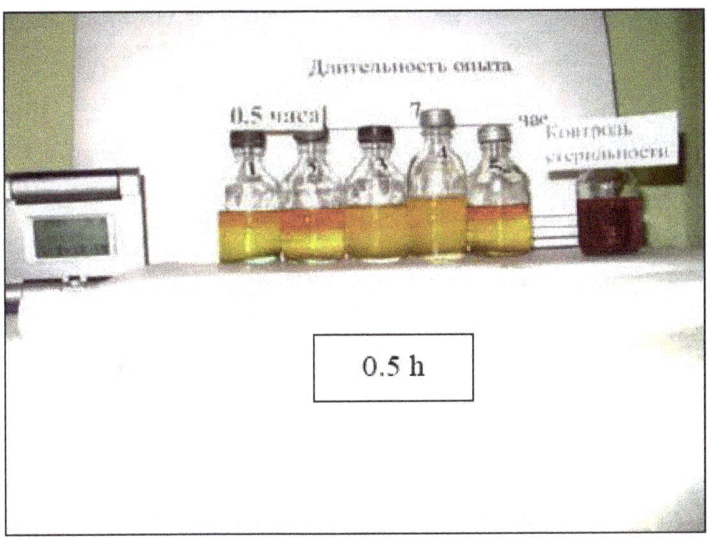

Figure 5.26. 7th hour of the experiment. The activation for 0.5 h. $R_{0.5} = 83$, $D_{0.5} = 0.24$, $K_{0.5} = 1.06$, $V_{0.5} = 0$.

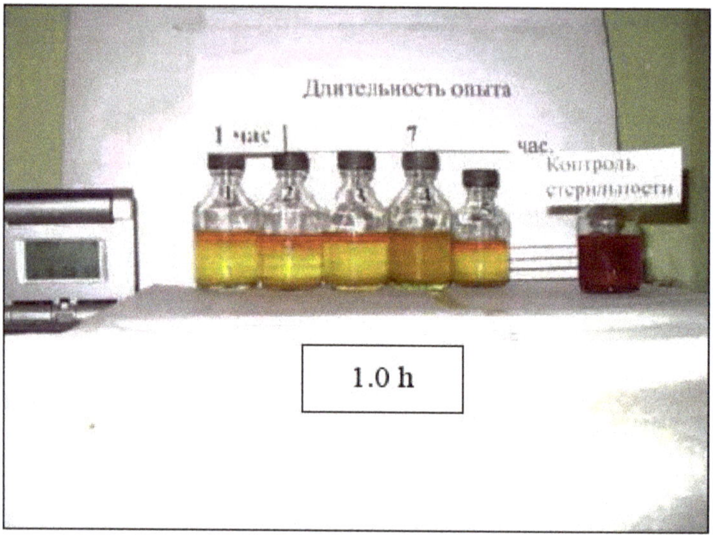

Figure 5.27. 7th hour of the experiment. The activation for 1.0 h. $R_{1.0} = 88$, $D_{1.0} = 0.04$, $K_{1.0} = 1.0$, $V_{1.0} = 0$.

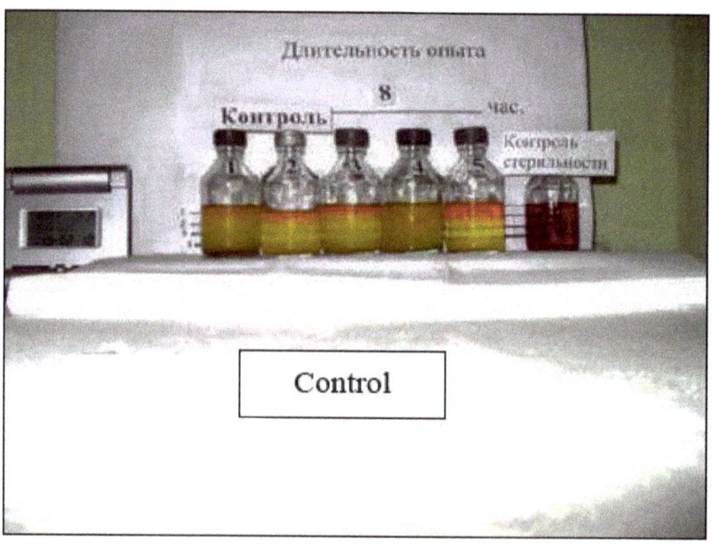

Figure 5.28. 8th hour of the experiment. Control. $R_c = 91$, $D_c = 0.52$, $V_c = 0$.

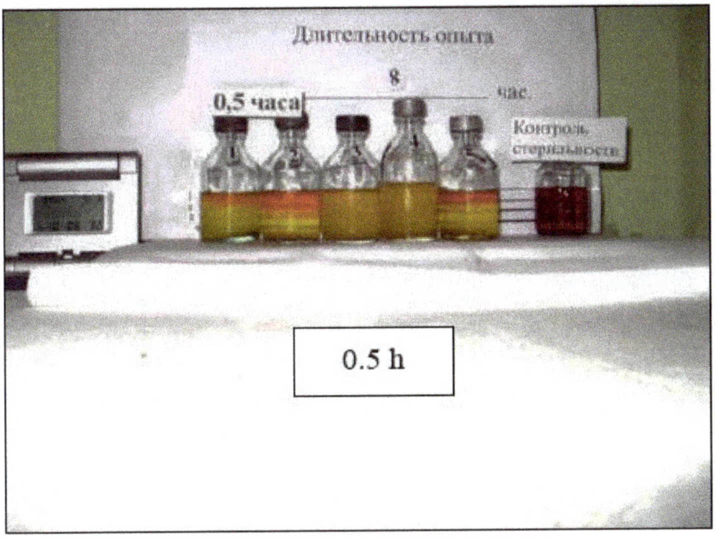

Figure 5.29. 8th hour of the experiment. The activation for 0.5 h. $R_{0.5} = 88$, $D_{0.5} = 0.54$, $K_{0.5} = 1.03$, $V_{0.5} = 0$.

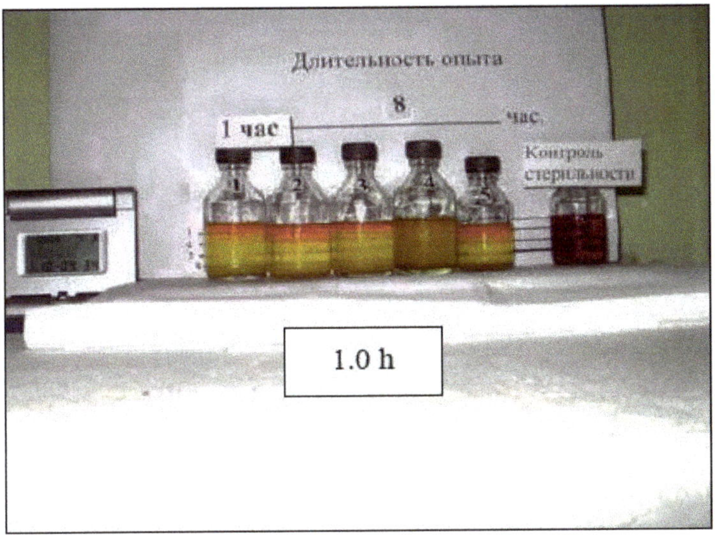

Figure 5.30. 8th hour of the experiment. The activation for 1.0 h. $R_{1.0} = 88$, $D_{1.0} = 0.48$, $K_{1.0} = 1.03$, $V_{1.0} = 0$.

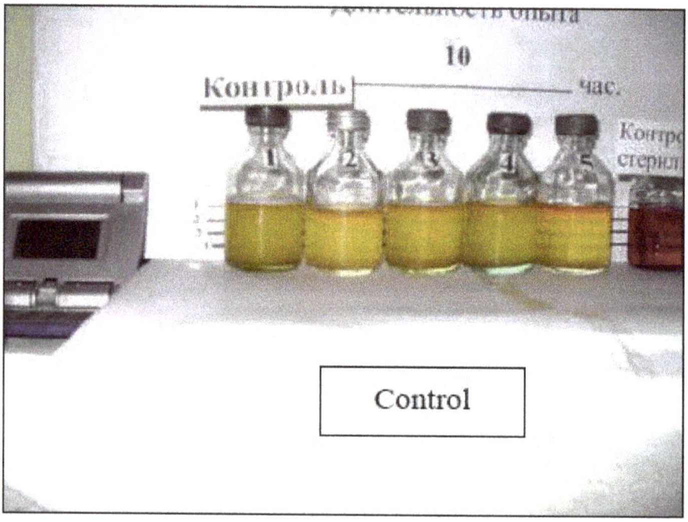

Figure 5.31. 10th hour of the experiment. Control. At the 10th hour of the cultivation for all variants of the experiment, glucose was introduced up to a final concentration of 2%. $R_c = 98$, $D_c = 0.82$, $V_c = 0$.

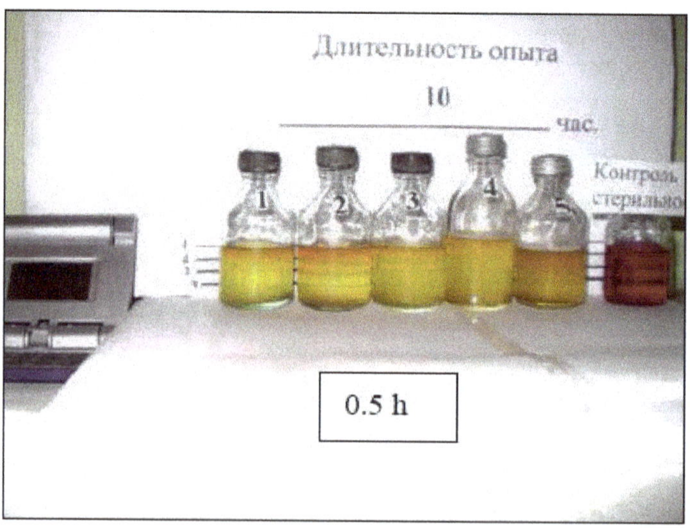

Figure 5.32. 10th hour of the experiment. The activation for 0.5 h. $R_{0.5} = 98$, $D_{0.5} = 0.7$, $K_{0.5} = 1.0$, $V_{0.5} = 2.2$.

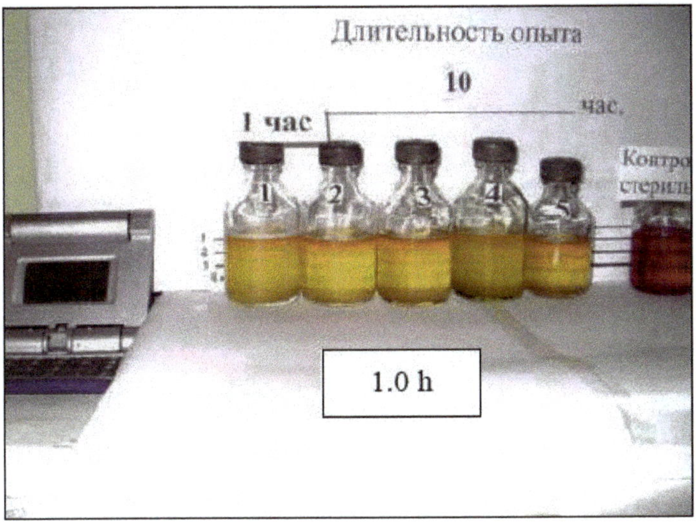

Figure 5.33. 10th hour of the experiment. The activation for 1.0 h. $R_{1.0} = 98$, $D_{1.0} = 0.88$, $K_{1.0} = 1.0$, $V_{1.0} = 2.0$.

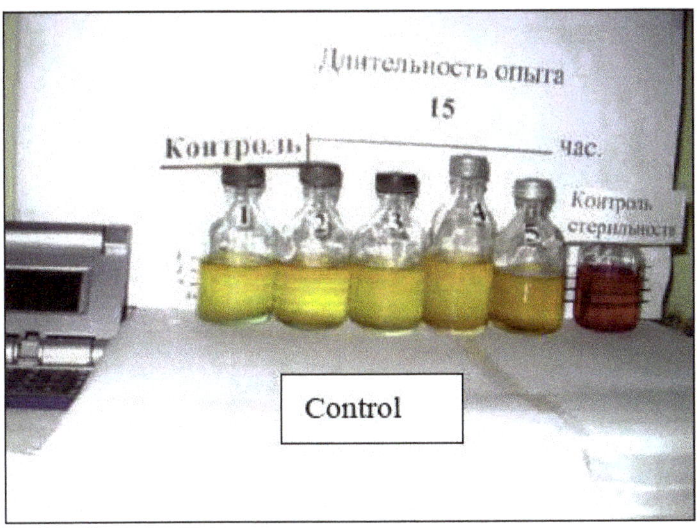

Figure 5.34. 15th hour of the experiment. Control. $R_c = 97$, $D_c = 0.9$, $V_c = 0.8$.

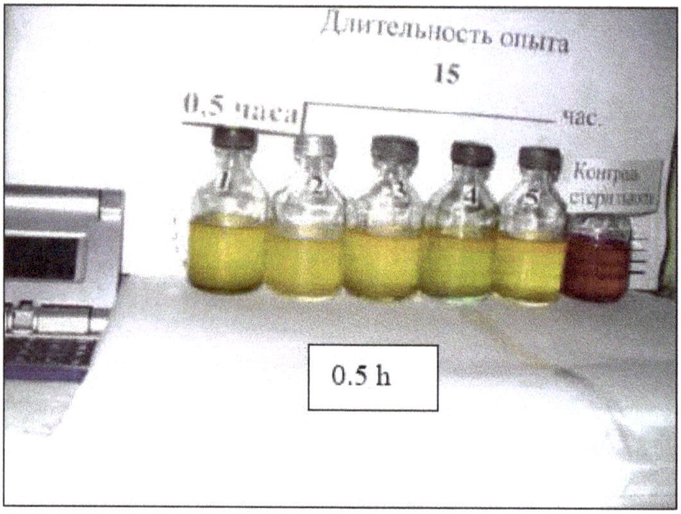

Figure 5.35. 15th hour of the experiment. The activation for 0.5 h. $R_{0.5} = 98$, $D_{0.5} = 0.9$, $K_{0.5} = 0.99$, $V_{0.5} = 0.4$.

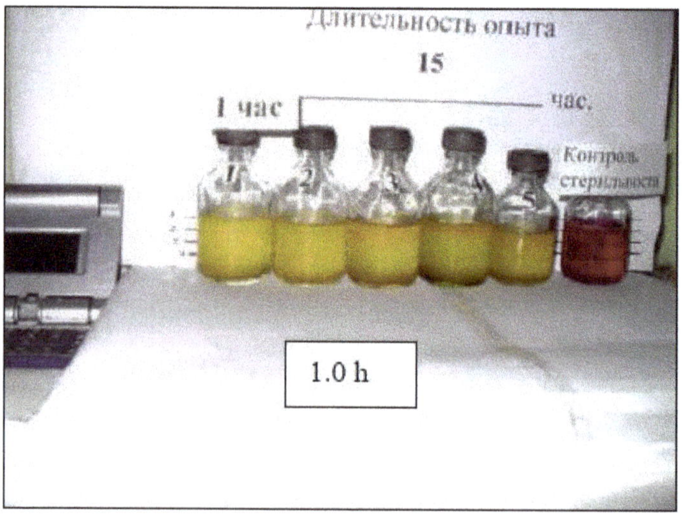

Figure 5.36. 15th hour of the experiment. The activation for 1.0 h. $R_{1.0} = 99$, $D_{1.0} = 0.9$, $K_{1.0} = 0.98$, $V_{1.0} = 1.2$.

In Figs. 5.15–5.36, you can see one flask with the nutrient medium (MPB) without *Escherichia coli* culture, but with resazurin added for the sterility control, near 15 flasks containing the nutrient medium on the basis of the different fractions of activated water and *Escherichia coli* culture. During all the experiments (all Figs. 5.15–5.36), the color of the control flask was invariable. This fact confirms the sterility maintenance in the course of the whole experiment.

As seen, resazurin is almost colorless ($R \approx 100$) at the 10th hour of the cultivation in all variants of the experiment. Therefore, starting from the 11th hour of the experiment, *E. coli* culture was under anaerobic conditions: redox-potential was $Eh < -100$ mV, and the free oxygen O_2 concentration tends to zero.

After 10 h of the experiment, culture *E. coli* was under strictly anaerobic conditions in all variants of the experiment (with different fractions of activated water). The redox-potential of the medium was very low ($Eh \ll -100$ mV). At the 16th hour of the experiment, the synthesis rate of the hydrogen–carbon-dioxide mixture was 2.2–3.4 ml/h, which testifies to the high reductase activity of the culture in that time period. Therefore, the strong oxidizing agent (which contained the highly toxic metal Cr) K_2CrO_4 was introduced into the medium up to a final concentration of 20 mg/L Cr(VI) at the 17th hour of the experiment (Figs. 5.37 and 5.38).

Anion $CrO_4{}^{2-}$ is the high-potential redox-buffer.

Since the reaction

$$CrO_4^{2-} + (n-1) \cdot H_2O + 5H^+ + 3e = Cr(OH)_3 \cdot nH_2O$$

has the standard redox-potential equal to $E_0' = +555$ mV, and resazurin has the standard potential of the transition to the colorless form (leukoform) equal to $E_0' = -100$ mV, the medium after the introduction of chromate becomes red almost instantly.

In 15 min after the introduction of chromate, the Cr(VI) concentration in the different variants of the experiment was as follows:

(i) control — $10 \cdots 12$ mg/L
(ii) the activation for 0.5 h — $11 \cdots 14$ mg/L, and
(iii) the activation for 1 h — $13 \cdots 16$ mg/L.

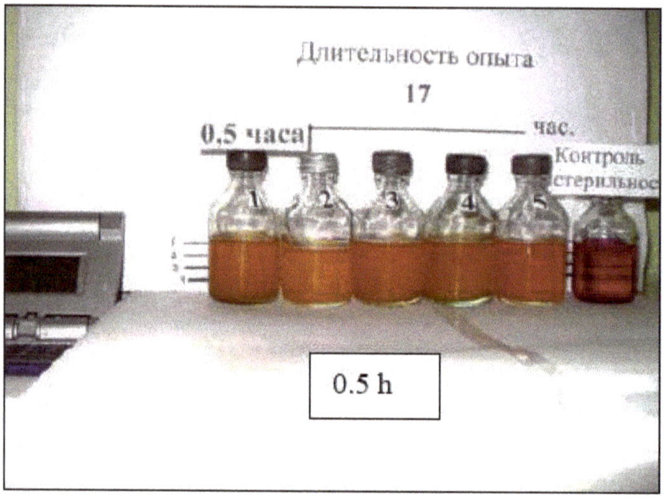

Figure 5.37. 17th hour of the experiment. The activation for 0.5 h. $R_{0.5} = 70$, $D_{0.5} = 1.0$, $K_{0.5} = 1.07$, $V_{0.5} = 0$. The reduction of chromates correlates with the discoloration of resazurin: sterility control [20 mg/L Cr(VI)] — is scarlet (SC), but the evident decrease of the medium coloration intensity occurs in the flasks in 15 min after the introduction of Cr(VI).

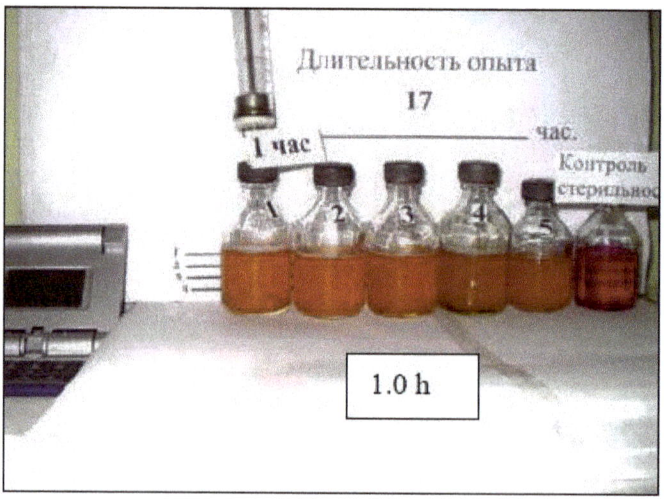

Figure 5.38. 17th hour of the experiment. The activation for 1.0 h. $R_{1.0} = 65$, $D_{1.0} = 1.0$, $K_{1.0} = 1.15$, $V_{1.0} = 0$.

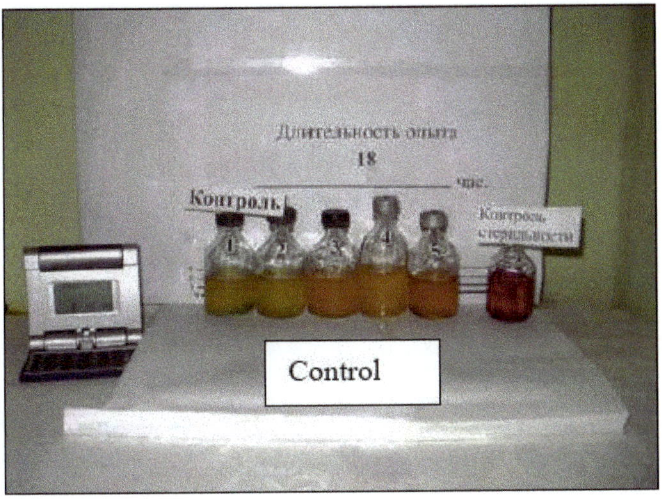

Figure 5.39. 18th hour of the experiment. Control. $R_c = 79$, $D_c = 1.0$, $V_c = 0$.

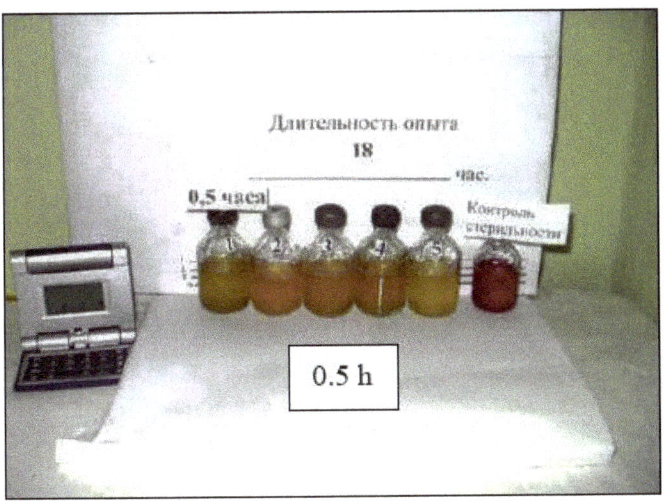

Figure 5.40. 18th hour of the experiment. The activation for 0.5 h. $R_{0.5} = 79$, $D_{0.5} = 1.0$, $K_{0.5} = 1.03$, $V_{0.5} = 0$.

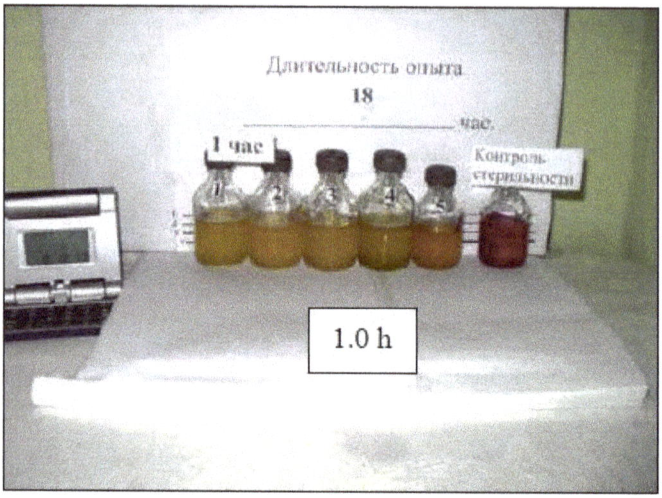

Figure 5.41. 18th hour of the experiment. The activation for 1.0 h. $R_{1.0} = 81$, $D_{1.0} = 1.0$, $K_{1.0} = 0.98$, $V_{1.0} = 1.2$.

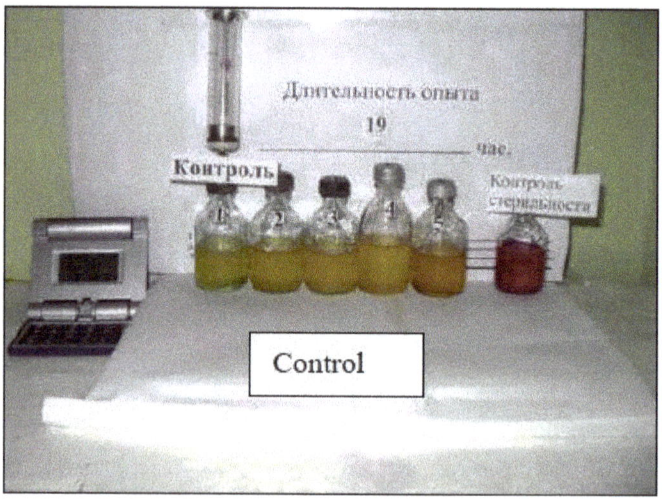

Figure 5.42. 19th hour of the experiment. Control. $R_c = 100$, $D_c = 1.0$, $V_c = 0$. In the figure, the sampling method of a gas from a closed flask by perforation through a rubber plug with a syringe is shown. It is shown that the excess pressure of a gas was absent (the syringe piston did not rise).

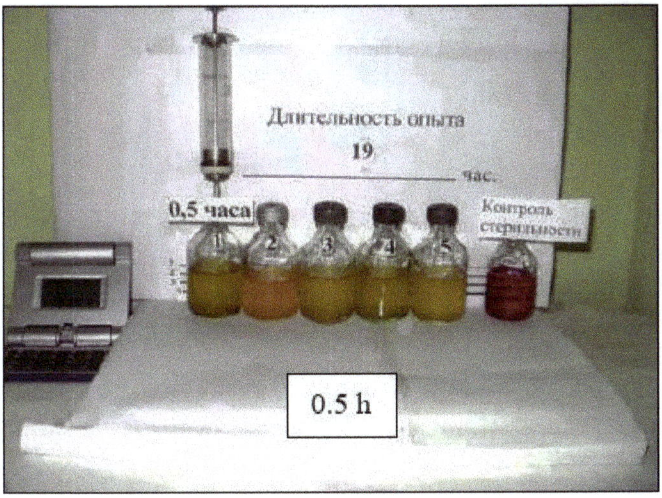

Figure 5.43. 19th hour of the experiment. The activation for 0.5 h. $R_{0.5} = 100$, $D_{0.5} = 1.0$, $K_{0.5} = 1.0$, $V_{0.5} = 0$.

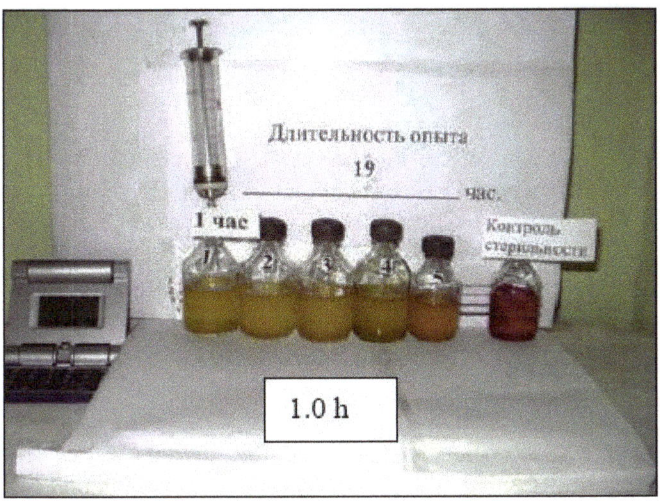

Figure 5.44. 19th hour of the experiment. The activation for 1.0 h. $R_{1.0} = 100$, $D_{1.0} = 1.0$, $K_{1.0} = 1.0$, $V_{1.0} = 0$.

In 30 min after the introduction of chromate, the Cr(VI) concentration in the different variants of the experiment was as follows:

(i) control — 3 · · · 4 mg/L,
(ii) 0.5 h of activation — 5 · · · 6 mg/L, and
(iii) 1.0 h — 7 · · · 8 mg/L.

In 60 min after the introduction of chromate, the redox-balance in the cultural medium restored little by little. In all three variants of the experiment (control, the activation for 0.5 h and 1.0 h), the gradual discoloration of the medium due to the decrease in redox-potential was observed (Figs. 5.39–5.41).

At the 19–20th hour of the experiment, the redox-balance in the cultural medium was restored completely. In all three variants of the experiment (control, the activation of water for 0.5 h and 1.0 h), almost complete discoloration of the medium took place (Figs. 5.42–5.44).

After the introduction of chromate, the synthesis of a gas by the culture did not resume.

In Fig. 5.45, we show the summary dependence of the reductase activity of the culture in three water-fractions grown under anaerobic conditions.

The obtained data testify that water activation has no effect on the metabolic parameters of *Escherichia coli* culture grown under anaerobic conditions during the whole experiment (20-h culture).

5.3. Cultural-Physiological Parameters of *Escherichia coli* Culture Grown on the Activated Meat-Peptone Agar Under Aerobic Conditions

5.3.1. *Effect of different fractions of activated water on the survivability of cells and the growth of colonies on meat-peptone agar under aerobic conditions*

The study of the cultural-physiological parameters of *Escherichia coli* culture during its growth on the activated and nonactivated agarized medium was designed with the purpose to determine the effect of the activation on the survivability of cells, the growth rate of colonies on the base of survived cells, and the shape and the size of grown cells.

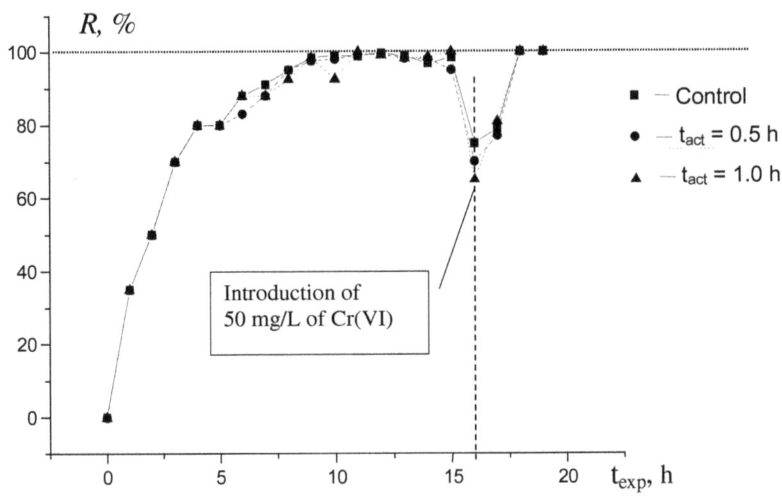

Figure 5.45. Reductase activity (R) of *Escherichia coli* culture grown under anaerobic conditions.

In the experiment, glass Petri dishes were used. Each dish was filled with 15 ml of melted meat-peptone agar. Then, with the purpose to test the sterility, the dishes were held for 24 h in a thermostat at 30°C.

During the investigation of the morphology of colonies and cells of *Escherichia coli*, we prepared the tenfold dilutions of the inoculum (1 ml of active 24-h culture of *Escherichia coli*), and then the corresponding aliquots were spread on the agarized medium by a Drigalsky spatula to obtain isolated cells, from which it was expected to obtain the isolated colonies as a result of their growth. The inoculation into dishes were made in 3, 4, and 5 dilutions: we spread uniformly 0.1 ml of the cell suspension over the surface of the agarized medium, using a glass spatula.

Such a method provides a comparatively low concentration of initial cells on the surface of the agarized medium, which allows visual control over the number of colonies of the culture separately.

The bacteria culture was incubated at a temperature of 20°C under aerobic conditions.

The effect of the activated medium on the growth of *Escherichia coli* on the agarized nutrient medium (meat-peptone agar) is quantitatively expressed as the collection of the coefficients which describe the morphological and physiological parameters of the culture.

Let us enumerate the most important coefficients which sufficiently describe the growth of colonies.

(1) $I_1 = (N_c/N_{t(\text{act})}) \times 100\%$ — index of total survivability:
The index of total survivability characterizes the influence of the duration of activation of the medium on the total number of survived cells (grown colonies) in comparison with the control (the medium prepared with the use of nonactivated water). In this formula, N_c and $N_{t(\text{act})}$ are the number of cells of the culture in 1 ml of the initial nonactivated medium (control) and the activated medium, respectively.

(2) $K_6 = N_c/N_{t(\text{act})}$ — the shape change coefficient of colonies:
In this formula, N_c and $N_{t(\text{act})}$ are the numbers of colonies with a changed shape in the control experiment and in the experiment with activated water, respectively.

(3) $K_8 = d_{t(\text{act})}/d_c$ — the size change coefficient of colonies:
In this ratio, $d_{t(\text{act})}$ and d_c are the average sizes of colonies grown in activated and control media, respectively.

(4) $I_9 = (N_a/N) \times 100\%$ — the anomalous division index of cells:
In this ratio, N_a are N are the numbers of anomalous and normal cells in a specific experiment with a specific fraction of the activated or nonactivated nutrient medium.

The cultural-morphological properties of *Escherichia coli* culture (the size, shape, and color of colonies, and the size and shape of cells) are the important criteria of the effect of activated water on microorganisms. The change of the mentioned parameters provides evidence for the essential influence of the activation on the cell division processes, the synthesis of the structural components of cells, pigments, the "programming" of the formation of a stereometry of colonies, etc.

Regarding the growth of *Escherichia coli* culture on the agarized medium, we took account of characteristics such as the color, size, and shape of colonies. To study the influence of the activated medium on the shape and size of cells under microscope, we prepared the cell suspensions of *Escherichia coli* culture which were grown on the media without activation and with the activation for 0.5 h and 1.0 h. These preparations were at first heated up to a high temperature, dyed with methylene blue, and then studied using a microscope with magnification $\times$ 2400. The preparations were photographed, and then the changes of the shape and size of cells depending on the duration of activation of the medium were calculated.

In Figs. 5.46–5.57, we show the photos of Petri dishes where the colonies of the microbiological culture were grown, at various time points

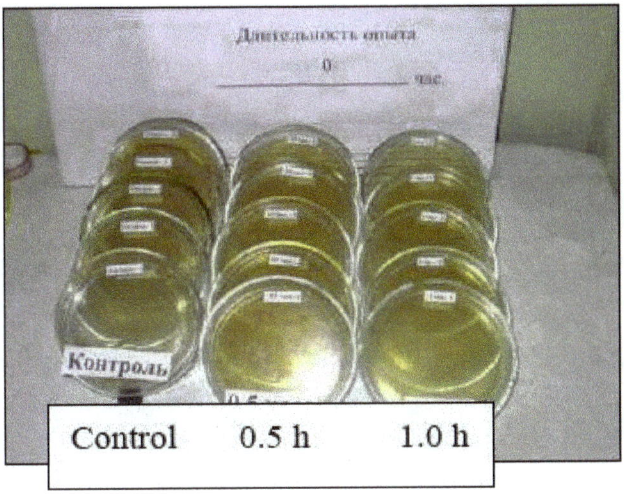

Figure 5.46. General view of Petri dishes. Start of the experiment. On the surfaces of three fractions of the agarized medium in every Petri dish, equal amounts of the cell suspension of tenfold dilution were inoculated. Each fraction corresponded to initial nonactivated water and water activated for 0.5 h or 1.0 h.

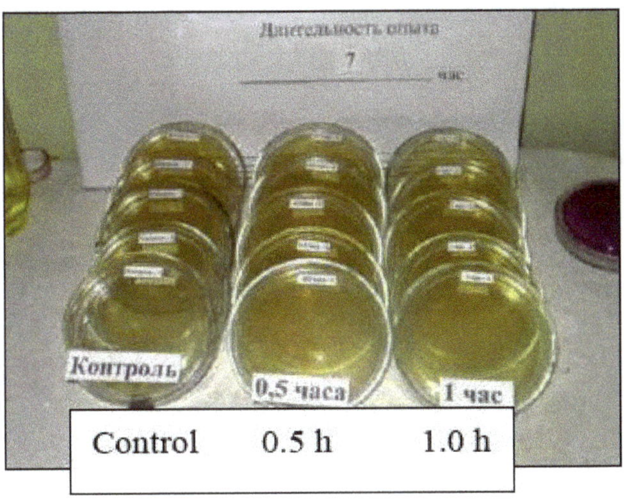

Figure 5.47. General view of Petri dishes. 7th hour of the experiment. There is no visual registration of the colonies of microbiological culture yet.

Figure 5.48. Control. 17th hour of the experiment. On the surface of the Petri dish, the initial growth and the appearance of hundreds of small colonies are seen. In the dishes with activated water, there are few very small colonies.

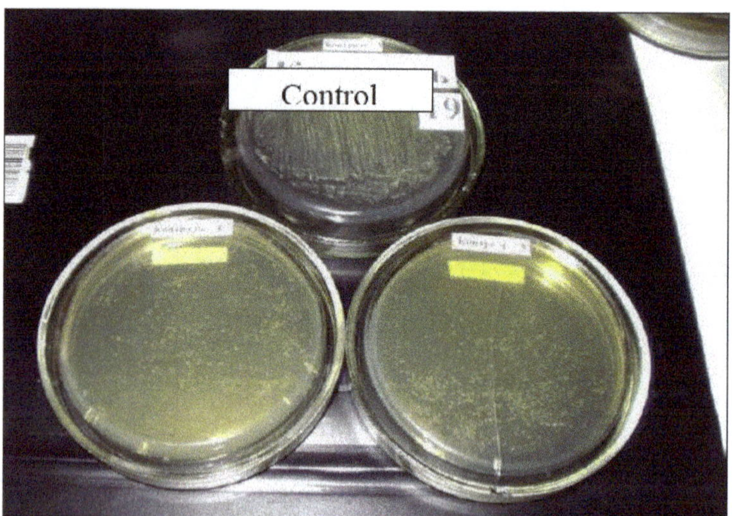

Figure 5.49. Control. 19th hour of the experiment. On the surface of the nonactivated nutrient medium, more than 100 colonies are registrated. Number of cells/ml in inoculate: $N = 1.57 \times 10^8$. Average diameter of colonies grown on the surface: $d = 0.8$ mm.

Figure 5.50. Medium with $t_{act} = 0.5$ h. 19th hour of the experiment. Number of cells/ml in inoculate: $N = 4.8 \times 10^6$. Average diameter of colonies grown on the surface: $d = 1.8$ mm.

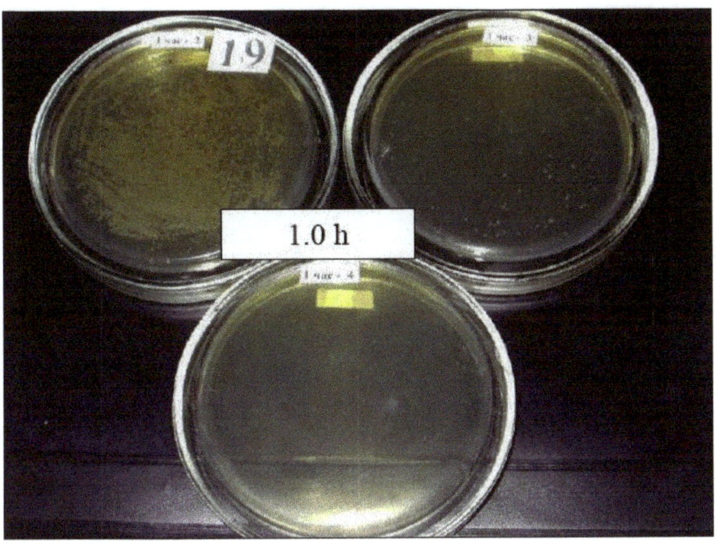

Figure 5.51. Medium with $t_{act} = 1.0$ h. 19th hour of the experiment. Number of cells/ml in inoculate: $N = 4.3 \times 10^5$. Average diameter of colonies grown on the surface: $d = 1.0$ mm.

Figure 5.52. Control. 23rd hour of the experiment. Number of cells/ml in inoculate: $N = 1.7 \times 10^8$. Average diameter of colonies grown on the surface: $d = 1.1$ mm.

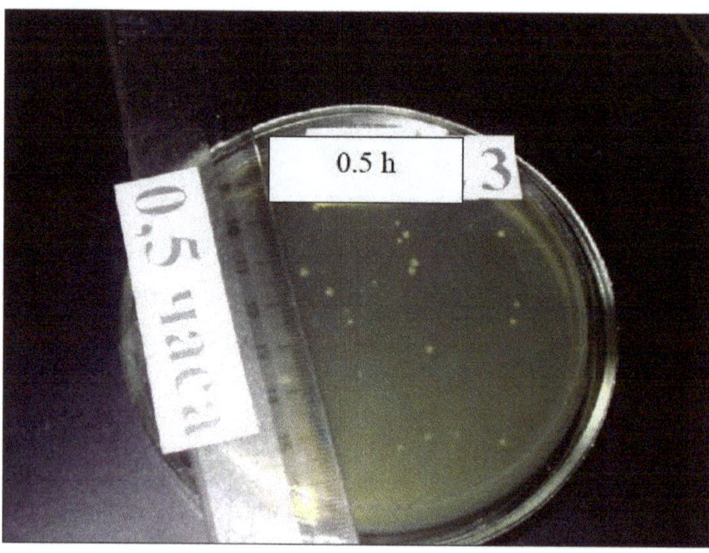

Figure 5.53. Medium with $t_{act} = 0.5$ h. 23rd hour of the experiment. Number of cells/ml in inoculate: $N = 6.4 \times 10^6$. Average diameter of colonies grown on the surface: $d = 1.8$ mm.

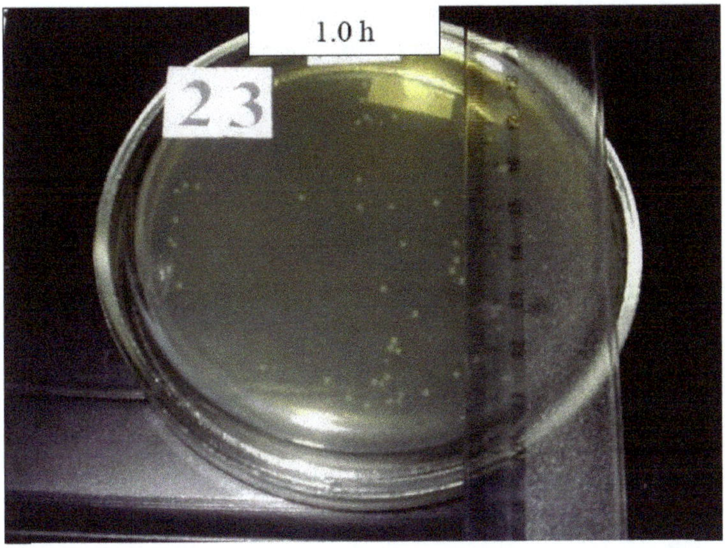

Figure 5.54. Medium with $t_{act} = 1.0$ h. 23rd hour of the experiment. Number of cells/ml in inoculate: $N = 5.2 \times 10^5$. Average diameter of colonies grown on the surface: $d = 1.5$ mm.

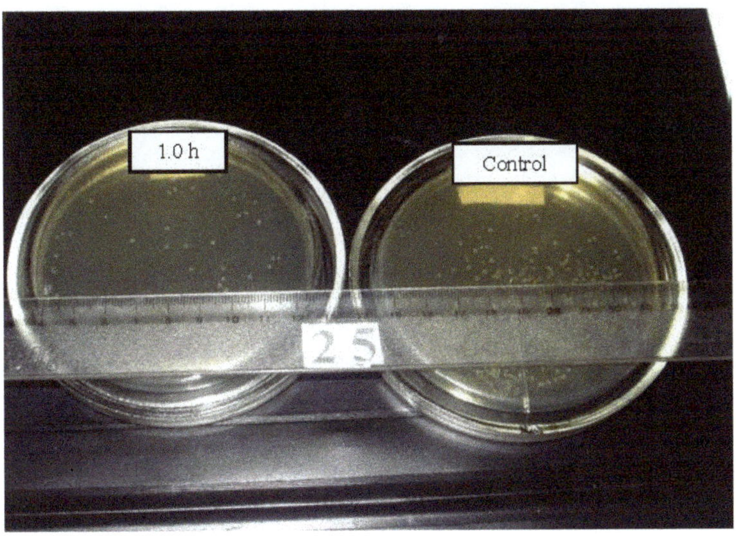

Figure 5.55. Control and medium with $t_{act} = 1.0$ h. 25th hour of the experiment. Control: $N_c = 1.7 \times 10^8$; $d = 1.2$ mm. Medium with $t_{act} = 1.0$ h: $N_{1.0} = 5.6 \times 10^5$; $d = 2$ mm.

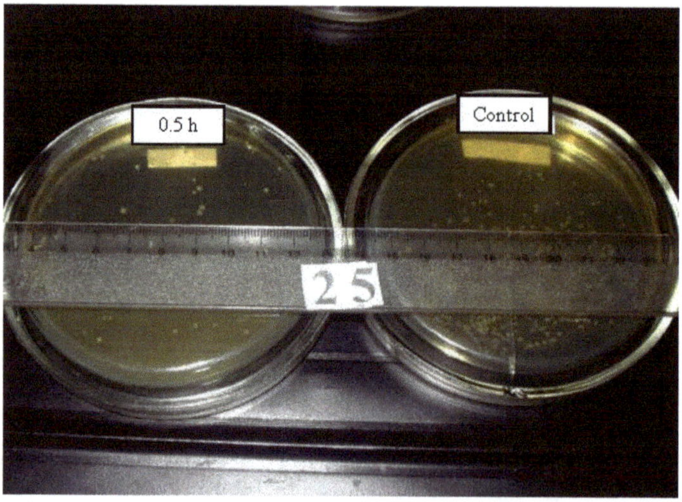

Figure 5.56. Control and medium with $t_{act} = 0.5$ h. 25th hour of the experiment. Control: $N_c = 1.7 \times 10^8$; $d = 1.2$ mm. Medium with $t_{act} = 0.5$ h: $N_{1.0} = 6.4 \times 10^6$; $d = 2.7$ mm.

(a)

Figure 5.57. Control and medium with $t_{act} = 0.5$ h and $t_{act} = 1.0$ h. (a) 29th and (b) 31st hours of the experiment. Control: $N_c = 1.7 \times 10^8$; $d = 1.2$ mm. $t_{act} = 0.5$ h: $N_{1.0} = 6.4 \times 10^6$; $d = 2.7$ mm. $t_{act} = 1.0$ h: $N_{1.0} = 5.6 \times 10^5$; $d = 2$ mm. Results for 29th and 31st hours of the experiment coincide with the data for the 25th hour of the experiment.

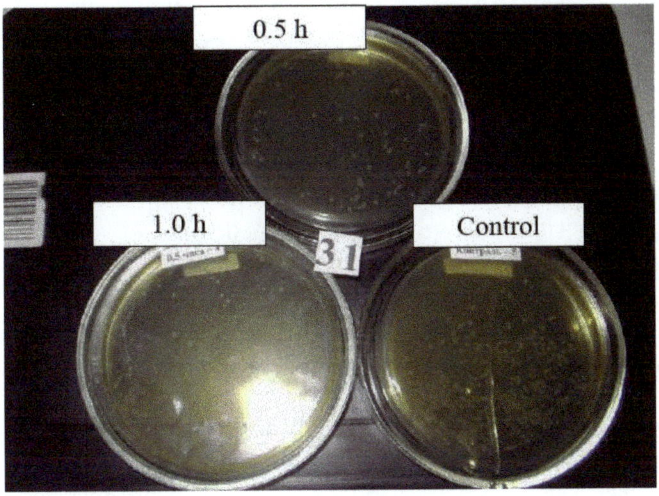

(b)

Figure 5.57. *(Continued)*

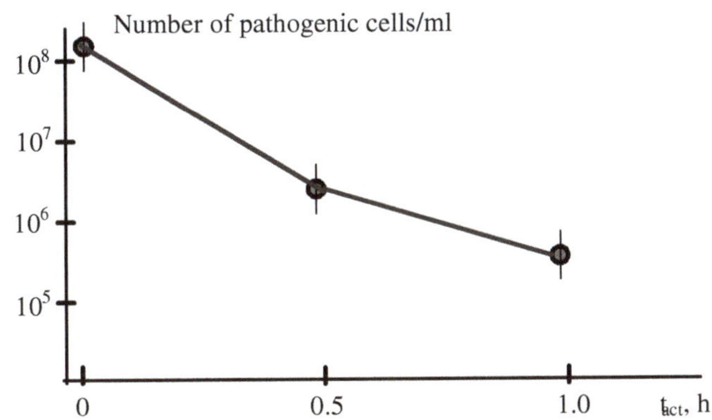

Figure 5.58. Bactericidal (bacteriostatic) action of different fractions of activated water on *Escherichia coli* cultured under aerobic conditions.

for different fractions of activated water. Below each photo, the necessary information describing the actual state of colonies at a given time point is presented with the corresponding comments.

The results of bactericidal (bacteriostatic) action of activated water on *Escherichia coli* are shown in Fig. 5.58.

Conclusions

The results of experiments on the study of the effect of the activated medium on the growth parameters of the microbiological culture *Escherichia coli* grown on MPA under aerobic conditions can be formulated as follows:

(1) The index of total survivability for the microbiological culture *Escherichia coli* is equal to:

- $I_{1(0.5\,h)} = 3.7$ (for the medium activated for 0.5 h);
- $I_{1(1.0\,h)} = 0.3$ (for the medium activated for 1.0 h).

Such values of the coefficients I_1 testify that the medium activation leads to a significant bactericidal effect:

- For the 0.5-h activation of the medium, the amount of survived cells was only 3.7% in comparison with the control.
- For the 1.0 h activation of the medium, the amount of survived cells was only 0.3% in comparison with the control.

(2) All colonies, both in experimental variants (in the media with the duration of activation of 0.5 and 1.0 h) and in control (nonactivated medium), were identical and had round hemispherical shape.

The value $K_6 = 1$ in both experimental variants testifies that the water activation does not cause a change of the shape of colonies of *E. coli* on the agarized medium.

(3) During the growth of colonies, their size was changed.

The shape-change coefficient of colonies is equal to:

- $K_{8(0.5h)} = 1.8$ (for the medium activated for 0.5 h measured at the 19th hour of the experiment).
- $K_{8(1.0h)} = 1.0$ (for the medium activated for 1.0 h measured at the 19th hour of the experiment).
- $K_{8(0.5h)} = 2.3$ (for the medium activated for 0.5 h measured at the 25th hour of the experiment).
- $K_{8(1.0h)} = 0.8$ (for the medium activated for 1.0 h measured at the 25th hour of the experiment).

These data provide evidence for the following:

(i) Activation of the agarized medium for 0.5 h leads to an increase of the diameter of colonies by ≈ 2 times.

(ii) Activation of the agarized medium for 1.0 h does not lead to an increase of the diameter of colonies, or is accompanied by its slight decrease.

(4) During the growth of colonies, the significant change of the shape of cells was observed.

At the end of the growth, the values of the anomalous division index of cells are equal to, respectively:

- $I_{9(c)} = 4$ for the nonactivated medium (control);
- $I_{9(0.5h)} = 280$ for the medium activated for 0.5 h; and
- $I_{9(1.0h)} = 220$ for the medium activated for 1.0 h.

(*Continued*)

(*Continued*)

Values of the coefficients I_9 testify that the activation of the agarized medium leads to an increase of the number of anomalous cells of the microbiological culture *Escherichia coli* by 220–280%.

In the control experiment, i.e. the culture grown on the nonactivated agarized medium, the number of anomalous cells (4%) does not exceed the usual stochastic deviation.

The morphological peculiarities of the cells that were grown in the activated agarized medium and some conclusions about the mechanism and consequences of the effect under study will be considered below.

5.3.2. *Peculiarities of the morphology and division of cells of Escherichia coli in activated meat-peptone agar under aerobic conditions*

It was shown above that the activation of a nutrient medium leads to the abrupt suppression of the survivability of cells of the microbiological culture grown on meat-peptone agar under aerobic conditions. One of the causes for such bactericidal effect is the influence of activated water on cell division. This problem will be discussed in details in Chap. 8.

The anomaly of cell division is manifested in this case through the creation in the colonies of *Escherichia coli* culture of the so-called "squibs" which are "blown" cells, whose division is violated but the growth is preserved.

In Figs. 5.59–5.61, we present the photos of fragments of the system of cells that were grown on the surfaces of the activated and nonactivated agarized nutrient media. These fragments correspond to Fig. 5.57 and are obtained from a large magnification of separate colonies on the surfaces of the corresponding Petri dishes by a microscope.

As seen from the fragment shown in Fig. 5.61, the main part of cells on the surface of the nonactivated medium corresponds to the "standard" shape for *Escherichia coli* culture (like elongate linear sticks), and the number of anomalous (nonseparated) cells is few. The estimations given above show that the concentration of anomalous cells is at most 4%, which is within the statistical fluctuations.

The basically different pattern of morphology of cells is observed for the culture grown on the activated medium.

In the case of the medium activated for 0.5 h, a great number of cells which do not separate after the division at the termination of the growth stage

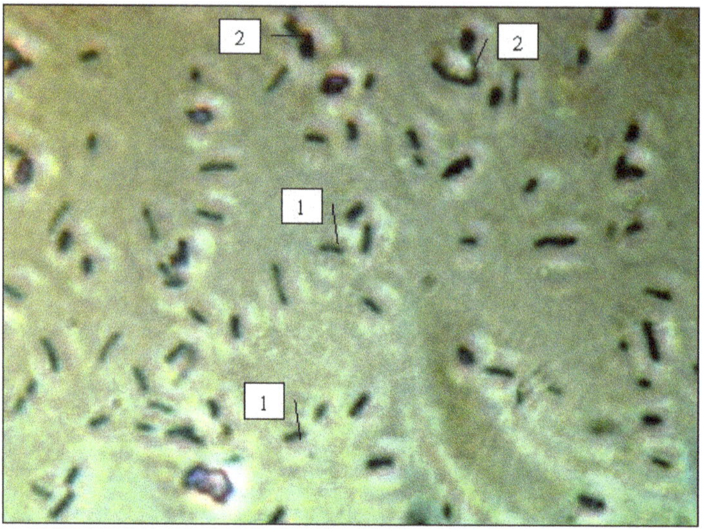

Figure 5.59. General view of a surface fragment of the nonactivated agar medium with grown cells. 1 — cells with normal division, 2 — cells with anomalous division.

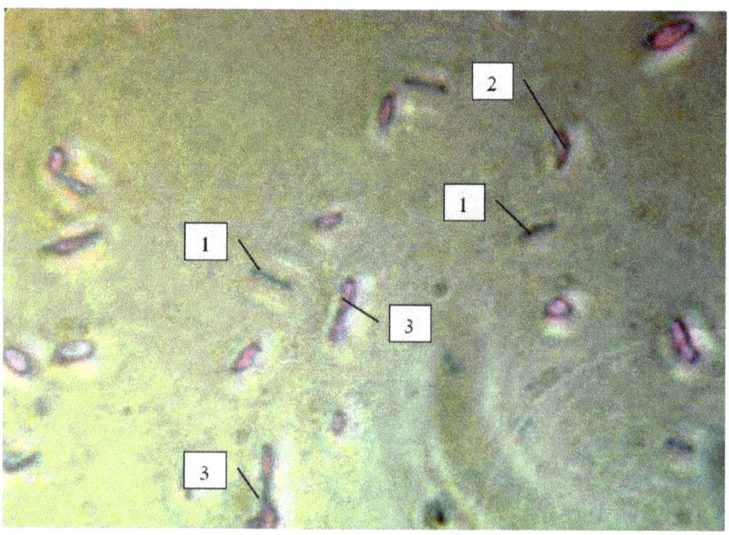

Figure 5.60. General view of a surface fragment of the agarized medium with grown cells activated for 0.5 h. 1 — cells with normal division, 2 — cells with anomalous division, 3 — non-separated linear chains of cells.

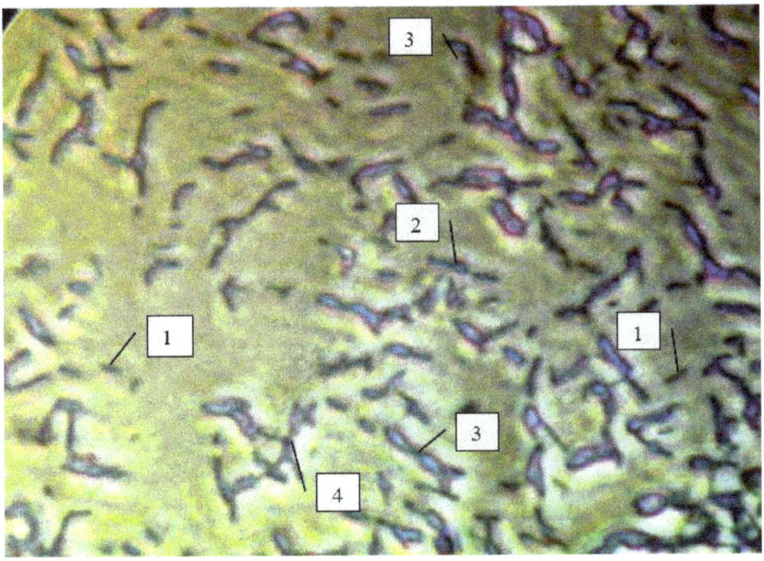

Figure 5.61. General view of a surface fragment of the agarized medium with grown cells activated for 1.0 h. 1 — cells with normal division, 2 — cells with anomalous division, 3 — non-separated linear chains of cells, 4 — nonseparated branched chains of cells.

and form the linear chains that contain two to three cells connected in series is observed. This result is shown in Fig. 5.60. Although the question about the ability of such cells for a further division remains unclear, it is evident that such a division will be strongly inhibited in any case. The quantitative estimates show that the concentration of such anomalous cells exceeds that of normal cells by a factor of 2.2.

The even greater influence of the activation on the morphology of cells of *Escherichia coli* corresponds to the medium activated for 1 h. As follows from the magnified fragment of the surface of such nutrient medium shown in Fig. 5.61, the anomalous cells become the dominating component of the process of division of cells. Their relative concentration exceeds that of normal cells by a factor of 2.7. It is worth to note that, in this case, not only nonseparated linear chains containing two to three cells are formed. We also observe the great number of nonseparated branched chains of cells with a complicated branched structure that contains many (up to 10–20) nonseparated cells in some cases. With regard for the fact that the cell division process in such anomalous systems is to be suppressed severely, one of the natural reasons for the formation of these superstructures is, apparently, the mutual joining of individual linear nonseparated fragments.

It is seen that, although the cell division process occurs most likely in accordance with the natural laws, the division rate of cells sharply decreases during the last stage of their growth. It seems almost evident that just the change of the properties of activated water plays the key role in this last stage. This is conditioned by the fact that the energetic characteristics of the cell division process, which leads to the increase of both the surface and the total surface energy of dividing cells, are directly connected to the influence of the "external" (with respect to cells) water which is the base of the nutrient medium on their surface energy. Saying simply, the full separation of a pair of cells after the division in activated water becomes energetically disadvantageous because of the excessive increase of the surface energy of fully separated cells. The indirect confirmation of this conclusion is the phenomenon observed, namely the very abrupt decrease of the viscosity coefficient of activated water at its low velocity. This effect testifies unambiguously to the weakening of intermolecular attracting forces on the boundary that isolates activated water from walls bounding the water localization zone. In this meaning, such properties of water can be identified with the phenomenon of hydrophobic effect. It is evident that, in the presence of this phenomenon, the growth of the boundary surface becomes energetically disadvantageous (and therefore impeded). It is very probable that such a phenomenon describes nicely the inhibition of the cell division process in the last stage, which leads to the appearance of nonseparated, branched many-link cell systems that were observed in the experiment. This question will be considered in details in Chap. 8.

The authors understand that this effect is registered yet only for *Escherichia coli* culture but they think that it has to be manifested in other cell systems (including callus tissue and cancer cells) because of the common physicomolecular characteristics. Two mentioned systems will be considered in details in what follows.

5.4. Influence of Activated Water on the Stability of the Microbiological Culture *Escherichia coli* to the Action of Antibiotics Under Aerobic Conditions

One of the most interesting aspects of the action of activated water on the biological objects is related to the modification of the efficiency of the

Table 5.1. Representatives of the investigated classes of antibiotics.

No	Name of antibiotic	Antibiotic class
1	Clindamycin	Lincosamides
2	Kanamycin	Aminoglicosides
3	Cephalexin	1st generation of cephalosporin
4	Ceftriaxone	3rd generation of cephalosporin
5	Chloromycetin	
6	Ciprofroxacin	Fluoroquinols
7	Ampicillin	
8	Tetracycline	
9	Cephaclor	2nd generation of cephalosporin
10	Carbenicillin	Penicillin

influence of various antibiotics on these objects in the presence of such water. For this purpose, we studied the influence of the activation degree of water, the base of nutrient media, on the efficiency of the influence of 10 different antibiotics on the pathogenic *Echerichia coli* culture. These antibiotics are listed in Table 5.1.

In each Petri dish on the surface of the agarized medium with uniform lawn of bacteria, five sterile paper disks containing five different antibiotics were placed. The labeled number of each antibiotic corresponded to its name from Table 5.1. The same antibiotics correspond to the photos shown in Fig. 5.62.

Violet resazurin was presented in the medium as the indicator of metabolic activity. Microorganisms during their growth on the medium surface decreased the redox-potential. As a result, resazurin was successively transformed into resorufin (red) and then into the colorless form (leukoresazurin).

The technique of the evaluation of the redox-potential by the indicator color is considered in Sec. 5.1.

If there was no growth, the redox-potential remained high, and the color of resazurin was not changed. Inside the growth inhibition zone (GIZ) (around the disk with the antibiotic), the reductase activity was absent. As a result, this zone was colored in blue-violet or red. The stability to the action of antibiotics (SA) was determined by the diameter (D_0, mm) of GIZ. The quantitative characteristics were determined in 24 h and 36 h of the growth of cultures in the presence of antibiotics.

The investigations were made for three water fractions — initial nonactivated water, water activated for 30 min (0.5 h), and water that was influenced with an activator for 60 min (1.0 h).

The investigations were made under aerobic conditions at a temperature of 20°C.

The results are shown in Figs. 5.62–5.64.

In Fig. 5.62, we show the general view of Petri dishes with the corresponding fractions of the agarized medium and in the presence of specific antibiotics. The registration was made after the 15-h presence of antibiotics on the surface of the medium layer, where *Escherichia coli* culture was grown.

During the measurements, it was discovered that, for the first 15 h, the GIZs around each disk with an antibiotic became partly contacting. Therefore, we were unable to make quantitative measurements. Inside this initial time interval, only qualitative estimations are correct. According to such estimations, the maximal increase of SA was observed in the medium based on the water activated for 0.5 h, and the minimal one — for the activation for 1 h. During the further growth period of *Escherichia coli*, the significant dispersion of the characteristics defining SA was observed.

After 24 h of the course of the experiment, GIZs for different antibiotics decreased significantly. These zones ceased to overlap, and the quantitative registration of SA by the diameters D_0 of GIZs became possible. These data are shown in Fig. 5.63. Let us discuss these results briefly.

The higher SA (the least diameter D_0) for all 10 antibiotics was observed in the medium with the 0.5-h activation, and the least SA was observed at the activation for 1 h. In all variants during the entire experiment, *Escherichia coli* culture has maximal stability to the action of clindamycin (GIZ was absent). The culture that grew in the nonactivated medium (control) was mostly inhibited by ceftriaxone ($D_0 = 36$ mm), chloromycetin ($D_0 = 32$ mm), and cephaclor ($D_0 = 31$ mm).

In the case of the medium activated for 0.5 h, the stability of *Escherichia coli* culture to all antibiotics increased in comparison with the control. The range of the increased SA is between 1.13 (ceftriaxone) and 4 (chloromycetin), i.e. the stability increased from 13% to 400%.

The 1.0-h duration of the medium activation resulted in an ambiguous effect on SA of the culture. For kanamycin, cephalexin, and tetracycline, SA increased in comparison with the control and the medium activated for 0.5 h by 2.07–9 times. For chloromycetin and cephaclor, SA increased in comparison with the control (by 1.03–1.2 times), but it decreased in

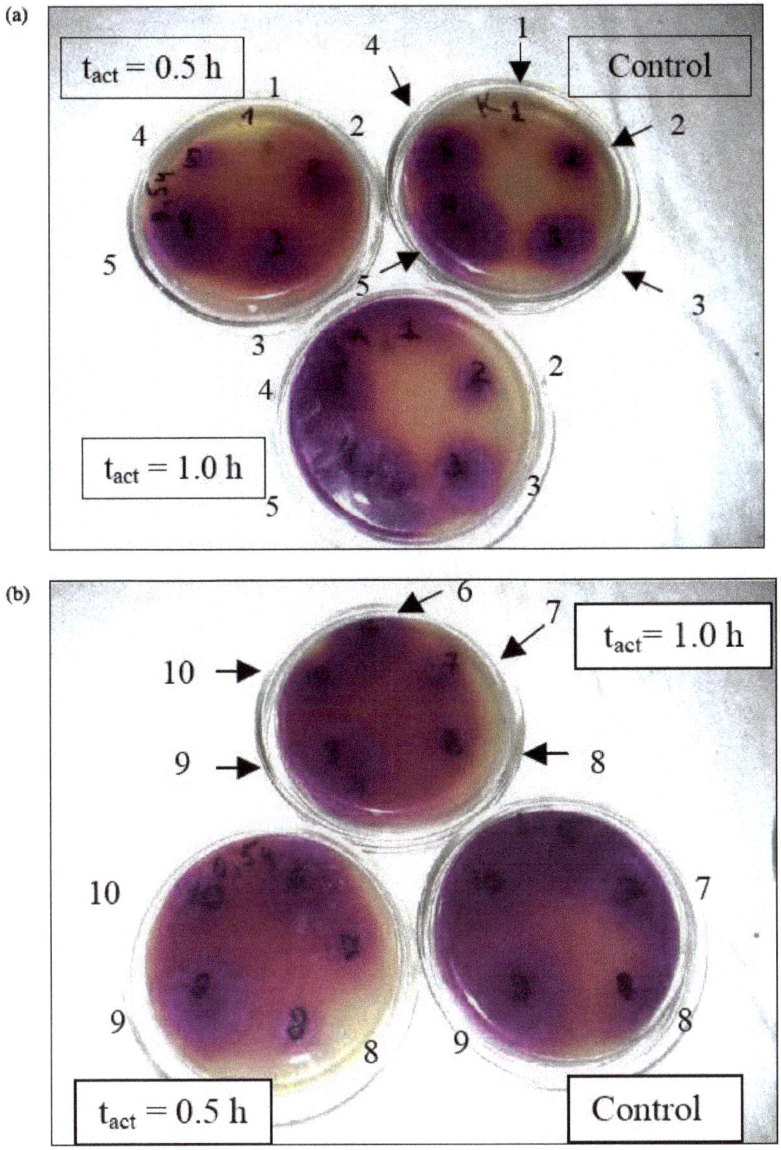

Figure 5.62. General view of Petri dishes with *Escherichia coli* culture for three fractions of water after 15 h of the experiment to study the influence of antibiotics on the culture: (a) clindamycin (1), kanamycin (2), cephalexin (3), ceftriaxone (4), chloromycetin (5); (b) ciprofroxacin (6), ampicillin (7), tetracycline (8), cephaclor (9), carbenicillin (10).

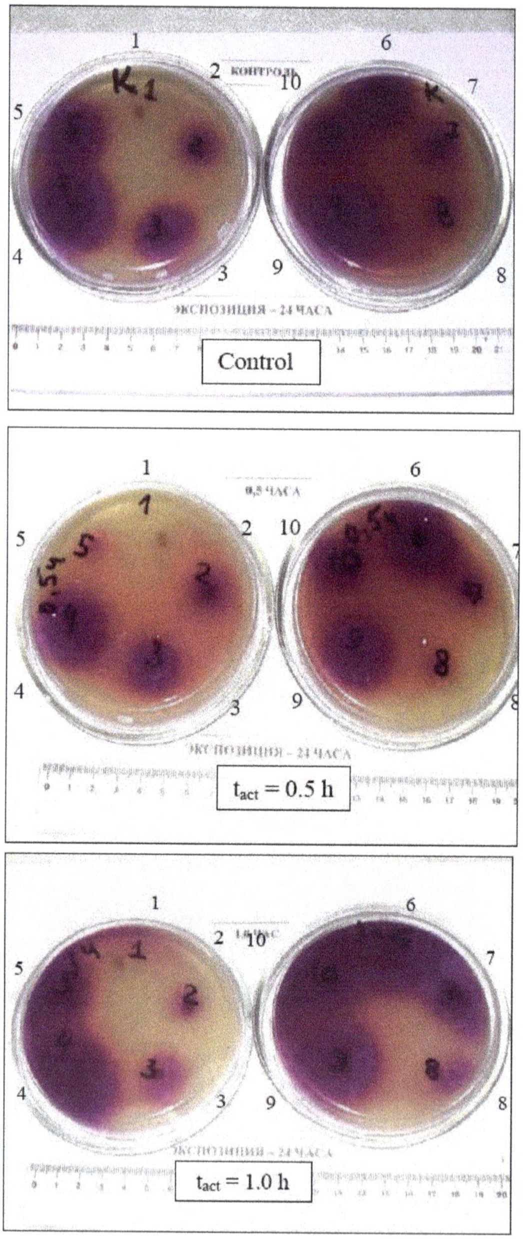

Figure 5.63. General view of Petri dishes with *Escherichia coli* culture for three fractions of water after 24 h of the experiment. The list of antibiotics corresponds to that for Fig. 5.62.

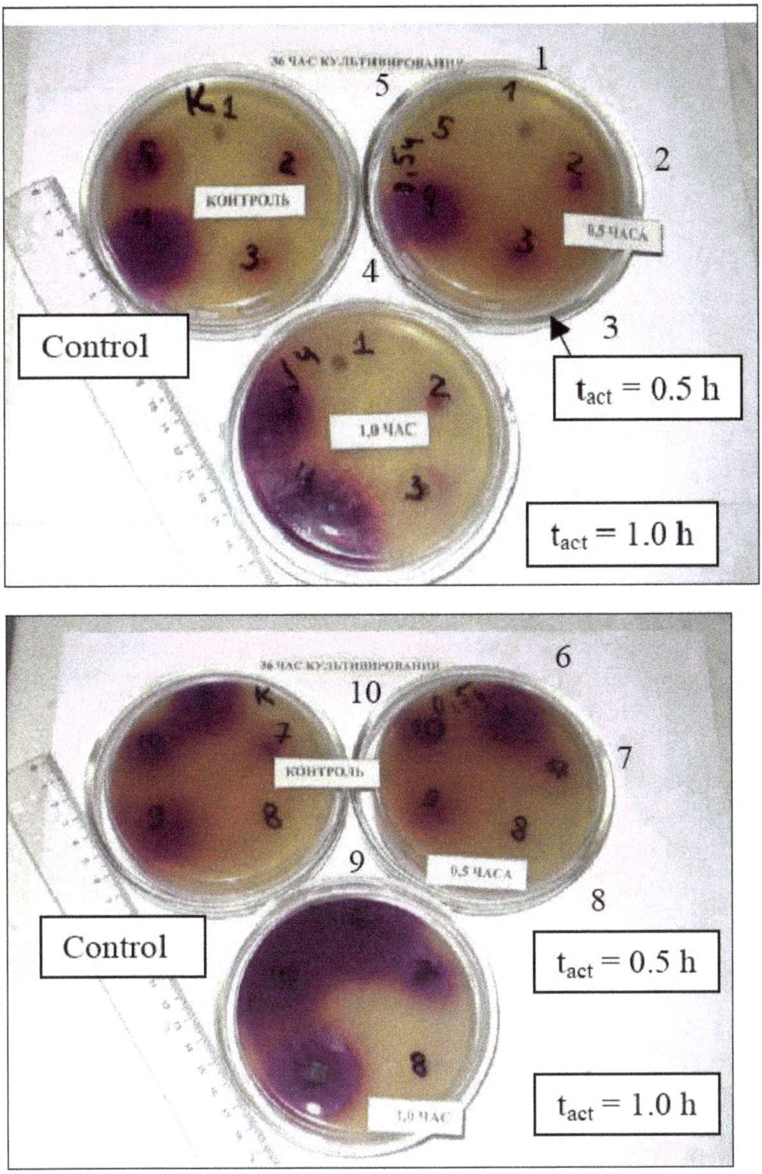

Figure 5.64. General view of Petri dishes with *Escherichia coli* culture for three fractions of water after 36 h of the experiment. The list of antibiotics corresponds to that for Fig. 5.62.

comparison with the variant where the medium was activated for 0.5 h. With the activation for 1.0 h, *Escherichia coli* culture is more sensitive to ciprofroxacin, ampicillin, and carbenicillin in comparison with both the control (by a factor from 0.64 to 0.84) and the variant with the activation for 0.5 h (by a factor from 0.53 to 0.74).

Another series of quantitative measurements was made in 36 h after the introduction of antibiotics onto the growing culture surface. The general view of Petri dishes with *Escherichia coli* culture for three fractions of water after 36 h of the experiment is shown in Fig. 5.64. Below, we note some main peculiarities of the action of antibiotics for this time period.

In the variant of the experiment with the activation for 0.5 h, the stability to chloromycetin increased in comparison with the control by 19 times, but it decreased to 0.9 for ceftriaxone.

In the variant of the experiment with the activation for 1.0 h, the stability to ciprofroxacin, carbenicillin, and cephaclor decreased in comparison with the control by 0.65–0.76 times. The stability to ampicillin decreased in comparison with the control by 0.44 times and it decreases to 0.9 for ceftriaxone. At the same time, the stability to kanamycin and cephalexin increases by 12–13 times.

The obtained results for both series of measurements and different modes of activation are shown in Table 5.2.

The obtained dependence determining the influence of the water activation duration on the efficiency of the action of various antibiotics are shown in the separate figures (Fig. 5.65).

The obtained results demonstrate the very strong and ambiguous influence of activated water on the efficiency of the action of different antibiotics on the microbiological culture. The probable interpretation of this effect will be made in Chap. 8.

5.5. Effect of the Nutrient Medium Activation on the Metabolic Parameters of Microbial Associations

The above-considered peculiarities of the growth of the pure microbiological culture *Escherichia coli* in activated water-based nutrient medium demonstrate the strong influence of the activation of water on the metabolic parameters, reductase activity, growth and shape of cells, and other characteristics of this culture. These results are of great importance for biotechnology.

Table 5.2. Stability indices of the microbiological culture *Echerichia coli k 12* to various antibiotics in activated media.

Water activation mode (duration of activation) and the marking of disks with antibiotics	Duration of experiment (hours)	Antibiotic on a disk, dose in μg	Diameter of GIZ D_0, mm
Control	24		
1		clindamycin, 30	0
2		kanamycin, 30	21
3		cephalexin, 30	29
4		ceftriaxone, 30	36
5		chloromycetin, 30	32
6		ciprofroxacin, 5	26
7		ampicillin, 10	16
8		tetracycline, 30	9
9		cephaclor, 30	31
10		carbenicillin, 100	24
0.5 h	24		
1		clindamycin, 30	0
2		kanamycin, 30	12
3		cephalexin, 30	22
4		ceftriaxone, 30	32
5		chloromycetin, 30	8
6		ciprofroxacin, 5	23
7		ampicillin, 10	13
8		tetracycline, 30	8
9		cephaclor, 30	22
10		carbenicillin, 100	19
1.0 h	24		
1		clindamycin, 30	0
2		kanamycin, 30	10
3		cephalexin, 30	14
4		ceftriaxone, 30	32
5		chloromycetin, 30	26
6		ciprofroxacin, 5	31
7		ampicillin, 10	25
8		tetracycline, 30	0
9		cephaclor, 30	30
10		carbenicillin, 100	29

(*Continued*)

Table 5.2. (*Continued*)

Water activation mode (duration of activation) and the marking of disks with antibiotics	Duration of experiment (hours)	Antibiotic on a disk, dose in μg	Diameter of GIZ D_0, mm
Control	36		
1		clindamycin, 30	0
2		kanamycin, 30	12
3		cephalexin, 30	13
4		ceftriaxone, 30	29
5		chloromycetin, 30	19
6		ciprofroxacin, 5	23
7		ampicillin, 10	11
8		tetracycline, 30	0
9		cephaclor, 30	22
10		carbenicillin, 100	19
0.5 h	36		
1		clindamycin, 30	0
2		kanamycin, 30	12
3		cephalexin, 30	15
4		ceftriaxone, 30	32
5		chloromycetin, 30	0
6		ciprofroxacin, 5	23
7		ampicillin, 10	11
8		tetracycline, 30	0
9		cephaclor, 30	20
10		carbenicillin, 100	18
1.0 h	36		
1		clindamycin, 30	0
2		kanamycin, 30	0
3		cephalexin, 30	0
4		ceftriaxone, 30	30
5		chloromycetin, 30	26
6		ciprofroxacin, 5	31
7		ampicillin, 10	25
8		tetracycline, 30	0
9		cephaclor, 30	30
10		carbenicillin, 100	29

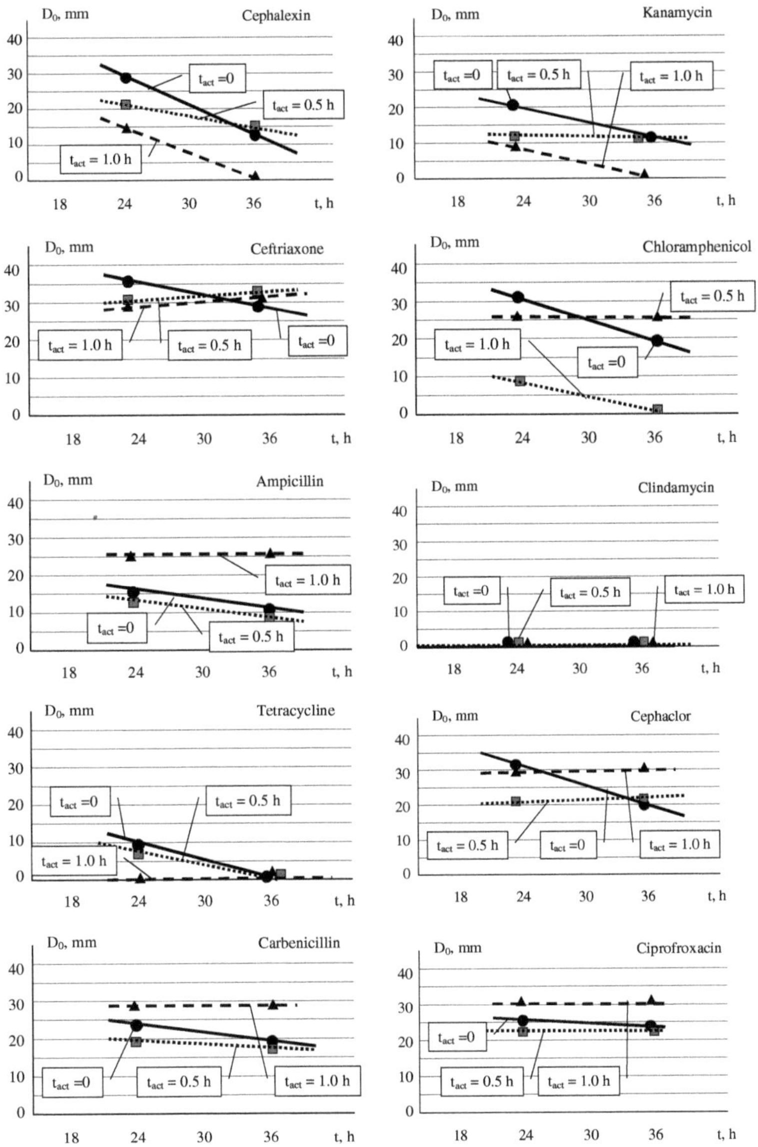

Figure 5.65. Influence of different fractions of activated water on the action of 10 different antibiotics on the microbiological culture *Escherichia coli k 12*. The decrease or increase of the diameter D of GIZ corresponds to the increase or decrease of the efficiency of specific antibiotic action in different fractions of activated medium. D_0 is diameter of GIZ, mm.

At the same time, we note that such experiments reflect, in essence, an idealized picture of the influence of activated water on animals and human beings. The matter is that the natural state of any complicated biological object is connected with a great complex (associations, cenoses) of different microbiological structures. They do not form a mechanical mixture, but they are in the complicated synergetic connection, by forming the multicomponent–multifunctional symbiotic system. Each of these cultures presents different physiological groups, and together they create a distinctive superorganism which is characterized by stable intra- and inter-group connections.

In this section, we study the peculiarities of the influence of activated water on the parameters of syntrophic microbial associations, whose composition and physiological properties are close to those of the human ones. These associations include the maximal variety of species and physiological groups of microorganisms (both prokaryotes and eukaryotes).

As a source of syntrophic microbial associations, we used the granulated preparation including microorganisms taken from the active sludge of an aerotank and the fermented sediment of a methane-tank, which together contained the maximal variety of representatives of the physiological groups of microorganisms, and correspondingly represented the variety of the variants of metabolic pathways. In the initial state, the microbial associations are a dry granulated preparation that includes syntrophic microbial associations, the collection of necessary macro- and micro-elements, and an agglutinative substance that keeps all the complex both in the dry state and in a water-based medium.

During the cultivation of associations, we introduced five grams of dry microbial granulated preparation and the 30-ml portion of MPB into each flask containing activated or nonactivated water; the flasks were closed by rubber plugs and screwed with metalic hoods with a thread.

To determine the effect of a state of water on the reductase activity, we used resazurin which is nontoxic for microbes and allows one to unambiguously visually determine the main redox-indices of a cultural medium. The main peculiarities of the application of resazurin to the reductase activity diagnostics were considered above.

In addition, the use of resazurin allows us to determine the threshold of the transition from aerobic to anaerobic conditions in a given series of experiments. At the start of each experiment, the state of a microbiological association corresponds to aerobic conditions even in a closed flask. The relation and the conditional boundary between aerobic and anaerobic conditions are determined by the ratio between the inflow rate of oxygen and the rate of its binding in a specific biochemical reaction. The anaerobic conditions hold when the latter exceeds the former. The abrupt decrease of the redox-potential becomes the evidence for the transition to the anaerobic mode, which is reflected in the discoloration of resazurin.

During the experiment, the following parameters were measured:

K_{ster} — coefficient of sterility that was determined by the coloration intensity of resazurin in the control medium (the discoloration of resazurin, in %);

D_c, $D_{0.5}$, and $D_{1.0}$ — average values of the optical density of a medium in the control experiment and in the media that were activated for 0.5 h and 1.0 h (a change of the optical density characterizes the biomass growth for a cultivated syntrophic microbial association);

V_c, $V_{0.5}$, and $V_{1.0}$ — average values of the gas volume in the control experiment and in the media that were activated for 0.5 h and 1.0 h. The samples of gas were taken with a syringe by piercing the rubber plug without violation of the medium hermiticity in a flask;

R_c, $R_{0.5}$, and $R_{1.0}$ — average values of the reductase activity index (the discoloration of resazurin, in %) in the control and in the media activated for 0.5 and 1.0 h;

$K_{0.5} = R_c/R_{0.5}$, $K_{1.0} = R_c/R_{0.5}$ — total inhibition coefficients for the reductase activity on the medium activated for 0.5 and 1.0 h; and

$K_{0.5;1.0}$ ($K_{0.5;1.0} = K_{0.5}/K_{1.0}$) — the relative efficiency coefficient of reductase activity inhibition.

In Figs. 5.66–5.79, we show the photos which illustrate the process of modification of the metabolic parameters of the syntrophic microbial association depending on the cultivation duration and the composition of water. Under each figure, the necessary information related to the time point of the measurement is provided.

As shown, starting from the 8th hour, we observed the abrupt increase of the reductase activity in the medium prepared on the base of the water activated for 0.5 h. This corresponds to the fact that the growth takes place under anaerobic conditions from this time moment. At the same time, the growth of associations in other flasks corresponds to aerobic conditions as before.

Starting from the 29th hour of experiment, the gas evolution began in all samples.

The gas has evolved till the end of the experiments (up to the 76th hour).

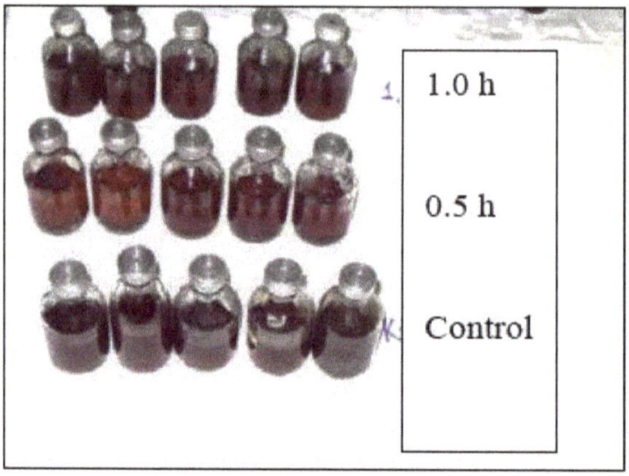

Figure 5.66. Start of the experiment.

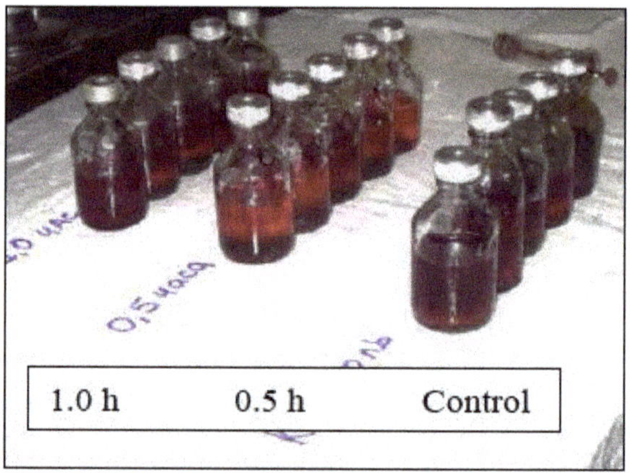

Figure 5.67. 1st hour of the experiment. $R_c = 1$, $R_{0.5} = 20$, $R_{1.0} = 10$, $K_{0.5} = 20$, $K_{1,0} = 10$, $K_{0.5;1.0} = 2.0$, $V_c = 0$, $V_{0.5} = 0$, $V_{1.0} = 0$.

At the 44th hour of experiment, we started the investigation of the influence of the duration of activation of a nutrient medium on the reduction rate of chromates by microbial associations.

The introduction of 50mg/L of Cr(VI) into the medium is accompanied by the abrupt increase of the redox-potential that leads to the appearance of a coloration in flasks (Eh $\geq +50$ mV). The time necessary for the execution of the analytic determination of Cr^{6+} in the medium (5 min) did not allow us to catch the difference in the reductase activities in different fractions

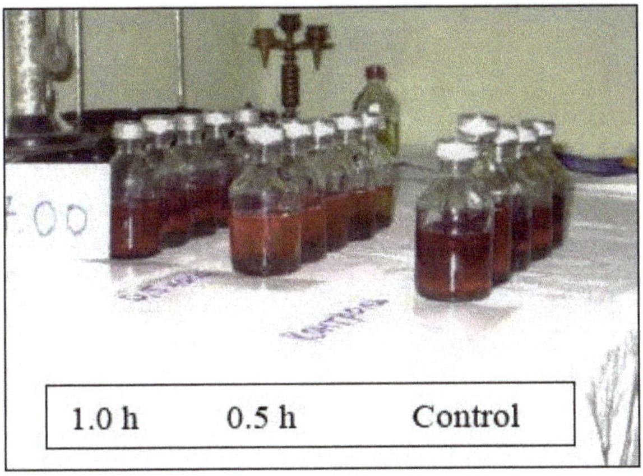

Figure 5.68. 7th hour of the experiment. $R_c = 5$, $R_{0.5} = 25$, $R_{1.0} = 15$, $K_{0.5} = 5$, $K_{1.0} = 3$, $K_{0.5;1.0} = 1.66$, $V_c = 0$, $V_{0.5} = 0$, $V_{1.0} = 0$.

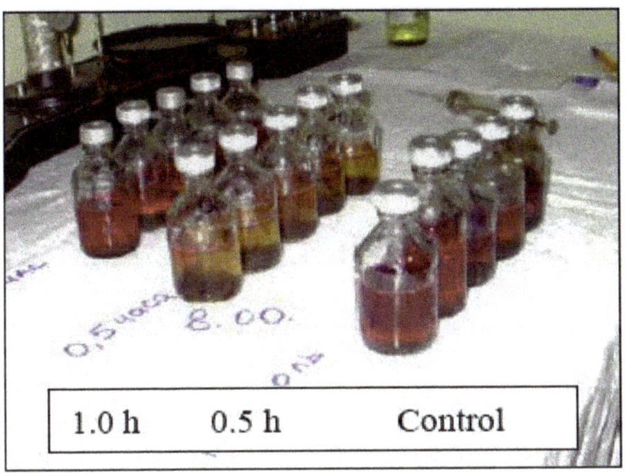

Figure 5.69. 8th hour of the experiment. $R_c = 30$, $R_{0.5} = 70$, $R_{1.0} = 15$, $K_{0.5} = 2.33$, $K_{1.0} = 0.5$, $K_{0.5;1.0} = 4.6$, $V_c = 0$, $V_{0.5} = 0$, $V_{1.0} = 0$.

of activated water and in the control. But it is shown by the discoloration of the medium in the flasks (Fig. 5.78) that the reduction rate of Cr^{6+} is a little higher in the control (nonactivated water) than in the variants with the activation for 0.5 h and 1.0 h: after 50–60 s from the introduction of chromates, the media in all flasks are colorless in the control (Eh ≤ -100 mV) and colored (Eh $\geq +50$ mV) in the variants with the activation for 0.5 h and 1.0 h.

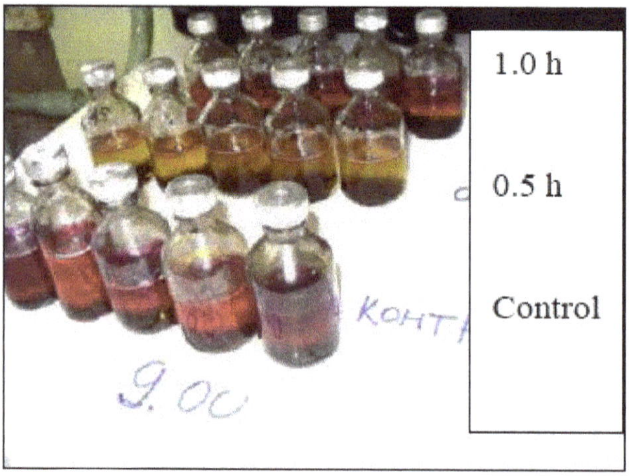

Figure 5.70. 9th hour of the experiment. $R_c = 30$, $R_{0.5} = 70$, $R_{1.0} = 15$, $K_{0.5} = 2.33$, $K_{1.0} = 0.5$, $K_{0.5;1.0} = 4.6$, $V_c = 0$, $V_{0.5} = 0$, $V_{1.0} = 0$.

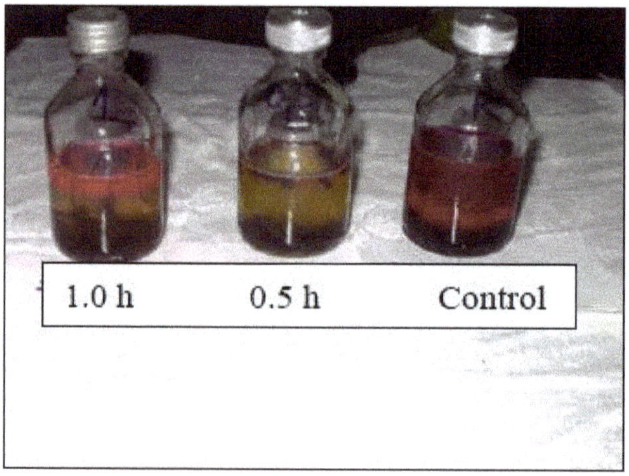

Figure 5.71. 9th hour of the experiment. The typical redox-activity is shown in the variants of the experiment at the 9th hour: (i) activation for 0.5 h almost: full discoloration of resazurin; (ii) activation for 1.0 h: discoloration by 80%, 20% of the upper layer is colored; (iii) in the control (nonactivated water): the medium is completely (100%) colored.

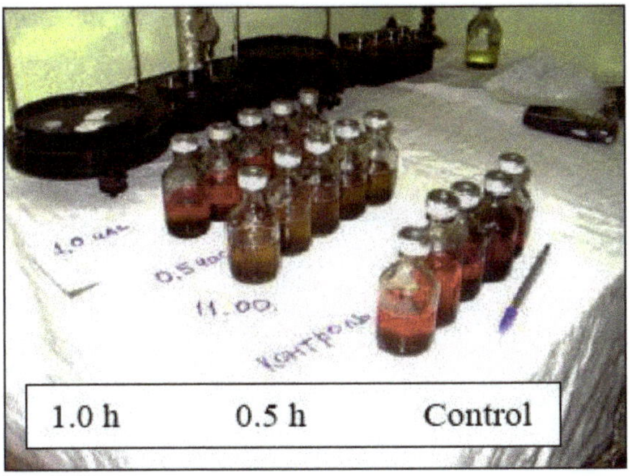

Figure 5.72. 11th hour of the experiment. $R_c = 55$, $R_{0.5} = 90$, $R_{1.0} = 50$, $K_{0.5} = 1.63$, $K_{1.0} = 0.9$, $K_{0.5;1.0} = 1.8$, $V_c = 0$, $V_{0.5} = 0$, $V_{1.0} = 0$.

Figure 5.73. 12th hour of the experiment. $R_c = 65$, $R_{0.5} = 95$, $R_{1.0} = 55$, $K_{0.5} = 1.46$, $K_{1.0} = 0.85$, $K_{0.5;1.0} = 1.72$, $V_c = 0$, $V_{0.5} = 0$, $V_{1.0} = 0$.

The same studies were made at the 45th and 69th hours of the experiment. The results are analogous to those described.

Let us consider the results of experiments presented above.

In Fig 5.80, we present the summary data which characterize the influence of different fractions of activated water on the metabolic activity of syntrophic microbial associations under anaerobic conditions.

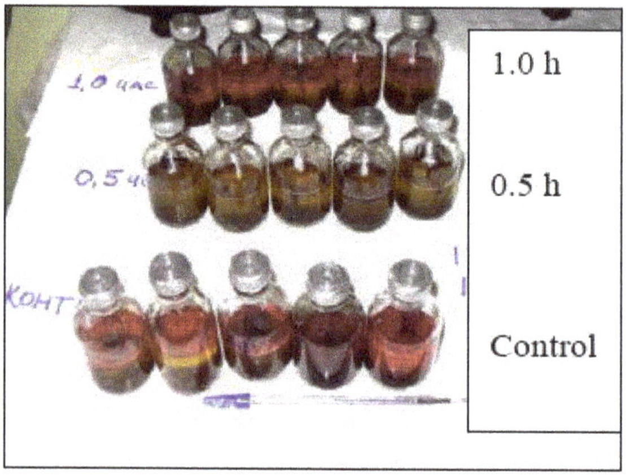

Figure 5.74. 22nd hour of the experiment. $R_c = 87$, $R_{0.5} = 100$, $R_{1.0} = 82$, $K_{0.5} = 1.15$, $K_{1.0} = 0.94$, $K_{0.5;1.0} = 1.22$, $V_c = 0$, $V_{0.5} = 0$, $V_{1.0} = 0$.

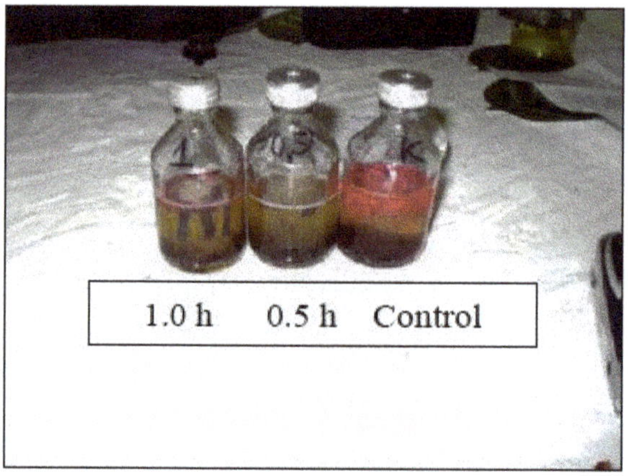

Figure 5.75. 23rd hour of the experiment. $R_c = 87$, $R_{0.5} = 100$, $R_{1.0} = 95$, $K_{0.5} = 1.15$, $K_{1.0} = 1.1$, $K_{0.5;1.0} = 1.05$, $V_c = 0$, $V_{0.5} = 0$, $V_{1.0} = 0$.

It is seen that the use of the medium activated for 0.5 h leads to the very abrupt increase of the metabolic activity for the first 10–20 h. This result is clearly demonstrated by Fig. 5.81, where we give the dependence of the relative reductase activity of microbial associations under anaerobic conditions in the medium activated for 0.5 h relative to the reductase activity of the same associations in the nonactivated medium. The maximal excess of the absolute value of the metabolic activity in the presence of such a

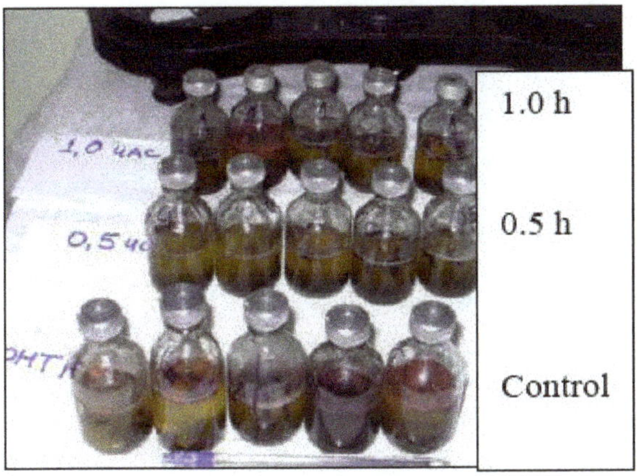

Figure 5.76. 26th hour of the experiment. $R_c = 95$, $R_{0.5} = 100$, $R_{1.0} = 95$, $K_{0.5} = 1.05$, $K_{1.0} = 1.0$, $K_{0.5;1.0} = 1.05$, $V_c = 0$, $V_{0.5} = 0$, $V_{1.0} = 0$.

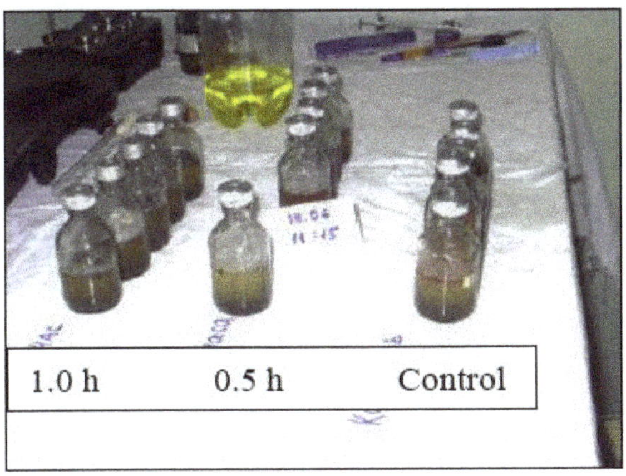

Figure 5.77. 44th hour of the experiment. $R_c = 99$, $R_{0.5} = 100$, $R_{1.0} = 99$, $K_{0.5} = 1.01$, $K_{1.0} = 1.0$, $K_{0.5;1.0} = 1.01$, $V_c = 60$ ml, $V_{0.5} = 90$ ml, $V_{1.0} = 50$ ml.

type of activated water (relative to the medium prepared with nonactivated water) reaches 40–50% in 10 h after the start of the experiment.

In contrast to this, the use of the medium activated for 1.0 h leads to a small but reliable decrease of the metabolic activity relative to the control medium. The value of this decrease in the same time interval corresponds to 10–15%.

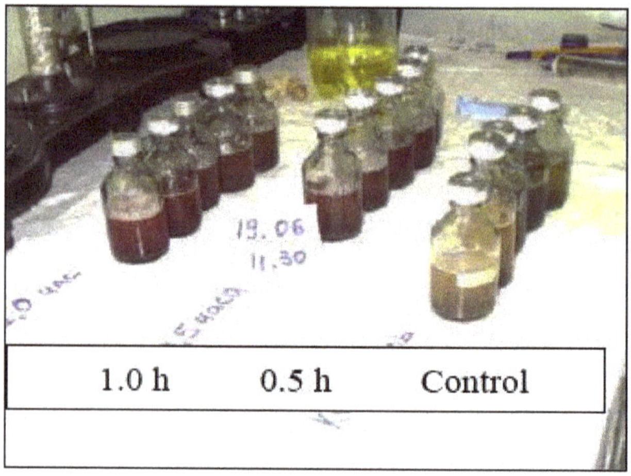

Figure 5.78. 44th hour of the experiment. The investigation of the influence of the duration of activation of the nutrient medium on the reduction rate of chromates by the microbial associations. The photo was taken in 60 s after the introduction of 50 mg/L of Cr(VI) into all samples.

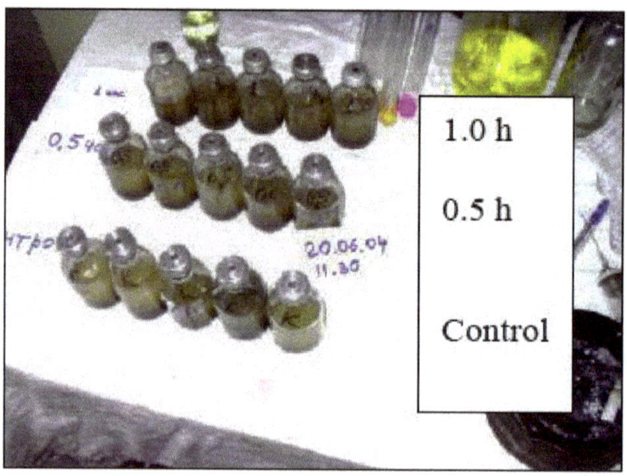

Figure 5.79. 68th hour of the experiment. $R_c = 100$, $R_{0.5} = 100$, $R_{1.0} = 100$, $K_{0.5} = 1.0$, $K_{1.0} = 1.0$, $K_{0.5;1.0} = 1.01$, $V_c = 180\,\text{ml}$, $V_{0.5} = 200\,\text{ml}$, $V_{1.0} = 160\,\text{ml}$. In all variants of the experiment, resazurin was completely discolored. This demonstrates the high reductase activity of the microbial association.

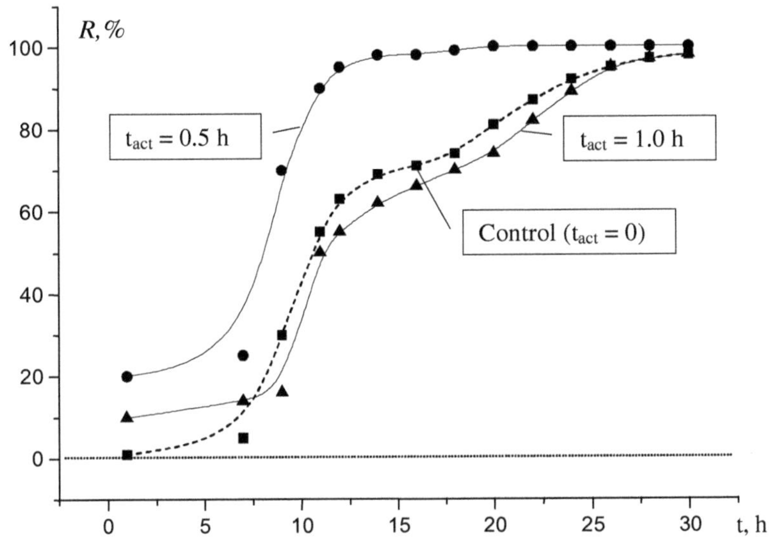

Figure 5.80. Effect of the medium activation duration on the reductase activity of microbial associations under anaerobic conditions.

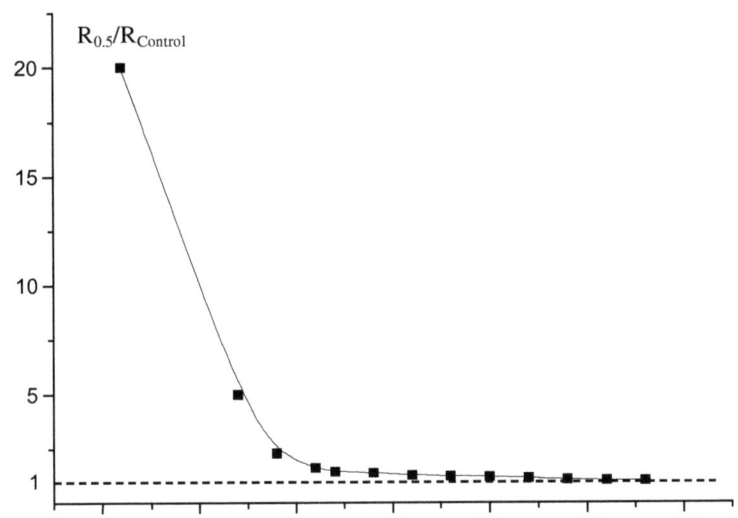

Figure 5.81. Effect of the medium activation duration on the relative reductase activity of microbial associations under anaerobic conditions in the medium activated for 0.5 h relative to that in the nonactivated medium.

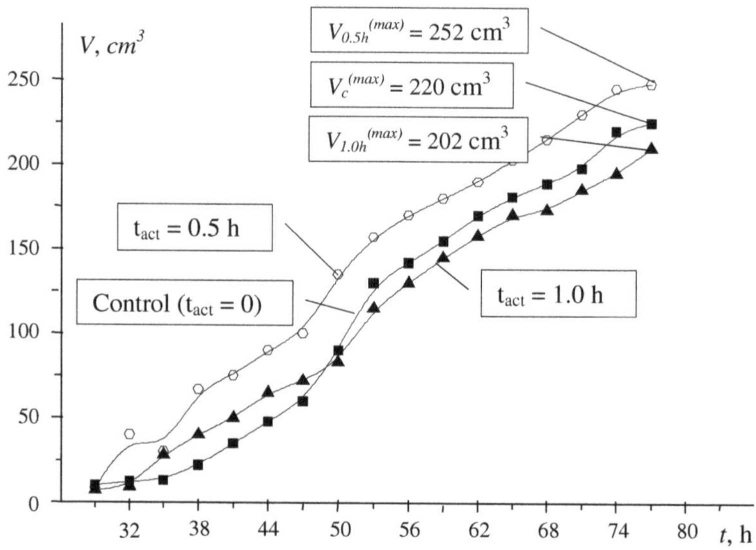

Figure 5.82. Effect of the medium activation duration on the gas synthesis by microbial associations. The quantity V refers to the total volume of the evolved gas.

After the first 30 h of the experiment, the indices of metabolic activity for all the three investigated water-fractions are approximately the same.

In Fig 5.82, we present the data on the gas evolution during the growth of microbial associations in the media with the different water fractions.

As was mentioned above, the intensive gas evolution began from the 29th hour of the experiment. It was discovered that the gas evolution was more intense in the medium prepared with water activated for 0.5 h, which yielded the great summary volume of the gas. In the following stages (starting from the 35th hour of the experiment), the gas evolution rate was approximately the same in all fractions of water. These data correlate well with the results shown in Fig. 5.80, according to which the reductase activity of all fractions of water became the same starting from the 33–35th hours of the experiment.

CHAPTER 6

Examination of the Influence of MRET Activated Water on Prophylaxis and Treatment of Oncology

6.1. Procedures of Examinations of the Influence of Activated Water on Oncology, Objects of Investigation, and Facilities

The results of studies showed that activated water has a number of anomalous and unique physicochemical properties that are distinctly different from those of ordinary water (see Chap. 3).

As shown in Chaps. 4–6, activated water shows potent effects on the growth of plant cells and bacterial cultures. In particular, certain fractions of activated water substantially modify the efficiency of some antibiotics, inhibit the growth of opportunistic pathogenic bacteria, change the reductase activity of microbial associations, and suppress the growth of callus tissue.

The next important step of study was to investigate the effects of different fractions of activated water on various transformed cells and cells of the immune system (Vysotskii *et al.*, 2006). The methods of studies and the results obtained are presented in details in what follows. To study how the different fractions of activated water affect the antitumor resistance of organism, the following experimental approaches and techniques were used:

- study of the possible antitumor efficiency of the prophylactic administration of different fractions of activated water; to this end, mice received activated water before the tumor cell transplantation ("prophylactic treatment" mode);
- study of the possible antitumor efficiency of the therapeutic administration of different fractions of activated water; to this end, mice received activated water after the tumor cell transplantation ("therapeutic treatment" mode);

- investigation of the functional activity of lymphocytes with natural killer cell cytotoxicity isolated from spleens of normal (without tumors) mice which received activated water.

Five different fractions of activated water were prepared to elucidate the efficiency of the antitumor effects of activated water depending on the duration of its activation. Four water fractions were obtained after water activation for 15 min, 30 min, 45 min, and 60 min. The samples of freshly activated water were prepared from fresh distilled water everyday and used at once after the activation was finished. Moreover, before the beginning of studies, a large volume of water was activated for 30 min and stored at 4°C for about 45 days. This water was also used in the study. The everyday required volume of this water was used for a specific cycle of continued studies. This fraction of activated water was named "old activated water". The studies with "old activated water" were aimed to research the influence of the "aging" of activated water on its antitumor and immunostimulatory effects *in vitro* and *in vivo*.

Moreover, the analogous control studies using nonactivated distilled water were carried out.

In the study, inbred adult male BALB/c mice aged 11 weeks with 23–24 g corporal weight were used. BALB/c mice (the genetic formula "bbcc", H-2^a is the major histocompatibility complex allele) are white mice which are very susceptible to various oncogenous viruses. BALB/c mice have a very low frequency of the development of mammary gland tumors (no more than 5%) and do not have the "milk factor". Lung tumors develop in 25–30% of mice. The animals were housed in standard facilities with free access to water and food. The animals did not show any signs of malignancy or other pathological processes. All animals were maintained under strict ethical conditions according to the International recommendations.

Experimental studies were carried in R. E. Kavetsky Institute of Experimental Pathology, Oncology, and Radiology of the National Academy of Sciences of Ukraine under supervision and participation of Dr. Yu. V. Yanish and Dr. S. Olishevsky.

The general scheme of studies for each tumor strain is presented in Fig. 6.1. This scheme was equally used to study the influence of different fractions of activated water on tumors (*Ehrlich carcinoma* and *sarcoma 37*).

To study the influence of activated water on the tumor growth, all mice were randomly divided into 11 groups with 20 animals in each group. These groups are showed in the form of column on the left side in Fig. 6.1.

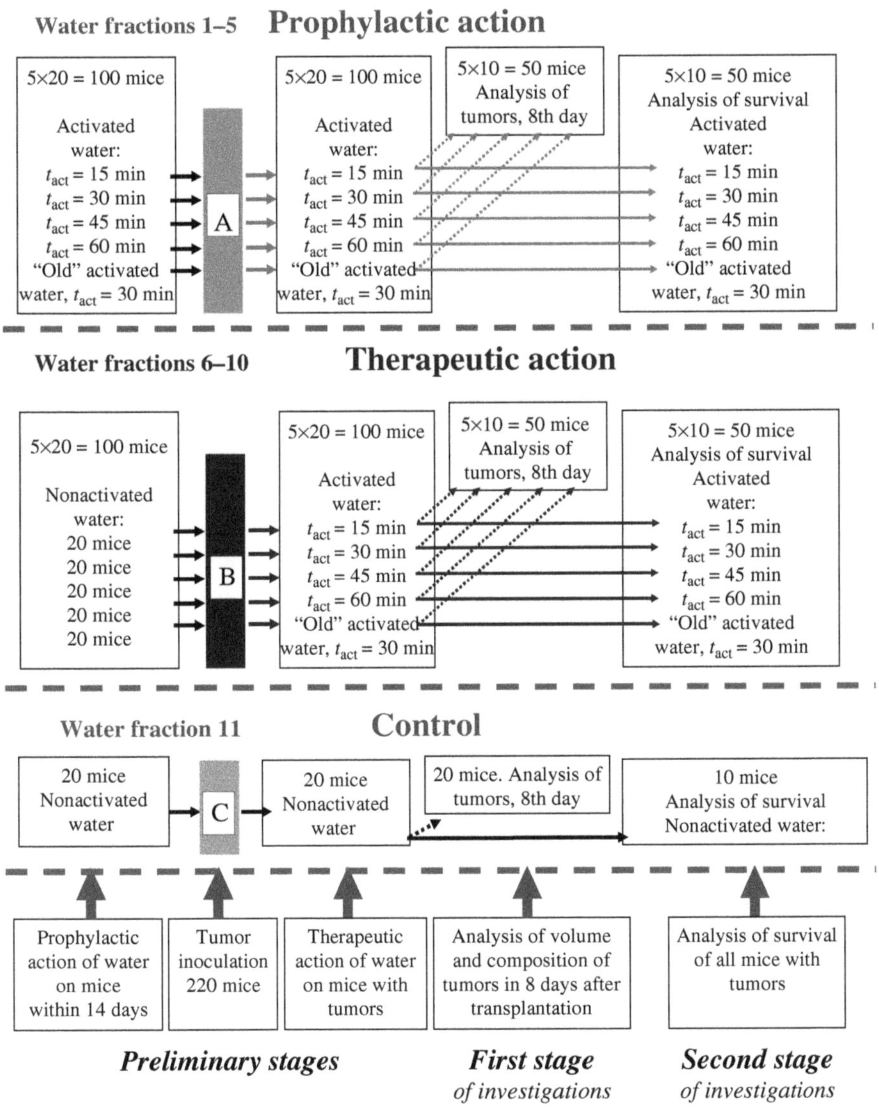

Figure 6.1. General scheme of studies of the antitumor effect of different fractions of activated water (fractions 1–5 and 6–10) on mice with transplanted tumors. Fraction 11 ("ordinary" nonactivated) water was used in the control experiments. Tumor transplantation A corresponds to the study of the prophylactic treatment with activated water; tumor transplantation B corresponds to the study of the therapeutic treatment of activated water; C corresponds to control investigations.

Five experimental groups (from the 1st to 5th group) comprised mice which received activated water in the "prophylactic treatment" mode, the other five groups (from the 6th to 10th group) were treated with activated water in the "therapeutic treatment" mode. One of the 11 groups served as a control and comprised mice which received nonactivated distilled water.

A set of 100 mice from groups 1–5 ("prophylactic treatment" mode) received the appropriate fraction of activated water (according to the daily rate of water intake) for 14 days before the tumor transplantation. After the tumor cell inoculation, all mice from these groups continued to receive activated water.

At this time, 100 mice from groups 6–10 ("therapeutic treatment" mode) and 20 control mice received ordinary nonactivated distilled water for a fortnight before the tumor transplantation. In 14 days after the experiment began, mice of all groups were inoculated intraperitoneally with an equal number of viable cells of a certain tumor strain. The next day after the tumor cell inoculation, all mice from groups 6–10 ("therapeutic treatment" mode) began to receive the appropriate fractions of activated water.

Control mice were administered with nonactivated distilled water before and after the tumor transplantation.

The first stage of studies was finished on the 8th day after the tumor cell inoculation, when 10 mice from each group were sacrificed and ascitic-fluids containing tumor cells were obtained from peritoneal cavities.

Furthermore, the average volume of ascitic-fluid, the number of tumor cells presented in one millimeter of ascitic-fluid, and the total number of tumor cells in the peritoneal cavity of a tumor-bearing mouse were determined.

The efficiency of the application of activated water was assessed in the following way.

(1) The average volume of ascitic-fluid was calculated as

$$\langle V \rangle = \sum_{n=1}^{10} V_n/10,$$

where $V_1, V_2, V_3, \ldots, V_{10}$ are the volumes of ascitic fluids obtained from peritoneal cavities of 10 tumor-bearing mice in each group.

(2) The average number of viable tumor cells in one milliliter of ascitic-fluid ($\langle C \rangle$/1 ml) was estimated according to the routine technique. The number of tumor cells was counted using a microscope and a hemocytometer with magnification $\times 160$. The viability of cells was determined

by the *Trypan blue exclusion test*, and uncolored cells were considered as viable.

(3) The average number of viable tumor cells in the peritoneal cavity of one mouse was calculated as

$$\langle C \rangle = (\langle C \rangle / 1\ \text{ml}) \times \langle V \rangle.$$

(4) The index of tumor growth inhibition was evaluated according to the obtained data and calculated as

$$D = [(\langle V \rangle_{\text{exp}} - \langle V \rangle_{\text{control}}) / \langle V \rangle_{\text{control}}] \times 100\%,$$

where $\langle V \rangle_{\text{control}}$ — average ascitic-fluid volume of control mice, and $\langle V \rangle_{\text{exp}}$ — average ascitic fluid volume for mice from the experimental group.

Other set of mice (10 in each group) continued to receive the same activated water fraction. The survival of these animals was daily monitored in order to study the effects of certain activated water fractions on the dynamics and the survival indices of tumor-bearing mice.

On the basis of the data, several additional quantitative parameters which characterize the effect of the application of activated water on the survival of experimental tumor-bearing animals were determined.

(5) The average survival time $\langle t \rangle$ was calculated on the basis of mortality data as

$$\langle t \rangle = \sum_{n=1}^{10} N_n / 10,$$

where $N_1, N_2, N_3, \ldots, N_n$ are the number of days survived by the mouse group after the tumor transplantation.

(6) The median survival time characterizes the survival time of 50% of animals in a group.

(7) Percentage increase in the lifetime was calculated as

$$K = [(\langle t \rangle_{\text{exp}} - \langle t \rangle_{\text{control}}) / \langle t \rangle_{\text{control}}] \times 100\%,$$

where $\langle t \rangle_{\text{control}}$ and $\langle t \rangle_{\text{exp}}$ are the average survival times in the control and experimental groups of mice, respectively.

Statistical analysis was performed using “GraphPad Instat”. Data are expressed as means ± standard error.

6.2. Study of the Antitumor Effects of Different Fractions of MRET Activated Water *in vivo* in the Modes of Prophylactic and Therapeutic Treatment Tested on the Tumor Model of Ascitic *Ehrlich carcinoma*

6.2.1. *Materials, methods, and experimental results*

The first series of studies was performed using experimental tumor growth model such as ascitic *Ehrlich carcinoma*.

A spontaneous mammary gland tumor appeared in an underbred female mouse was further maintained as an experimental strain of solid *Ehrlich adenocarcinoma*. Ascitic *Ehrlich carcinoma* was obtained from the inoculation of tumor cells in the peritoneal cavities of mice.

In the present work, cells of ascitic *Ehrlich carcinoma* were obtained from peritoneal cavities of underbred white mice on the 7th–8th day of tumor growth. All mice from 11 groups were inoculated intraperitoneally with 200,000 viable cells/mouse of ascitic *Ehrlich carcinoma* in accordance with the general scheme presented in Fig. 6.1.

In seven days after the tumor transplantation, ascitic-fluid from the peritoneal cavities of half of all animals was obtained.

Figure 6.2 shows test-tubes with obtained ascetic-fluids of five mice from the "prophylactic treatment" group (water was activated for 30 min) and control mice in experiments with the ascitic *Ehrlich carcinoma* model.

The photo in Fig. 6.2 clearly demonstrates the dramatic difference in the volumes of ascitic-fluids from mice which received activated water before and after the tumor cell inoculation ("prophylactic treatment" group) as compared with control mice which received ordinary water ("control" group).

The volume of ascitic-fluid can be compared to the volume of a tumor which developed for seven days. In this case, the application of activated water decreased the tumor volume more than twice as compared with the control results (data in Table 6.1).

The correlation between the ascitic-fluid volumes and the tumor cell number of mice from the prophylactic treatment, therapeutic treatment, and control groups allows us to determine the effects of applications of different activated water fractions on the growth and size of tumors in tumor-bearing mice. The results of experimental measurements of the ascitic-fluid average volume and both the total and unit-volume average cell numbers of ascitic

Figure 6.2. Ascitic fluid samples obtained from peritoneal cavities of mice with transplanted ascitic *Ehrlich carcinoma.* On the left side, five samples from mice after the prophylactic administration ("prophylactic treatment" group) of water activated for 30 min; on the right side, 5 samples obtained from control mice ("control" group). Every test-tube contains ascitic fluid obtained from one mouse.

Ehrlich carcinoma are presented in Table 6.1. In this table, we present the data corresponding to the resumptive parameter (the index of tumor growth inhibition — *D*) that characterized the efficiency of the influence of activated water on the tumor growth.

Table 6.1 shows that the highest index of tumor growth inhibition exceeded 50% and was achieved when mice were treated with water activated for 30 min.

These characteristics indicating the effects of activated water on the tumor parameters in tumor-bearing mice are clearly presented in Figs. 6.3–6.5.

The more detailed analysis of the results obtained will be described below.

When the analysis of the volume and cellular content of ascitic-fluid was completed, each group consisted of 10 mice. The study of the animal survival dependence on the time interval after the tumor transplantation allows us to determine the efficiency of different fractions of activated water on the lifetime of tumor-bearing mice.

Moreover, this study allows the determination of the difference in the effects of freshly activated water and "old activated water" of the same type.

Table 6.1. Influence of different fractions of activated water on the investigated parameters of ascitic *Ehrlich carcinoma*.

Type of treatment and fraction of water used	Investigated parameters			Index of tumor growth inhibition, D, %
	Average volume of ascitic-fluid in one tumor $\langle V \rangle$, ml	Average number of viable tumor cells in 1 ml of ascitic-fluid $\langle C/V \rangle$, ml^{-1} ($\times 10^3$ cells)	Average number of viable tumor cells in peritoneal cavity of one mouse $\langle C \rangle$, ($\times 10^6$ cells)	
Control, $t_{act} = 0$	2.85 ± 0.2	235724 ± 44915	672	—
Prophylactic, $t_{act} = 15$ min	1.6 ± 0.1	117354 ± 35134	188	43.6
Prophylactic, $t_{act} = 30$ min	1.4 ± 0.07	111268 ± 23714	156	50.9
Prophylactic, $t_{act} = 45$ min	1.5 ± 0.1	120068 ± 11711	180	47.4
Prophylactic, $t_{act} = 60$ min	2.2 ± 0.08	181868 ± 36784	400	22.8
Prophylactic, "Old activated water", $t_{act} = 30$ min	1.9 ± 0.1	161166 ± 13774	306	33.3
Therapeutic, $t_{act} = 15$ min	2.2 ± 0.1	193231 ± 32144	425	22.8
Therapeutic, $t_{act} = 30$ min	2.0 ± 0.07	151283 ± 30561	303	30.0
Therapeutic, $t_{act} = 45$ min	2.3 ± 0.1	150014 ± 11301	345	19.3
Therapeutic, $t_{act} = 60$ min	2.5 ± 0.08	210067 ± 23677	525	12.3
Therapeutic, "Old activated water", $t_{act} = 30$ min	2.2 ± 0.1	166541 ± 23454	366	22.8

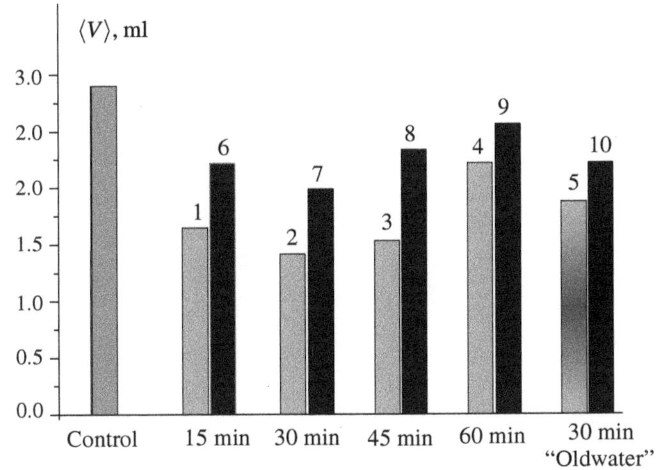

Figure 6.3. Effect of the prophylactic (1–5) and therapeutic (6–10) applications of activated water on the average volume of ascitic-fluid obtained from mice inoculated intraperitoneally with tumor cells of *Ehrlich carcinoma.*

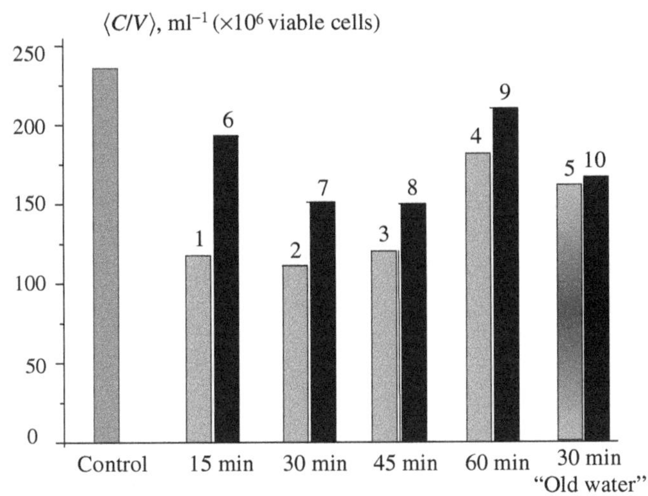

Figure 6.4. Effect of the prophylactic (1–5) and therapeutic (6–10) applications of activated water on the average number of viable cells in 1 ml of ascitic-fluid obtained from mice inoculated intraperitoneally with tumor cells of *Ehrlich carcinoma.*

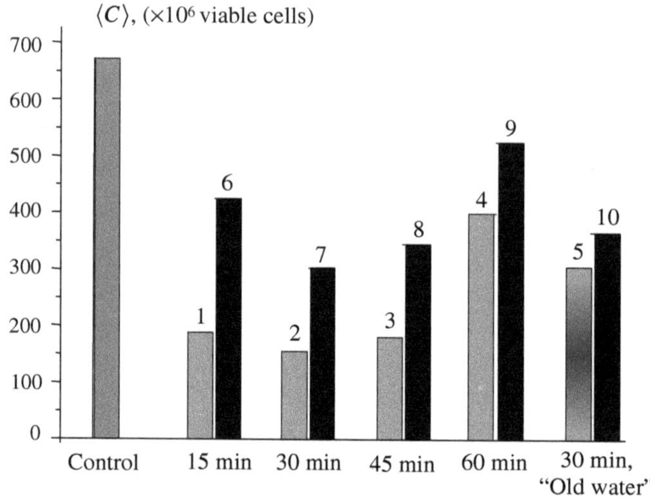

Figure 6.5. Effect of the prophylactic (1–5) and therapeutic (6–10) applications of activated water on the average number of viable cells in an ascitic tumor obtained from mice inoculated intraperitoneally with tumor cells of *Ehrlich carcinoma.*

Figure 6.6 shows the appearance of mice from the "control" and "prophylactic treatment" groups in 18 days after the ascitic *Ehrlich carcinoma* transplantation [$t_{act} = 30$ min (b)]. The unbiased registration of the appearance of mice in the terminal stage of tumor growth allows presentation of the clear difference in the effects of both activated and nonactivated water on the tumor size.

Figure 6.7 shows the whole view of cages with mice from two different "prophylactic treatment" groups. Mice received the different types of water activated for 30 min. On the left side of the photo, mice received freshly activated water are placed; on the right side, animals which were treated with earlier activated water ("old activated water") are shown. The photo was taken on the 19th day after the ascitic *Ehrlich carcinoma* transplantation.

Figure 6.8 shows the photos of dead animals of the "control" group (on the left side) and live mice of the "prophylactic treatment" group (on the right side). Mice of those groups received water activated for 45 min. The photo was taken on the 21st day after the tumor cell inoculation. It is evident that dead mice show an enlargement of the abdomen, which can be explained by growing tumors in peritoneal cavities.

(a) (b)

Figure 6.6. The appearance of mice from the (a) "control" and (b) "prophylactic treatment" groups (the duration of activation was 30 min) on the 18th day after the ascitic *Ehrlich carcinoma* cell inoculation.

Table 6.2 shows the statistical data of experimental results that characterize the survival of mice with transplanted *Ehrlich carcinoma* after different applications of activated water in the "prophylactic treatment" and "therapeutic treatment" modes.

The data on the survival dynamics of tumor-bearing mice which received different types of activated water in the "prophylactic treatment" and "therapeutic treatment" modes are presented in Figs. 6.9 and 6.10.

The relative number of mice (the number of mice before the tumor transplantation was taken as 100%) survived to a definite day after the tumor transplantation was shown on the vertical axis.

The data on the lifetime of tumor-bearing mice for both application modes and different types of activated water are presented in Fig. 6.11.

All figures clearly demonstrates the rapid increase in the lifetime of animals which received water activated for 30 min in the "prophylactic treatment" mode.

Figure 6.7. Mice from "prophylactic treatment" groups that received water activated for 30 min. Mice treated with freshly activated water are placed on the left side of the photo, whereas mice received "old activated water" is on the right side. The photo was taken on the 19th day after the inoculation of mice with cells of ascitic *Ehrlich carcinoma.*

6.2.2. *Results and discussion*

The results of experimental studies suggest that the applications of optimal types of activated water in the optimal mode show the substantial positive effects resulted in the inhibition of tumor growth observed in mice with transplanted ascitic *Ehrlich carcinoma.* In spite of the fact that the prophylactic application of activated water did not affect the rate of tumor transplantation in mice (it was equal to 100% in both cases of control animals and animals treated with activated water), the positive antitumor effect was displayed in the decrease of both the volume of ascitic-fluid in the peritoneal cavities of tumor-bearing mice and the content of viable tumor cells (Table 6.1, Figs. 6.3–6.5). In particular, the values of the ascitic-fluid volume were twice decreased in animals of the "prophylactic treatment" group ($t_{act} = 30$ min) in comparison with control animals. A similar tendency and approximately the same efficiency were observed in other groups with the prophylactic application of activated water ($t_{act} = 30$ min, 15 min, and 45 min). In contrast to the results mentioned above, the effect of activated

Figure 6.8. Dead mice of the "control" group (on the left side) and live mice from the "prophylactic treatment" group (water activated for 45 min was used). The photo was taken on the 21st day after the inoculation of mice with cells of ascitic *Ehrlich carcinoma*.

water (t_{act} = 60 min) on the tumor growth in mice of the "prophylactic treatment" group was significantly weaker (the average ascitic-fluid volume of each mouse from this group was 2.2 ml as compared with 2.85 ml in control).

In Table 6.1, we present values of the indices characterizing tumor growth together with values of the index characterizing the inhibition of tumor growth (D): 43.6, 50.9, 47.4, and 22.8 for the "prophylactic treatment" groups which received activated water with t_{act} = 15 min, 30 min, 45 min, and 60 min, respectively.

Table 6.2. Effect of different fractions of activated water on the survival of mice with ascitic *Ehrlich carcinoma*.

		Observational data		
Type of treatment and fraction of water used	Number of animals	Average lifetime $\langle t \rangle$, days	Median survival time, days	Percentage increase in lifetime, K,%
Control, $t_{act} = 0$	10	14.9 ± 0.5	15	—
Prophylactic, $t_{act} = 15$ min	10	21.7 ± 0.7	21	45.6
Prophylactic, $t_{act} = 30$ min	10	24.6 ± 1.1	26	61.7
Prophylactic, $t_{act} = 45$ min	10	21.6 ± 0.9	20	45.0
Prophylactic, $t_{act} = 60$ min	10	18.8 ± 1.4	17	26.2
Prophylactic, "Old activated water", $t_{act} = 30$ min	10	20.2 ± 0.5	23	35.6
Therapeutic, $t_{act} = 15$ min	10	18.6 ± 0.5	18	24.8
Therapeutic, $t_{act} = 30$ min	10	21.2 ± 1.0	19	42.3
Therapeutic, $t_{act} = 45$ min	10	21.2 ± 1.1	19	22.8
Therapeutic, $t_{act} = 60$ min	10	14.8 ± 0.9	14	≈ 0
Therapeutic, "Old activated water", $t_{act} = 30$ min	10	18.3 ± 1.3	19	22.8

It is important to note that long-term storage of activated water decreased its antitumor activity, but did not abrogate it. Such water serves as a sufficiently efficient antitumor substance. For instance, the tumor inhibition index in mice which received "old activated water" ($t_{act} = 30$ min) in the "prophylactic treatment" mode was approximately equal to 33.3%. It was substantially better as compared with mice from the "prophylactic

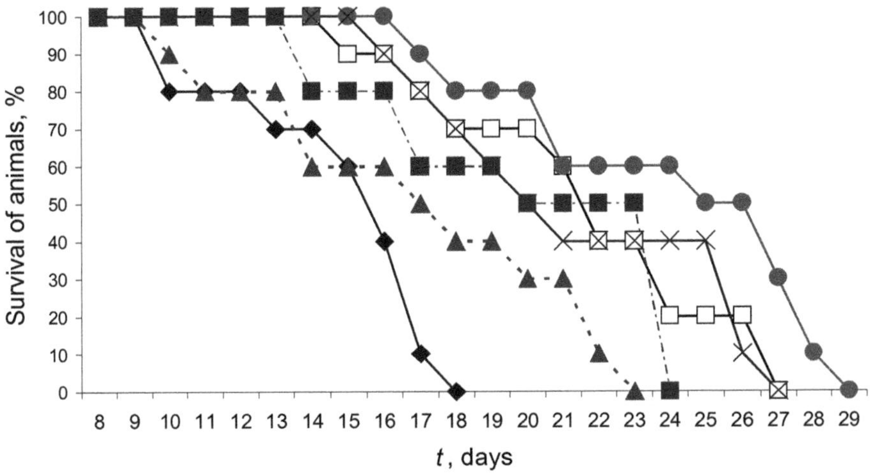

-◆- control ($t_{act} = 0$); -□- $t_{act} = 15$ min; -●- $t_{act} = 30$ min; -×- $t_{act} = 45$ min;
- ▲ - $t_{act} = 60$ min; -■- $t_{act} = 30$ min ("old activated water").

Figure 6.9. Survival dynamics of tumor-bearing mice which received different types of activated water in the "prophylactic treatment" mode.

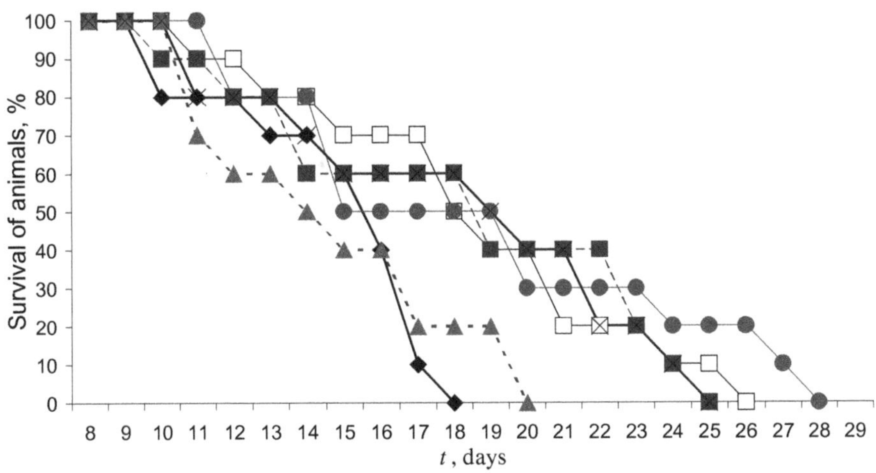

-◆- control ($t_{act} = 0$); -□- $t_{act} = 15$ min; -●- $t_{act} = 30$ min; -×- $t_{act} = 45$ min;
- ▲ - $t_{act} = 60$ min; -■- $t_{act} = 30$ min ("old activated water").

Figure 6.10. Survival dynamics of tumor-bearing mice which received different types of activated water in the "therapeutic treatment" mode.

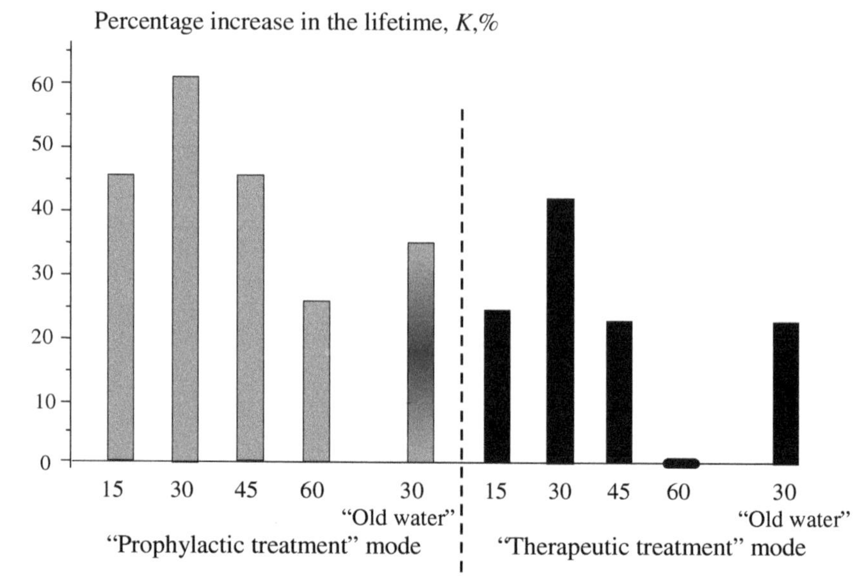

Figure 6.11. Percentage increase in the lifetime of tumor-bearing mice with ascitic *Ehrlich carcinoma* which received different types of activated water in the "prophylactic treatment" and "therapeutic treatment" modes. The numbers near the charts correspond to the water-activation duration in minutes.

treatment" group when animals received water freshly activated for 60 min. The percentage increase in the lifetime was 22.8% (Table 6.1).

We have observed the potent effect of activated water on the total tumor cell content. In particular, the total tumor cell content in mice of the "prophylactic treatment"group which received water activated in the most optimal mode ($t_{act} = 30$ min) was 4.2-fold decreased in comparison with control mice!

The therapeutic application of activated water was less efficient than the prophylactic treatment as suggested by values of the tumor growth inhibition index which arranged about 12–30%. The general tendency in the dependence of antitumor effects on the water-activation duration was similar. The application of water activated for 30 min was the most efficient.

It is important that activated and long-term stored water had significant antitumor effects when used for therapeutic treatment. As shown in Table 6.1, water activated for 30 min was sufficiently efficient in the therapeutic mode of application even after the storage for 14–45 days.

The efficiency of antitumor effects of such "old activated water" was equal to that for water freshly activated for 15 min.

The application of activated water has a substantial influence on the survival of tumor-bearing animals. When activated water was applied in the "prophylactic treatment" mode, the increase in the lifetime was observed in all groups of mice (Table 6.2). Water activated for 30 min has the most potent effect on the survival of mice with transplanted tumors. It has been shown that the average survival time of mice which received water ($t_{act} = 30$ min) in the "prophylactic treatment" mode increased to 61.7%. The very marked increase in the lifetime (about 45%) was observed when mice were treated with activated water in the "prophylactic treatment" mode at $t_{act} = 15$ min and $t_{act} = 45$ min.

When mice received activated water in the "therapeutic treatment" mode, the significant positive effects such as an increase in the lifetime were also observed. However, values of the estimated index were by 30–50% lower than that in the "prophylactic treatment" mode. It can be suggested that water activated for 60 min is not efficient for the therapeutic application.

In conclusion, the above data suggest that the prophylactic application of freshly activated water with $t_{act} = 30$ min is a very promising treatment approach directed to the inhibition of tumor growth, as was observed in the experiments with the ascitic *Ehrlich carcinoma* model.

The efficiency of the prophylactic application of activated water is very close to that of potent antitumor chemotherapeutic drugs! Furthermore, the application of such water aimed at the tumor growth prophylaxis had no adverse effects which are typical of chemotherapy of cancer.

6.3. Study of Antitumor Effects of the Application of Activated Water on the Experimental Tumor Model of Ascitic *Sarcoma 37*

6.3.1. *Materials and methods*

The results obtained suggest the antitumor efficiency of different types of activated water in the therapeutic or prophylactic treatment of mice with ascitic *Ehrlich carcinoma*. This very important result should be verified again using other tumor strains.

The ascitic *sarcoma 37* model was used as other histological tumor type for the studies of possible antitumor effects of the application of activated water. This tumor-growth model was chosen for the purpose of comparison of the antitumor effects of activated water because this tumor has connective-tissue histogenetic origin, whereas *Ehrlich carcinoma* belongs to epithelial tumors.

Ascitic sarcoma cells were obtained from the peritoneal cavities of unbred white mice on the 7–8th day after the tumor transplantation. All mice of 11 groups were inoculated intraperitoneally with 200,000 viable *sarcoma 37* cells according to the scheme presented in Fig. 6.1. The set of 220 mice was used in one series of studies. Modes of application of activated water (14 days before the tumor transplantation in the "prophylactic treatment" mode and on the next day after the tumor transplantation in the "therapeutic treatment" mode) and the division of mice into the "prophylactic treatment" and "therapeutic treatment" groups were almost similar to the above-mentioned studies using the *Ehrlich carcinoma* model.

In seven days after the ascitic *sarcoma 37* transplantation, the studies of the volume and cellularity of tumors were performed. Ten mice in each group were used. The photos of the ascitic-fluid volumes of several animals from the "control" and "prophylactic treatment" groups are presented in Figs. 6.12–6.13.

These mice received water activated for 15 min and 45 min. Each test-tube contained ascitic-fluid consisted of tumor cells corresponding to one

Figure 6.12. Measurement of the ascitic-fluid volume in mice with transplanted ascitic *sarcoma 37* of the "prophylactic treatment" group (on the right side) which received water activated for 45 min and "control" group (on the left side).

Figure 6.13. Measurement of the ascitic-fluid volume in mice with transplanted ascitic *sarcoma 37* of the "prophylactic treatment" group (on the left side) which received water activated for 15 min and "control" group (on the right side).

tumor obtained from one mouse. These photos show the tumor volumes of mice from different groups.

The data on the ascitic-fluid volume and the cell content are presented in Table 6.3 and Figs. 6.14–6.16. The more-detailed analysis of data will be described in what follows.

When the analysis of the volume and the cellular content of ascitic fluid was completed, every group consisted of 10 mice. Mice were under observation up to the experiment termination in order to evaluate the dependence of the survival on the type of activated water and the mode of treatment. Figures 6.17 and 6.18 show the whole view of cages with mice from two different groups. The number of mice in the "prophylactic treatment" group is larger than those in the "therapeutic treatment" and "control" groups. These photos correspond to the data obtained on the 17th and 20th days after the tumor transplantation and are shown in Figs. 6.19 and 6.20.

The photos clearly show the difference in the application efficiency of nonactivated and activated water in the "prophylactic treatment" and "therapeutic treatment" modes. Those photos were taken on the termination stage of the experiment when tumor burdens were approximately equal to critical burdens that cause death of animals.

Table 6.3. Effect of activated water on the parameters of transplanted ascitic *sarcoma 37*.

Type of treatment and fraction of water used	Investigated parameters			Index of tumor growth inhibition, D, %
	Average volume of ascitic-fluid in one tumor $\langle V \rangle$, ml	Average number of viable tumor cells in 1 ml of ascitic-fluid $\langle C/V \rangle$, ml^{-1} ($\times 10^3$ cells)	Average number of viable tumor cells in peritoneal cavity of one mouse $\langle C \rangle$ ($\times 10^6$ cells)	
Control, $t_{act} = 0$	3.1 ± 0.3	200454 ± 24908	621	—
Prophylactic, $t_{act} = 15$ min	2.2 ± 0.09	151324 ± 15330	333	29.0
Prophylactic, $t_{act} = 30$ min	2.0 ± 0.07	121161 ± 13742	242	35.5
Prophylactic, $t_{act} = 45$ min	2.1 ± 0.1	135030 ± 12219	284	32.3
Prophylactic, $t_{act} = 60$ min	2.8 ± 0.08	171266 ± 23713	479	9.7
Prophylactic, "Old activated water", $t_{act} = 30$ min	2.6 ± 0.2	141132 ± 30716	367	16.0
Therapeutic, $t_{act} = 15$ min	2.7 ± 0.08	183134 ± 23164	494	12.9
Therapeutic, $t_{act} = 30$ min	2.4 ± 0.05	160080 ± 32501	384	22.5
Therapeutic, $t_{act} = 45$ min	2.6 ± 0.15	169984 ± 21671	442	19.2
Therapeutic, $t_{act} = 60$ min	2.9 ± 0.1	220060 ± 33007	638	6.5
Therapeutic, "Old activated water", $t_{act} = 30$ min	2.7 ± 0.2	186842 ± 30059	504	12.4

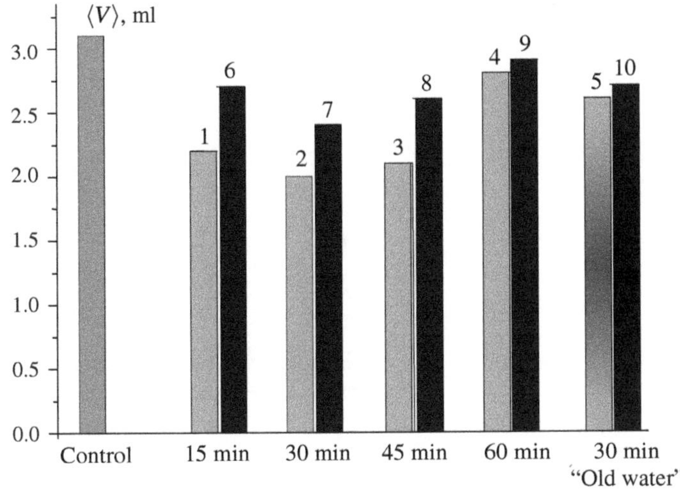

Figure 6.14. Effects of activated water on the average volume of one tumor obtained from mice transplanted intraperitoneally with *sarcoma 37* and treated with activated water in the prophylactic (1–5) and therapeutic (6–10) modes of application.

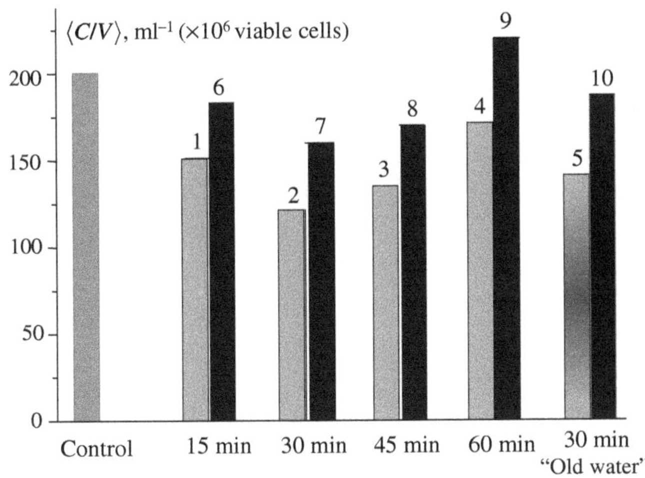

Figure 6.15. Effects of activated water on the average number of viable cells in 1 ml of a tumor obtained from mice transplanted intraperitoneally with *sarcoma 37* and treated with activated water in the prophylactic (1–5) and therapeutic (6–10) modes of application.

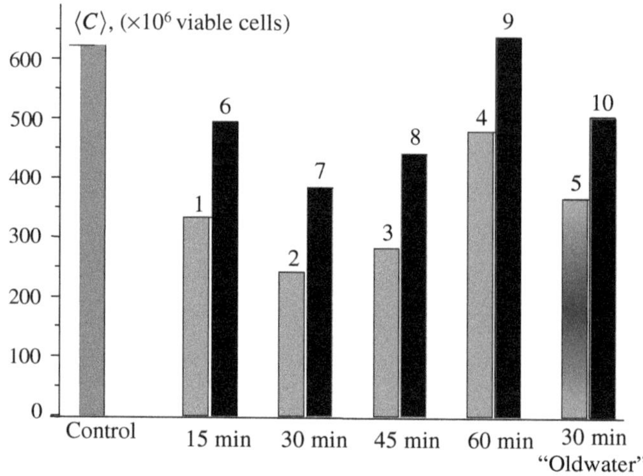

Figure 6.16. Effects of activated water on the average number of viable cells in one tumor obtained from mice transplanted intraperitoneally with *sarcoma 37* and treated with activated water in the prophylactic (1–5) and therapeutic (6–10) modes of application.

Figure 6.17. Tumor-bearing mice of the "prophylactic treatment" (on the left side; mice received water activated for 30 min) and "control" groups (on the right side). Mice were inoculated with cells of ascitic *sarcoma 37*. The photo was taken on the 17th day after the tumor cell inoculation.

Figure 6.18. Tumor-bearing mice of the "prophylactic treatment" (on the left side; mice received water activated for 15 min) and "therapeutic treatment" (on the right side; water was activated for 15 min) groups. Mice were inoculated with cells of ascitic *sarcoma 37*. The photo was taken on the 20th day after the tumor cell inoculation.

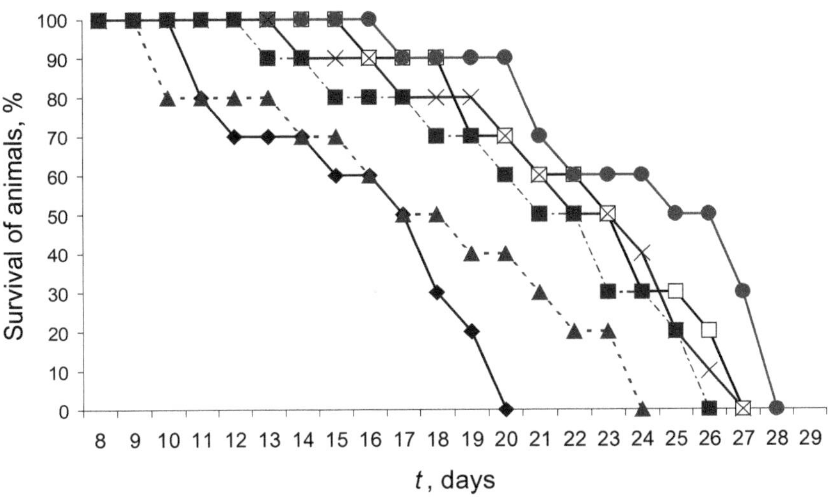

-◆- control ($t_{act} = 0$); -□- $t_{act} = 15$ min; -●- $t_{act} = 30$ min; -×- $t_{act} = 45$ min; - ▲ - $t_{act} = 60$ min; -■- $t_{act} = 30$ min ("old activated water").

Figure 6.19. Survival dynamics of tumor-bearing mice with ascitic *sarcoma 37* which received different types of activated water in the "prophylactic treatment" mode.

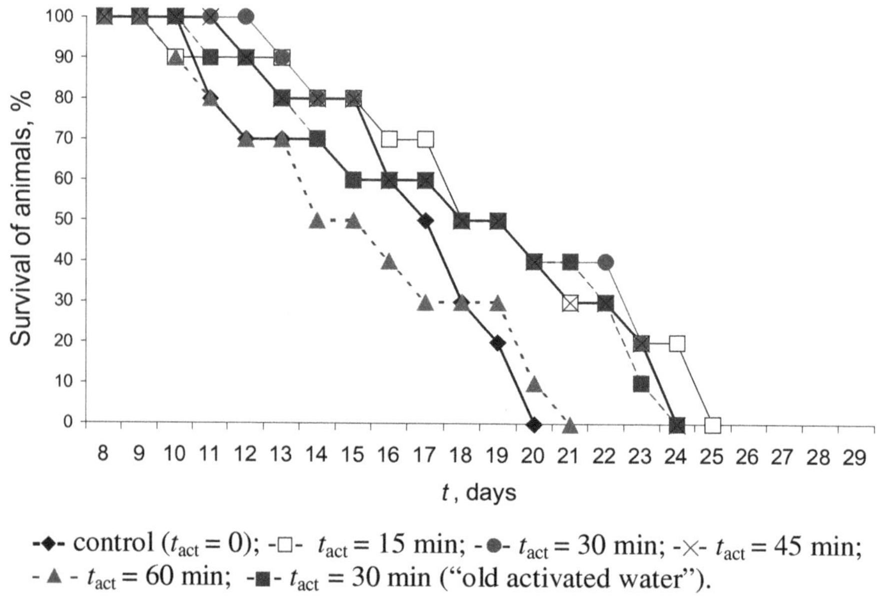

Figure 6.20. Survival dynamics of tumor-bearing mice with ascitic *sarcoma 37* which received different types of activated water in the "therapeutic treatment" mode.

The presented photos show that mice of the "prophylactic treatment" group have sufficiently greater vitality on the 17th and 20th days of the experiment, and they are moving (standing on pads). Mice of this group have smaller tumor burdens. At the same time, mice of the "control" group show sickliness and have very large volumes of ascitic fluid. It is well-known that good therapeutic effect can be achieved after the application of potent chemotherapeutic drugs. However, the chemotherapeutic treatment of cancer patients results in serious adverse effects.

The summary results of studies of the survival of mice with ascitic *sarcoma 37* after different modes of application of activated water are presented in Table 6.4.

Figures 6.19 and 6.20 show the data on the survival dynamics and the lifetime of mice with *sarcoma 37* after the prophylactic or therapeutic treatment with different types of activated water.

Figure 6.21 shows the summary data that characterize the dependence of the increase in the lifetime of tumor-bearing mice on the treatment mode and the type of received activated water.

Table 6.4. Effect of different fractions of activated water on the survival of mice with ascitic *sarcoma 37*.

Type of treatment and fraction of water used	Number of animals	Observational data		
		Average lifetime $\langle t \rangle$, days	Median survival time, days	Percentage increase in lifetime, K, %
Control, $t_{act} = 0$	10	16.1 ± 0.3	17	—
Prophylactic, $t_{act} = 15$ min	10	22.2 ± 0.5	23	37.8
Prophylactic, $t_{act} = 30$ min	10	24.4 ± 0.9	26	51.6
Prophylactic, $t_{act} = 45$ min	10	22.1 ± 0.7	23	37.3
Prophylactic, $t_{act} = 60$ min	10	17.8 ± 1.2	18	10.6
Prophylactic, "Old activated water", $t_{act} = 30$ min	10	21.0 ± 0.9	22	30.4
Therapeutic, $t_{act} = 15$ min	10	19.0 ± 0.5	19	18.0
Therapeutic, $t_{act} = 30$ min	10	19.1 ± 1.0	19	18.3
Therapeutic, $t_{act} = 45$ min	10	18.7 ± 1.0	19	16.1
Therapeutic, $t_{act} = 60$ min	10	15.6 ± 0.7	15	−3.1
Therapeutic, "Old activated water", $t_{act} = 30$ min	10	18.4 ± 1.3	19	14.3

6.3.2. *Results and discussion*

The studies indicate that the application of activated water resulted in the inhibitory effect on the growth of transplanted *sarcoma 37* in mice. However, such effects were less marked as compared with the ascitic *Ehrlich carcinoma* model. It is confirmed by values of the volume and the cellular

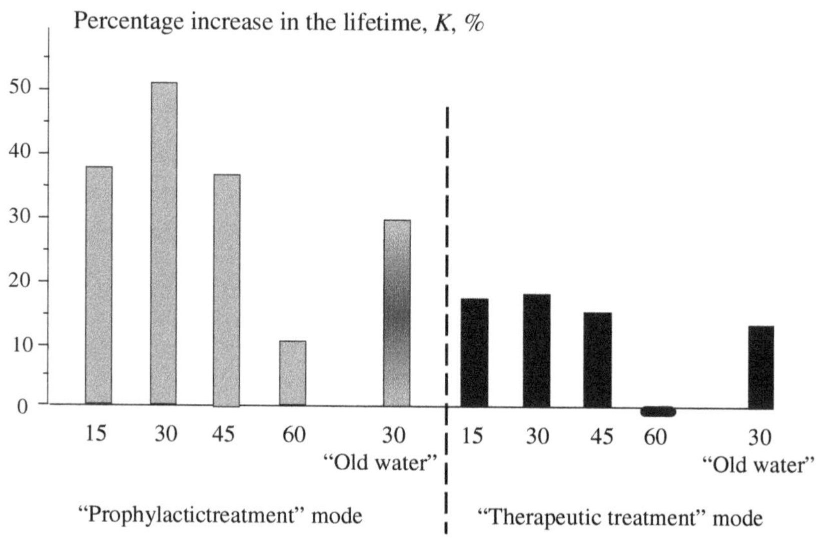

Figure 6.21. Percentage increase in the lifetime of tumor-bearing mice with ascitic *sarcoma 37* which received different types of activated water in the "prophylactic treatment" and "therapeutic treatment" modes. The numbers near the charts correspond to the water-activation duration in minutes.

content of ascitic-fluid obtained from the peritoneal cavities of tumor-bearing mice (Table 6.3 and Figs. 6.13–6.15). The application of optimally activated water (the duration of activation was 30 min) inhibits the tumor growth of *sarcoma 37* by 35.5%, which is a good result. The application of water activated for 15 min and 45 min inhibits the tumor growth by 29–32%, which is nearly similar as compared with the effects of the application of water activated for 30 min. At the same time, it is worth to note that activated water did not affect the tumor transplantability in mice that was equal to 100%.

The effect of activated water on the survival of mice with *sarcoma 37* was similar to that of the ascitic *Ehrlich carcinoma* model, but was less expressed (Table 6.4 and Figs. 6.19–6.21).

In this investigation, similarly to the *Ehrlich carcinoma* model, the greater values of the percentage increase in the lifetime were observed when mice received water activated for 30 min in the "prophylactic treatment" mode. The increase in the lifetime was 51.6%. When mice were treated with water activated for 15 min and 45 min, the increase in the lifetime was quite similar and approximately equal to 37–38%.

The important results were obtained when mice received "old activated water" (the duration of activation was 30 min). At the same time, the percentage increase in the lifetime was equal to that attained after the prophylactic and therapeutic applications of water activated for 45 min.

The "therapeutic treatment" mode of the application of activated water was less efficient than the "prophylactic treatment" mode. It should be noted that the application of water activated for 60 min resulted in an insignificant increase in the lifetime of mice which received activated water in the "prophylactic treatment" mode. The "therapeutic treatment" mode of the application of such water had no effect on the lifetime of tumor-bearing animals (in this case, the negative effect of the application of activated water was statistically insignificant).

In conclusion, we note that the prophylactic application of optimally activated water resulted in the significant tumor-growth inhibition that was shown in the *sarcoma 37* model. The increase in the lifetime of tumor-bearing animals was 50–55%.

At the same time, the therapeutic treatment of activated water was less efficient than the prophylactic application. Furthermore, the long-term storage of activated water did not cause a significant decrease of the antitumor efficiency, and this water can be successfully used for the experimental treatment of cancer.

Moreover, tumor growth model such as *sarcoma 37* was less susceptible to the application of activated water as compared with the *Ehrlich carcinoma* model.

6.4. Research of the Influence of Activated Water on the Cytotoxic Activity of Murine Lymphocytes

In order to understand the possible mechanism of antitumor effects of activated water, the studies of changes in the cytotoxic activity of lymphocytes of mice treated with different fractions of activated water were carried out.

Natural killer (NK) cells are the important cells of immune systems. Based on their defining function of spontaneous cytotoxicity without prior immunization, NK cells have been thought to play a critical role in immune surveillance and cancer therapy. NK cells that infiltrate tumors may protect against the spread of tumor. The second major role of NK cells is their production of cytokines, which may be critically important to eliminate infections. Currently, the active search and the screening of substances with potent immunostimulatory activity are carried out.

An augmentation of the NK cell activity caused by different modifying substances of natural origin is of great interest, e.g. for the correction of the NK cell functional activity in various pathological states and, especially, neoplastic malignancies. The results of studies of the effects of different activated water fractions on the NK cell activity are presented below. The study was aimed to evaluate the optimal modes of water activation and the application mode of activated water for the maximal stimulation of the cytotoxic activity of NK cells.

The following procedure and the scheme of investigation were used (Figs. 6.22 and 6.23).

In the first stage of investigation, mice of the experimental and control groups received activated water for different time intervals. Mice of the "prophylactic treatment" and "short prophylactic treatment" groups received activated water, respectively, for 21 days and for 14 days. When the treatment with activated water was finished, mononuclear lymphocyte fractions enriched with NK cells were isolated from spleens of mice of experimental groups.

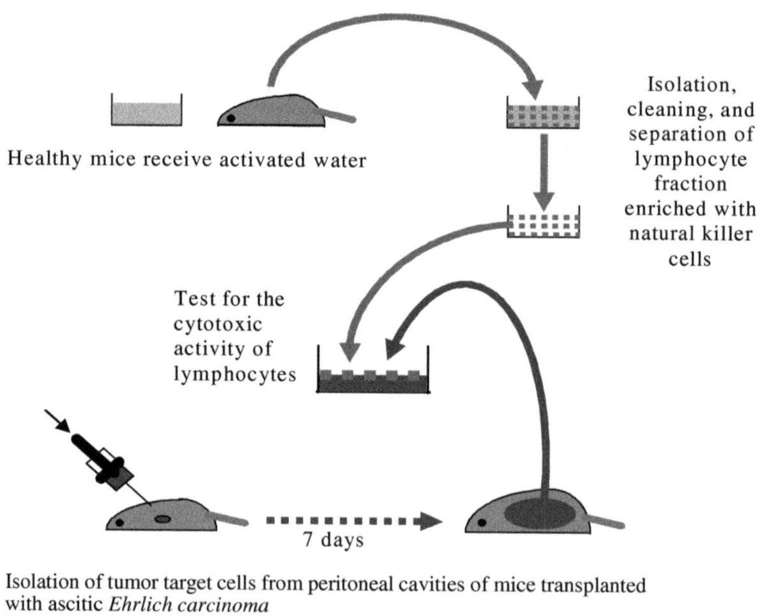

Figure 6.22. Procedure of the study of the influence of activated water on the cytotoxic activity of lymphocytes.

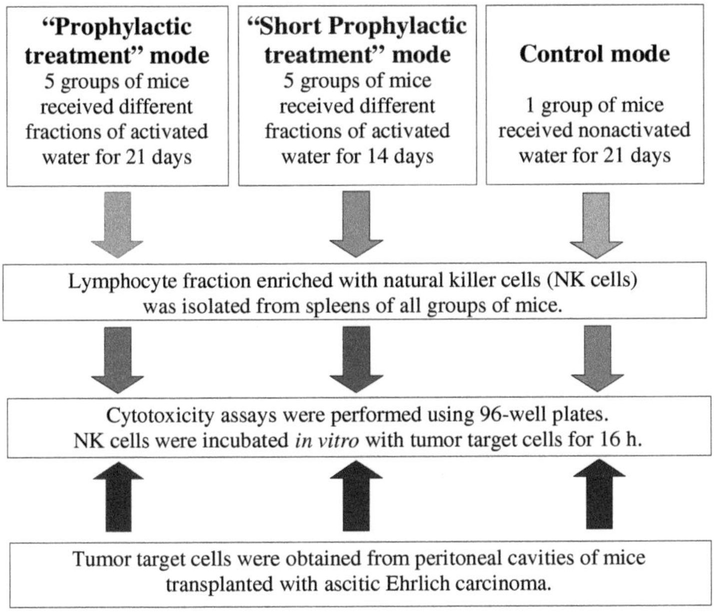

Figure 6.23. Scheme of studies of the effects of activated water on the cytotoxic activity of lymphocytes.

In the second stage, the cytotoxic activity of NK cells incubated with tumor target cells obtained from the peritoneal cavities of mice with transplanted ascitic *Ehrlich carcinoma* was studied.

In the study, inbred adult male BALB/c mice aged 12 weeks with 21–24 g corporal weight were used. All mice were randomly divided into 11 groups with five animals in each group. The application of activated water was as follows:

- mice of the "control" group received nonactivated distilled water for 14 days;
- "prophylactic treatment 15-min, 30-min, 45-min, and 60-min" groups of mice received water activated directly before the application for 15 min, 30 min, 45 min, and 60 min, respectively;
- "prophylactic treatment 30-min" ("old activated water") group of mice received for 21 days water activated for 30 min before the start of the experiment and stored in a cooler; that water was used up to the end of the experiment;

- mice of the "short prophylactic treatment 15-min, 30-min, 45-min, and 60-min" groups received water activated directly before the application for 15 min, 30 min, 45 min, and 60 min, respectively; and
- "short prophylactic treatment 30-min" ("old activated water") group of mice received water activated for 30 min before the start of the experiment and stored in a cooler for 14 days; that water was used up to the end of the experiment.

Animals of 10 experimental groups received the appropriate fraction of activated water (according to the daily rate of water intake), whereas the "control" group of mice received ordinary (nonactivated) water.

The subject of the investigation was the fraction of lymphocytes enriched with NK cells.

Because the obtained results are very important, a more detailed technique of isolation of mononuclear lymphocytes will be described below.

Splenocytes were obtained by the homogenization of spleens resected from mice in a Potter's homogenizer to obtain the single-cell suspension, which were then passed through a nylon mesh filter for the removal of clumps.

Mononuclear lymphocytes were isolated by the standard Ficoll-Verografin technique. To this end, splenocyte suspensions were centrifuged in the gradient of Ficoll-Verografin (1.077 g/cm^3; "Pharmacia Fine Chemicals") for 40 min (400 g) at 5°C. Then mononuclear lymphocytes were collected from the interface layer and twice washed in a sterile phosphate buffer solution. Lymphocytes were purified by further incubation into plastic Petri dishes for 40 min at 37°C in the humidified athmosphere with 5% CO_2 to deplete adherent cells and phagocytes. As a result of the applied techniques mentioned above, we obtained the lymphocyte fraction enriched with NK cells (about 25–30%). Isolated lymphocytes were suspended in RPMI-1640 medium ("Sigma", USA) supplemented with 10% heat-inactivated fetal calf serum ("Sigma", USA), penicillin (40 U/ml) ("Kyivmedpreparat", Ukraine), and streptomycin (40 μg/ml) ("Kyivmedpreparat", Ukraine). The final concentration of lymphocytes was 7.5×10^6 cell/ml.

Ascitic *Ehrlich carcinoma* cells were isolated from the peritoneal cavities of white mice aged 8–9 weeks on the 8th day after the tumor cell inoculation and were used in the cytotoxic test as NK-resistant target cells (TC). Tumor cells were suspended in the culture medium to a concentration of 2.5×10^6 cell/ml, and their number and viability were determined in the microscopic supravital test with trypan blue. The cell viability was about 98%.

All cytotoxicity tests were performed in 96-microwell round-bottomed plates using a target — effector cell ratio of 1:3 in the total volume of 200 μl/well. In controls, only TC and the nutrient medium, or TC and lymphocytes obtained from mice of the "control" group were added. To experimental wells, 0.1 ml of TC and 0.1 ml of lymphocytes isolated from spleens of mice treated with different types of activated water were added.

The general statistics of the experiments was as following: mononuclear lymphocytes obtained from each experimental mouse were placed into three wells of a plate. So, the effect of each fraction of activated water (and nonactivated control water as well) on the cytotoxic activity of NK cells was evaluated in 15 independent experiments and was considered as significant.

Similarly, 15 independent experiments were performed when only 0.1 ml of the TC suspension and 0.1 ml of the nutrient medium were added into one well. Those experiments allowed us to determine the average "basal" number of dead tumor target cells N_{TC} that could not be caused by the cytotoxic influence of effector cells. The "basal" number of dead tumor target cells was a "reference point" for the evaluation of the cytotoxic activity of NK cells obtained from mice after different types of the application of activated water.

After the incubation at 37°C for 18 h in the humidified atmosphere with 5% CO_2, microplates were gently centrifuged (400 g, 5 min). The numbers of viable and dead TC in control and experimental wells were determined using the microscopic test with supravital staining with trypan blue. The cytotoxic activity of NK cells was expressed as the cytotoxicity index (CI, %) and was calculated as follows:

$$\text{CI} = [(N_{TC} - N_{TT+NK})/N_{TC}] \times 100\%$$

where N_{TC} is the number of viable tumor cells in wells with only TC; and N_{TC+NK} refers to the number of viable tumor cells in wells with TC and NK cells.

Based on this definition, CI for "basal" experiments was equal zero.

In addition, to compare the activity of lymphocytes obtained from mice treated with activated water, we studied the immunostimulatory action of a reference chemical agent that belongs to the phorbol ether family. Phorbol myristate acetate (PMA; chemical formula: 2-O-tetradecanoylphorbol-13-acetate) in a concentration of 50 ng/ml was used as a standard stimulant of lymphoid-macrophage lines. PMA increases the ability of lymphocytes to generate superoxide radicals and activates protein kinase C which participates in many signal transduction pathways in cells.

The effects of the prophylactic application of different fractions of activated water on the levels of the cytotoxic activity of splenic mononuclear lymphocytes with NK-activity are shown in Table 6.5 and in Figs. 6.24 and 6.25. Values of the cytotoxic index were estimated on the basis of changes of the cytotoxic activity concerning the "basal" system, TC of which were not incubated with NK cells.

Figure 6.25 shows the changes of the cytotoxic activity of murine mononuclear lymphocytes stimulated during the treatment with activated water in comparison with the same activity values of lymphocytes isolated from spleens of mice treated with nonactivated water.

The data obtained demonstrate that the immunostimulatory potential of activated water is dependent on both the duration of activation and the duration of storage of activated water before the treatment of mice.

Figures 6.24 and 6.25 clearly demonstrate the changes of the cytotoxic activity of NK cells of mice after the period of treatment with activated water was prolonged.

In particular, when water activated for 30 min was applied in the prophylactic mode, a significant increase of the cytotoxic activity of lymphocytes was observed (Fig. 6.25). The application of this water in the "prophylactic treatment" mode resulted in the increase of the cytotoxicity index by 20% as compared with control values obtained after the application of nonactivated water (Fig. 6.24).

The analysis of the cytotoxic activity changes which depended on the duration of the application of activated water showed that the efficiency of the prophylactic application of water was increased when the period of treatment with activated water was prolonged.

The "short prophylactic treatment" mode of application of activated water (the duration of activation was 30 min) resulted in an insignificant increase of the natural cytotoxicity levels that did not exceed 6%. A similar immunostimulatory effect of the "prophylactic treatment" mode of application of "old activate water" (the duration of activation was 30 min) was observed. The data obtained suggested that the short prophylactic application of water activated for 15 min and 30 min resulted in a modest potentiation of the cytotoxic activity of NK cells. However, the prolongation of the application period of activated water from 14 days ("short prophylactic treatment" mode) to 21 days ("prophylactic treatment" mode) normalized the values of the cytotoxicity index which were equal to control data.

Table 6.5. Effects of activated water on the cytotoxic activity of splenic mononuclear lymphocytes with NK-activity.

Type of treatment and fraction of water used	Average number of viable tumor cells in microwells, $\times 10^3$	Cytotoxicity index (CI, %)	Change of cytotoxicity index (in relation to nonactivated water), %
"Basal" tumor cells	235.0	0	−16
Control, $t_{act} = 0$	197.5	16	0
Reference immunostimulant (PMA)	174.0	25.9	9.9
Prophylactic, $t_{act} = 15$ min	197.5	16	0
Prophylactic, $t_{act} = 30$ min	188.0	20	4.0
Prophylactic, $t_{act} = 45$ min	200.0	15	−1.0
Prophylactic, $t_{act} = 60$ min	200.0	15	−1.0
Prophylactic, "Old activated water", $t_{act} = 30$ min	195.0	17	1.0
Short Prophylactic, $t_{act} = 15$ min	202.0	14	−2.0
Short Prophylactic, $t_{act} = 30$ min	195.0	17	1.0
Short Prophylactic, $t_{act} = 45$ min	200.0	15	−1.0
Short Prophylactic, $t_{act} = 60$ min	202.5	13.8	−2.2
Short Prophylactic, "Old activated water", $t_{act} = 30$ min	200.0	15	−1.0

The application of other fractions of activated water did not cause statistically significant changes of the cytotoxic activity of NK cells.

Thus, the significant increase of the lymphocyte cytotoxicity levels was observed only when mice were treated with water activated for 30 min.

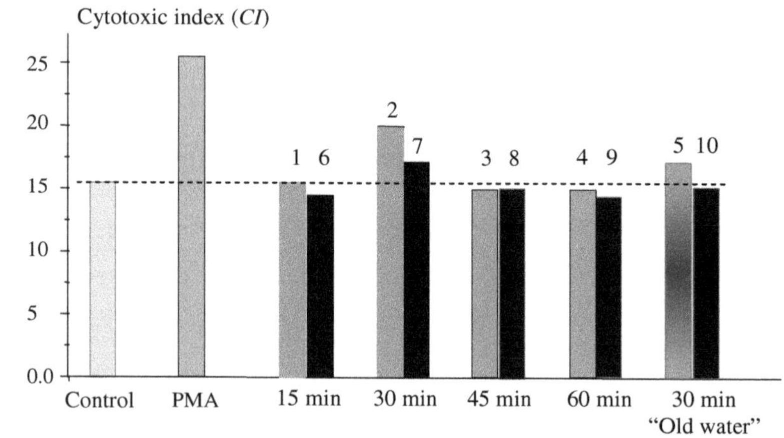

Figure 6.24. Effects of activated water on the cytotoxic activity of lymphocytes with natural killer cell activity. Activated water was applied in the prophylactic (1–5) and short prophylactic (6–10) mode. PMA was used as a standard stimulant of lymphoid cells.

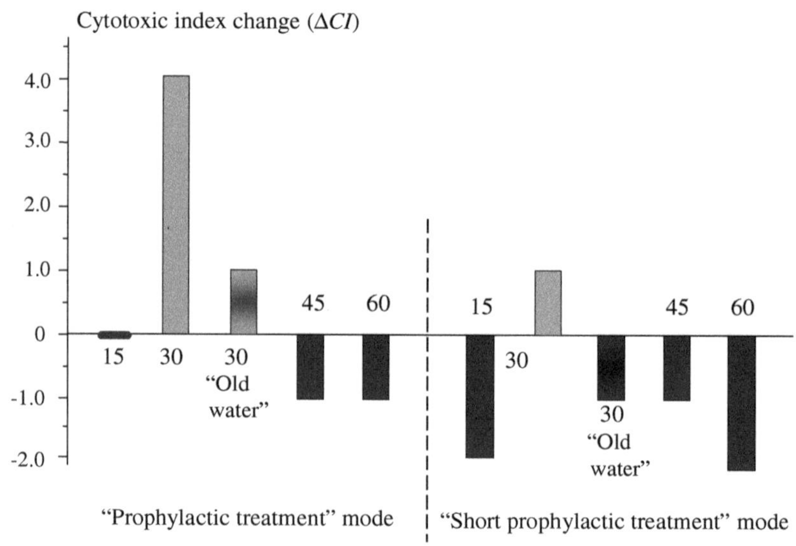

Figure 6.25. Changes of the cytotoxic activity of mononuclear lymphocytes of mice which received different types of activated water in comparison to the results of the application of nonactivated water. Numbers corresponds to the water-activation duration in minutes.

It must be noted that the immunostimulatory potential of activated water was less pronounced than that observed after the *in vitro* stimulation of lymphocytes with PMA. The CI values of PMA-stimulated NK cells were increased by 62%, whereas the application of activated water resulted in the increase of the natural cytotoxicity only by 20%. However, the augmentation of the NK cell activity caused by the application of activated water is quite a promising approach for the nondrug stimulation of immunity.

The usage of reference lymphocyte activity stimulants (e.g. PMA) can be used only in laboratory studies because they have significant toxicity or mutagenic activity. At the same time, the application of optimally activated water allows one to potentiate the cytotoxic activity of lymphocytes without any adverse effects.

It is possible that the prolongation of the treatment period of activated water will cause a more expressed augmentation of the NK cell cytotoxic activity.

In conclusion, the application of activated water can induce a significant activation of the NK cell cytotoxic potential. In this scenario, it is apparent that the application of activated water in tumor-bearing immunocompetent hosts would result in the cytotoxic activation of NK cells to destroy tumor cells. Thus, the obtained results may be important for the future therapeutic approaches that will implicate activated water. A more complete understanding of the antitumor and immunopotentiatory effects of activated water should promote the development of more rational and efficient therapeutic strategies of cancer treatment.

CHAPTER 7

Effect of MRET Activated Water on Staphylococcal Infection *in vivo* in Animal Model (on the Cells of Immune System) and *in vitro* on the Culture of *Staphylococcus aureus* Wood-46

7.1. Immune Response and the Purpose of Investigation

The phagocytic system is one of the main factors of natural nonspecific cellular resistance to infections and inflammations. It is the first line of protection of an organism against the penetration and reproduction of pathogenic microorganisms. The protective role of phagocytes is based on their capacity to identify, engulf and neutralize the alien agents penetrating into internal environment of a macro-organism. Phagocytosis is the main mechanism of natural resistance of the body especially at the first stage of contagious process. At the same time, it is a regular part of the process of formation of the specific immune response.

The infections induced by viruses and bacteria constitute the main category of infectious diseases subject to immune response. Staphylococcal infections are one of the most widespread infections. The microorganisms causing the staphylococcal infections are characterized by a number of pathogenic factors. They are capable of mutation and persist in microorganism-forming centers of the chronic infections periodically activated in case of absence of stable immunity. Following the penetration of pathogens in the human body, the primary nonspecific mechanisms of protection such as the cellular factors of natural resistance (mononuclear phagocytes/macrophages, polymorphonuclear leucocytes/neutrophils, and natural killer cells) are mobilized and activated.

Phagocytes (neutrophils, monocytes/macrophages) are the key elements of protection of an organism against infections and play the important role supporting homeostasis of the body. The phagocytes identify, engulf, destroy, and neutralize the alien substances. Macrophages combined with molecules of I and II classes of the main complex of histo-compatibility to T-lymphocytes constitute the antigens to the infectious microorganisms. Phagocytes are essential in the formation of the specific part of immune response: they neutralize the pathogens and contribute to the cessation of the development of the inflammatory process.

From the other hand, the infectious microorganisms are capable of suppressing immune-biological reactions of the body such as the production of immune-regulatory cytokines, the factors of cellular and humoralious immune response. The incapacity of natural and specific immune factors to neutralize pathogens can cause the following consequences:

- the equilibrium between pathogens and protective forces of the body (the chronic form of infection);
- the penetration of pathogens through protective barriers in the blood or in intracellular environment and then all over the body which can lead to death.

Considering the above, it is important to study the methods of enhancement of the factors of natural resistance of the body. The results of this investigation confirm that one of the perspectives is related to the effect of regular consumption of MRET Activated Water on the characteristics of immune response of biological organisms at cellular level.

In the process of the research, the effect of MRET Activated Water was studied in animal mice model on the characteristics of weight and cellularity of lymphoid organs of immune system, and on functional activity of phagocytes (peritoneal macrophages and neutrophils of the peripheral blood).

The investigation was conducted in two steps. The experiments *in vivo* were conducted on mice infected with *Staphylococcus* culture after preventive consumption of MRET water. In the experiments *in vitro*, the growth of identical staphylococcal culture was studied on meat-peptone agar (MPA) treated with MRET activator. These investigations were conducted under the supervision and participation of Prof. Lydia S. Kholodna from the Biological Department of Kiev National Shevchenko University.

The activation of water was conducted with the help of MRET Water Activator generating subtle composite low-frequency nonionizing electromagnetic field.

7.2. Functional Activity of Cells of the Immune System of Mice Infected with Staphylococcal Culture Following Preventive Consumption of MRET Water

7.2.1. *The methodology of investigation*

The investigation of the effect of MRET Activated Water was conducted in two steps: the evaluation of the immune-stimulatory effect following the ingestion of MRET water on the immune-competent cells in the model of mice infected with *Staphylococcus aureus Wood-46* (*in vivo*), and the evaluation of the inhibition of growth of *Staphylococcus aureus Wood-46* culture in MRET activated nutrient medium (*in vitro*). The *Staphylococcus aureus Wood-46* culture was received from the Czechoslovak collection of microorganisms.

The 400 male mice of line BALB in the age of 11–13 weeks and of the mass 18–21 g were used in the study *in vivo*. After preliminary experiments on the persistence of pathogen in homogenate of kidneys of mice conducted on five groups of mice (45 animals per each group), the optimal 30-min of MRET water activation was chosen for the main line of the investigation. The main line of experiments was conducted on three groups of mice with 50 animals in each group, and the following strategy of examinations was applied.

Prior to the inoculation of *Staphylococcus aureus Wood-46* culture, one group of mice (Group No. 1) consumed MRET activated distilled water for four weeks, another group (Group No. 2) consumed MRET water for two weeks, the control group consumed nonactivated ordinary distilled water. During the following two weeks of experiment, the first two groups continued to consume MRET water and the control group consumed ordinary distilled water. The view of an open-air cage with mice is shown in Fig 7.1.

In the process of the investigation, two types of staphylococcal infections were studied: the local inflammation and the intra-peritoneal infection. In order to induce a local inflammation, the culture of *Staphylococcus aureus Wood-46* was inoculated in the hind left paws of mice (Fig 7.2). For other series of experiments, the inoculation of culture of *Staphylococcus aureus Wood-46* was conducted intra-peritoneally in dose LD_{30} in order to spread the infection all over the body.

The second step of investigation was conducted *in vitro* based on the analysis of the growth of staphylococcal culture on meat-peptone agar

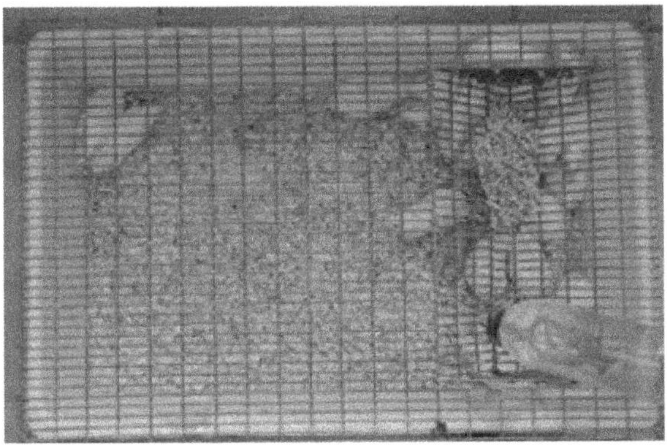

Figure 7.1. The view of an open-air cage with mice and the container with MRET water available to mice for unlimited consumption (on the right bottom part of the picture).

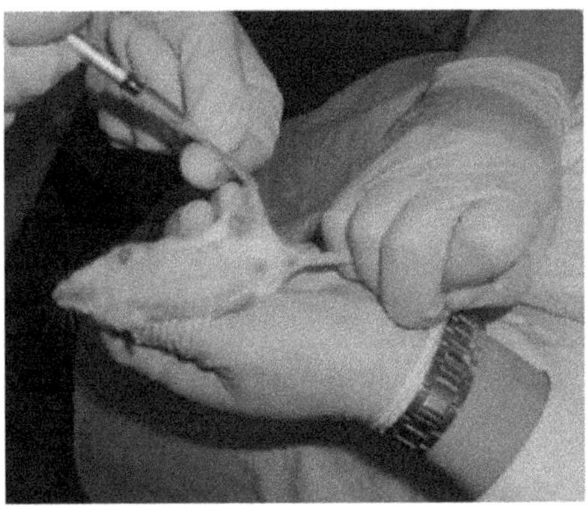

Figure 7.2. The procedure of inoculation of *Staphylococcus aureus* culture in the hind left paw of a mouse.

(MPA) at a temperature of 37°C during 18–24 h with different initial concentrations of cells (from 10 to 10^9 cells/ml). The samples were treated with the help of MRET activator during different periods of time (in the range of 15 to 60 min) right after the introduction of staphylococcal culture to MPA. The results of this investigation will be presented in the second part of this chapter.

7.2.2. *The examination of functional activity of cells of the phagocytic system*

7.2.2.1. *Characteristics of functional activity of the phagocytic system*

The phenomenon of phagocytosis depends on the characteristics of the functional state of cells of the phagocytic system, neutrophils and macrophages, and on informative modifications in their activity. The most common methodology applied in the studies of the functional activity of phagocytes is the examination of their phagocytic (engulfing of alien cells) and oxygen-dependent bactericidal activity.

Phagocytic activity of neutrophils and macrophages is estimated based on *Index of Phagocytosis* (relative quantity of the phagocytes which engulfed test-bacteria) and on *Phagocytic Number* (the number of test-bacteria engulfed by one phagocyte). The cultures of *Staphylococcus aureus* and *Latex* are usually used as test-bacteria.

The oxygen-dependent bactericidal activity of phagocytes is studied with the help of *NBT-test*: the formation of sediment on activated phagocytes in the solution of Nitro-Blue Tetrazolium. The principle of such testing is based on the fact that the soluble nitro-blue tetrazolium converts in diphormazan which spreads in the cytoplasma or on the surface of activated phagocytes in the form of dark blue granules. With the help of NBT-test, it is possible to distinguish the activated phagocytes from the nonactivated ones. The modifications in the characteristics of NBT-test (their significant increase or decrease) coincide with the changes in the mechanisms of oxygen-dependent bactericidal activity of phagocytes.

7.2.2.2. *The extraction of neutrophils of the peripheral blood*

Neutrophils were extracted from heparinized venous blood with a method of fractionation of cells based on a gradient of density of fractions of blood cells spread on phycol-verografin ($\rho = 1077$ g/ml). In order to prepare the solution of phycol-verografin with the density 1077 g/ml, 6.48 g of phycol were dissolved in 85 ml of hot distilled water and then the 17 ml of verografin were added gradually. The density of the solution was checked with an aerometer. The solution was sterilized with the help of autoclaving and kept at 4°C. Then, the heparinized venous blood was mixed with 0.15 M solution of NaCl in the ratio 1:4. The 2 ml of dilute blood were accurately spread on phycol-verografin. After that, the mixture was centrifuged for 45 min at 1500 rpm. After the process of centrifugation neutrophils are localized between plasma and a layer of phycol-verografin.

7.2.2.3. *The extraction of macrophages from the abdominal cavity*

In order to extract the peritoneal macrophages, the animals were inoculated intra-peritoneally with 4–5 ml of specific refrigerated fluid "medium 199". Then, the abdominal cavity was palpated for one minute. After that, peritoneal fluid was extracted in the test-tubes for further centrifugation. The suspension of cells was centrifuged at 1500 rpm.

7.2.2.4. *The phagocytic activity of neutrophils and macrophages*

Phagocytic activity of neutrophils and macrophages is estimated based on *Index of Phagocytosis* calculated as a relative quantity of the phagocytes which engulfed test-bacteria (in %), and also on *Phagocytic Number* calculated as an average of the number of test-bacteria engulfed by one phagocyte (in standard units). The cultures of *Staphylococcus aureus* (209G or 8325, Wood-46, Cowan-1) and *Latex* are usually used as test-bacteria.

Staphylococcal culture was grown for 24 h on a firm nutrient medium containing 10% of NaCl. After that, the biomass was picked up (at the same stage of cultivation) and washed with 0.15 M solution of NaCl for three times by centrifugation at 6–8 thousands rpm for 15 minutes. The concentration of test-bacteria was brought to 10^9 cells/ml using turbidity standard.

The suspension of phagocytes in the culture medium (concentration 5×10^6 cells/ml) together with the suspension of test-bacteria in 0.15 M solution of NaCl was spread over a slide-glass in the amount of 0.2 ml of each culture. Thus, the relation of phagocytes to test-bacteria was about 1:200. The slide-glasses were placed at the bottom of Petri dishes of small diameter. The cells were cultivated at 37°C for 30–45 min in water-saturated atmosphere with a fixed level of carbon dioxide (5%). Then, nonbonded bacteria were washed twice with Henk's solution from a monolayer of cells. The resulting preparations were dried.

After that, the cells were fixed by methanol for four minutes, dried and colored with azureosine for 15–20 min. The 100–200 cells were examined (concentration 5×10^6 cells/ml) under the microscope, and the Index of Phagocytosis and Phagocytic Number were calculated.

7.2.2.5. *The bactericidal activity of neutrophils and macrophages*

The oxygen-dependent bactericidal activity of phagocytes is estimated with the help of spontaneous and stimulated *NBT-test*. The testing is based on the conversion of the solution of *Nitro-Blue Tetrazolium* (NBT) into the dark-blue sediment on activated phagocytes. It is conducted following the cytomorphological methodology. The value of activated or NBT-POSITIVE phagocytes is calculated as a relative quantity of the phagocytes containing the dark-blue granules of diphormazan (in %).

Following the cytomorphological methodology, 0.2 ml of the suspension of phagocytes in the culture medium RPMI-1640 containing 5% fetal-veal serum (concentration 5×10^6 cells/ml) are spread on the slide-glass placed at the bottom of a Petri dish with a small diameter.

In spontaneous NBT-test, 0.2 ml of 0.15 M solution of NaCl are added to phagocytes. Cells are cultivated in an incubator at 37°C for 45 min in water-saturated atmosphere with a fixed level of carbon dioxide (5%).

In stimulated NBT-test, 0.2 ml of the suspension of test-bacteria *Staphylococcus aureus* in 0.15 M solution of NaCl (concentration $N_0 = 10^9$ cells/ml) are added to the incubated medium.

After the incubation in water-saturated atmosphere with a fixed level of a carbon dioxide (5%) for 10 min at 37°C, the 0.2 ml of the solution of NBT in 0.2% phosphate-solvate buffer are added to phagocytes.

After that, nonbonded bacteria are washed with Henk's solution from the slide-glasses with phagocytes (in the case of stimulated NBT-test). The preparations are dried and fixed for four minutes in methanol. Then, they are colored with safranine for 30 s. The value of *NBT-test* is calculated as a percentage of phagocytes containing the dark-blue granules of diphormazan (in %). The Functional Reserve of phagocytes is defined as a difference between the values of stimulated and spontaneous NBT-test (in %).

7.2.3. *Statistical calculations*

The data were calculated and evaluated with the help of the package of statistical software program "STATISTIRA for Windonws S.O.". The *p-values* were calculated for mean values of studied characteristics for the groups of mice which consumed MRET activated distilled water compared to the control group of mice which consumed nonactivated distilled water, following the criteria of Students' distribution.

7.3. Results and Discussion

7.3.1. *The effect of MRET water on the development of the local acute inflammation*

The local inflammation was induced with the help of the inoculation of *Staphylococcus aureus* culture in the hind left paws of mice. The ordinary inflammatory reaction was observed in the group of mice fed with nonactivated water: the intensive reddening of the hind left paw (Fig 7.3). Both groups of mice on MRET water (preventive consumption for four and two weeks respectively) did not develop any reddening of the hind left paw inoculated with *Staphylococcus aureus* culture (Fig 7.4).

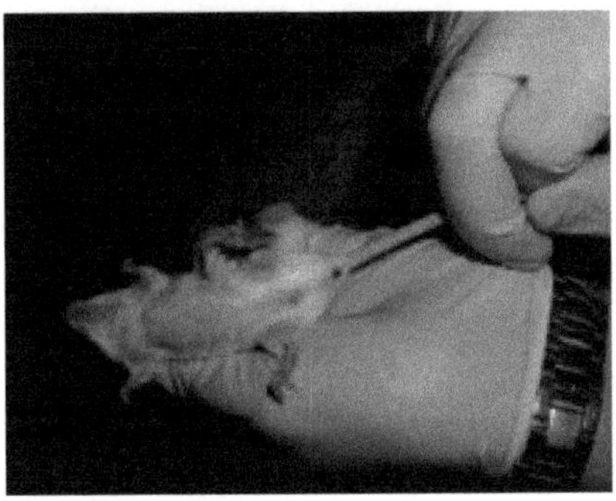

Figure 7.3. The view of paws of a mouse fed with nonactivated water (reddening of the injected paw) in 24 hours after the inoculation of *Staphylococcus* culture.

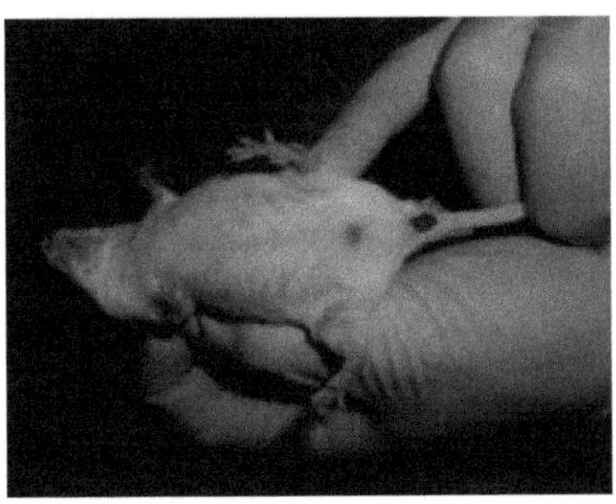

Figure 7.4. The view of paws of a mouse fed with MRET activated water (no reddening of the injected paw) in 24 hours after the inoculation of *Staphylococcus* culture.

The results of this experiment confirm the fact of the substantial inhibition of inflammatory infection in the case of regular consumption of MRET water.

7.3.2. *The effect of activated water on the death rate of animals in the case of intra-peritoneal staphylococcal infection*

There was no case of animal death in all investigated groups within the first 24 hours after intra-peritoneal inoculation of *Staphylococcus* culture, which is a pretty standard result. During the next eight days, 30% of animals died in control group which is an expected result for such experimental procedure. There was no death in both groups of mice that consumed MRET Activated Water, and this is a very unusual result. Thus, the consumption of MRET water reduced the death rate from 30% (control group) to 0% (MRET groups) during the first nine days of experiment.

Nevertheless, the main consequences of *Staphylococcus* infection do not manifest in death of animals as in the case of oncology diseases. *Staphylococcus* microorganisms affect the living systems and organs of the body. These pathogenic microorganisms cause inflammations, suppurations, abscesses, furuncles, quinsy, cepsical conditions, etc. That is why a detailed investigation of the process of stimulation by MRET water of phagocytes and lymphoid organs of the immune systems of animals infected with *Staphylococcus aureus* culture was conducted.

7.3.3. *The preliminary examination of the effect of activated water on staphylococcal infected mice*

For the investigation of protective properties of MRET Activated Water, the persistence of pathogens in the organisms of mice was analyzed on five groups of mice: Groups No. 1 (preventive consumption for four weeks) and Groups No. 2 (preventive for two weeks) which consumed 15-min and 30-min MRET activated distilled water, and Control group of mice which consumed nonactivated distilled water. The results of examination of homogenate of kidneys are presented in Table 7.1.

The significant protective properties of MRET water were confirmed by substantial decrease of *Staphylococcus* colony forming units (CFU) in homogenate of kidneys of mice which consumed MRET water, compared to control group of mice following the intra-peritoneal staphylococcal

Table 7.1. The effect of consumption of MRET activated water on the persistence of pathogens of staphylococcal infection in homogenate of kidneys of mice.

Groups of experimental animals	Period of activation, min	Number of CFU in 1 ml of homogenate of kidneys in 1 day	Number of CFU in 1 ml of homogenate of kidneys in 3 days	Number of CFU in 1 ml of homogenate of kidneys in 5 days
Group 1, N = 15	15	24266 ± 1330*	43227 ± 5600*	15160 ± 1310*
Group 1, N = 15	30	19316 ± 1460	29600 ± 1890*	14000 ± 1660*
Group 2, N = 15	15	23387 ± 2760*	42550 ± 4500*	14550 ± 1750*
Group 2, N = 15	30	24060 ± 870*	41760 ± 3090*	10600 ± 1200*
Control, N = 15	—	19000 ± 2620	76590 ± 4340	31250 ± 2220

CFU — colony forming units
* — marks statistically significant results with $p < 0.05$ compared to control.

infection after the first 24 hours. The analysis of data in the beginning of experiments leads to the conclusion that significant decrease of pathogen colonies in homogenate of kidneys of mice fed with MRET water begins only after 24 hours following the inoculation of *Staphylococcus* culture. The results on 30-min activated water were much better than on 15-min activated water, and all further experiments were conducted with 30-min activated water.

7.3.4. *The effect of activated water on the cellularity and the weight of lymphoid organs*

The results of experiments revealed that the consumption of MRET Activated Water significantly affected the cellularity (quantity of cells) and the weight of lymphoid organs such as spleen, thymus and lymph nodes after two weeks following the intra-peritoneal inoculation of *Staphylococcus aureus* culture. The appearance of the lymph node extracted from the body of a mouse is presented in Fig 7.5. The weight of lymph nodes was measured with the help of the torsion scales (Fig 7.6).

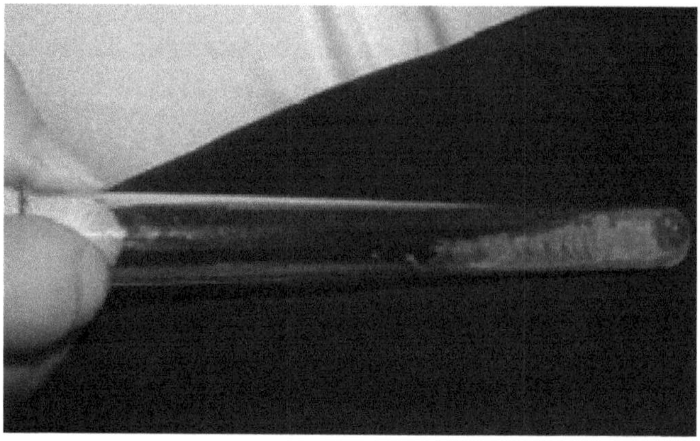

Figure 7.5. The view of a lymph node.

Figure 7.6. Measurement of the weight of a lymph node.

In the beginning of experiment following the intra-peritoneal inoculation of staphylococcal culture after the preventive application of MRET water for four weeks (Group No. 1) and for two weeks (Group No. 2), there was no distinct tendency in modifications of the weight of lymphoid organs in the groups of mice with MRET water compared to the control group of mice with nonactivated distilled water (Table 7.2). An insignificant increase of the cellularity of lymphoid organs of animals which consumed MRET water was observed in the beginning of experiment (Table 7.2).

In the first week of experiment, there was observed a tendency of the increase of the weight of lymphoid organs of mice which consumed MRET water, compared to control group (Table 7.3). A distinct tendency of increasing cellularity of lymphoid organs of mice which consumed MRET water was revealed. Most part of the results are statistically significant with $p < 0.05$ compared to the respective mean values of control group of mice (Table 7.3 and Fig. 7.7).

In two weeks of experiment, the distinct tendency of increasing weight and cellularity of lymphoid organs of mice which consumed MRET water was observed. The results for spleen and lymph nodes are statistically significant with $p < 0.05$ compared to the respective mean values of control group of mice (Table 7.4, Figs. 7.8 and 7.9).

Thus, after two weeks of experiments, in both groups of mice which consumed MRET water, it was observed that there was the statistically significant ($p < 0.05$) increase of the cellularity (quantity of cells) and the weight of spleen and lymph nodes, as well as the insignificant increase of the cellularity and the weight of thymus. These results confirm the fact of significant intensification of immune system response in animals fed with MRET water which were subjected to *Staphylococcus* infection.

The difference in studied parameters between the groups of mice which consumed MRET water (four weeks and two weeks of preventive consumption of MRET water) was insignificant, which confirms the fast and beneficial effect of MRET water on the immune activity of lymphoid organs.

In the beginning of experiment, the cellularity and the weight of lymphoid organs in MRET groups did not show the distinct tendency to modifications. It is reasonable to admit that the consumption of MRET water affects the weight and the cellularity of lymphoid organs only during the infection period.

7.3.5. *The effect of activated water on functional activity of cells of the phagocytic system*

The cells of the phagocytic system are essential in the process of formation of a nonspecific immune response of the body to infections. The role of cells of the phagocytic system (monocytes/macrophages, polymorphonuclear leucocytes/neutrophils, dendritic cells) in the process of protection

Table 7.2. The weight and the cellularity of lymphoid organs in the beginning of experiment.

Groups of experimental animals (N — number of animals)	Weight of lymphoid organs, mg			Cellularity of lymphoid organs		
	Spleen	Thymus	Lymph nodes	Spleen ($\times 10^8$)	Thymus ($\times 10^7$)	Lymph nodes ($\times 10^6$)
Control, N = 15	357.8 ± 6.3	44.3 ± 4.8	28.7 ± 4.4	2.10 ± 0.90	2.20 ± 0.30	3.10 ± 0.80
Group 1, N = 15	239.4 ± 3.6*	65.5 ± 8.1*	39.4 ± 3.9*	2.30 ± 0.80	2.50 ± 0.40	3.30 ± 0.19
Group 2, N = 15	314.9 ± 9.1*	68.4 ± 3.3*	41.3 ± 5.1*	2.70 ± 0.13	2.3 ± 0.60	3.40 ± 0.17

* — marks statistically significant results with $p < 0.05$ compared to control.

Table 7.3. The weight and the cellularity of lymphoid organs in one week of experiment.

Groups of experimental animals (N — number of animals)	Weight of lymphoid organs, mg			Cellularity of lymphoid organs		
	Spleen	Thymus	Lymph nodes	Spleen ($\times 10^8$)	Thymus ($\times 10^7$)	Lymph nodes ($\times 10^6$)
Control, N = 7	298.4 ± 7.8	54.0 ± 4.9	41.1 ± 5.3	1.70 ± 0.08	2.10 ± 0.13	1.90 ± 0.10
Group 1, N = 7	328.9 ± 6.5*	64.3 ± 5.2	40.7 ± 4.3	3.70 ± 0.12*	1.60 ± 0.11*	2.02 ± 0.08
Group 2, N = 7	478.1 ± 4.3*	56.3 ± 7.2	54.8 ± 5.6*	2.80 ± 0.14*	2.20 ± 0.08	2.27 ± 0.06*

* — marks statistically significant results with $p < 0.05$ compared to control.

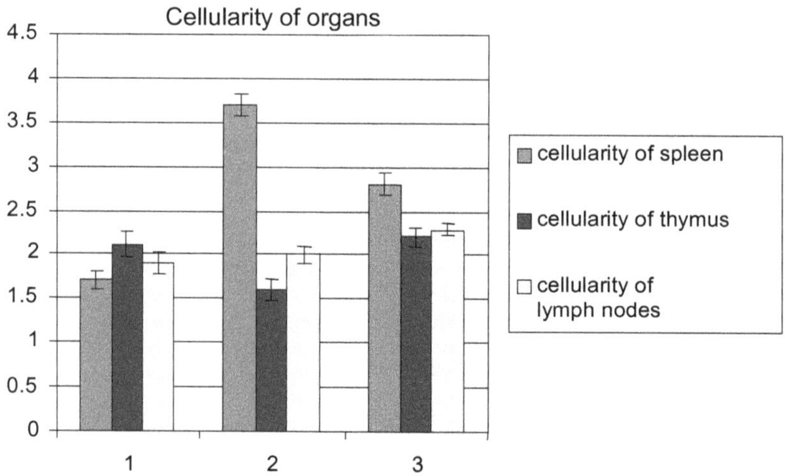

Figure 7.7. The cellularity of lymphoid organs of mice in one week of experiment: 1 — Control group; 2 — Group No. 1 which consumed MRET water (preventive for four weeks); 3 — Group No. 2 which consumed MRET water (preventive for two weeks).

of an organism against infectious diseases consists of engulfing, isolation and inactivation of genetically alien substances. The activated phagocytes form a number of cytokines which perform important regulatory functions. Besides, the mononuclear phagocytes combined with the molecules of classes I and II of MHCC (*major histo-compatibility complex*) to T-lymphocytes constitute antigens, and are essential in the process of formation of specific immune response affecting the production of immune-globulins, cytokines, and the activity of T-lymphocytes.

The investigation conducted in animal mice model revealed that the consumption of MRET Activated Water led to the stimulation of functional activity of cells of the phagocytic system: peritoneal macrophages and neutrophils of the peripheral blood. The consumption of MRET Activated Water significantly increased the intensity of engulfing function of phagocytes (phagocytic activity) and their oxygen-dependent bactericidal activity after two weeks following the intra-peritoneal inoculation of *Staphylococcus aureus* culture.

In the beginning of experiment following the intra-peritoneal inoculation of staphylococcal culture after the preventive application of MRET water for 4 weeks (Group No. 1) and for 2 weeks (Group No. 2), there was no distinct

Table 7.4. The weight and the cellularity of lymphoid organs in two week of experiment.

Groups of experimental animals (N — number of animals)	Weight of lymphoid organs, mg			Cellularity of lymphoid organs		
	Spleen	Thymus	Lymph nodes	Spleen ($\times 10^8$)	Thymus ($\times 10^7$)	Lymph nodes ($\times 10^6$)
Control, N = 7	78.5 ± 4.5	50.4 ± 5.1	21.0 ± 5.1	1.70 ± 0.07	2.19 ± 0.1	2.09 ± 0.06
Group 1, N = 7	219.1 ± 5.5*	54.7 ± 5.5	40.5 ± 4.9*	4.19 ± 0.10*	2.28 ± 0.10	3.31 ± 0.12*
Group 2, N = 7	155.1 ± 5.5*	54.7 ± 5.5	46.8 ± 5.4*	3.12 ± 0.10*	2.39 ± 0.05*	4.00 ± 0.10*

* — marks statistically significant results with $p < 0.05$ compared to control.

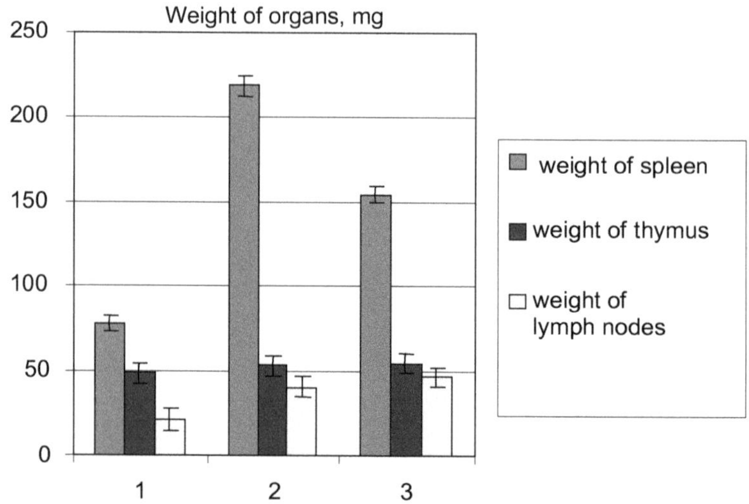

Figure 7.8. The weight of lymphoid organs in two weeks of experiment: 1 — Control group; 2 — Group No. 1 which consumed MRET water (preventive for four weeks); 3 — Group No. 2 which consumed MRET water (preventive for two weeks).

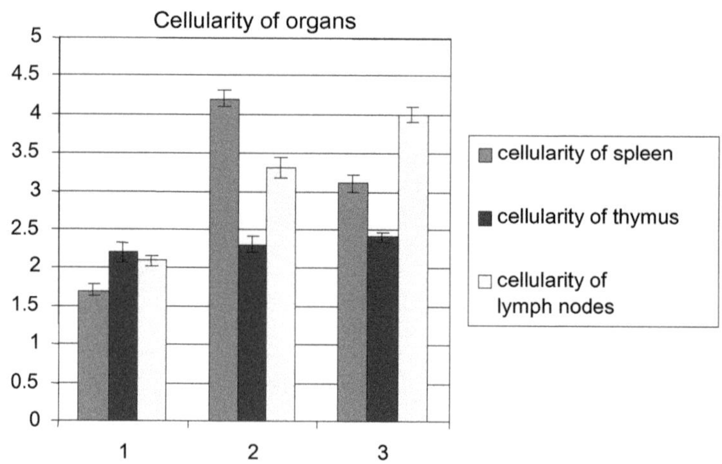

Figure 7.9. The cellularity of lymphoid organs in two weeks of experiment: 1 — Control group; 2 — Group No. 1 which consumed MRET water (preventive for four weeks); 3 — Group No. 2 which consumed MRET water (preventive for two weeks).

tendency in modifications of the phagocytic activity in the case of *Staphylococcus aureus* as an object of phagocytosis, or in oxygen-dependent bactericidal activity in the groups of mice which consumed MRET water compared to the control group of mice which consumed nonactivated distilled water (Table 7.5).

In the case of *Latex* as an object of phagocytosis, the parameters of phagocytic activity (Index of Phagocytosis and Phagocytic Number) of neutrophils and macrophages increased in most cases (Table 7.6, Figs. 7.10 and 7.11). The spontaneous NBT-test does not require any object of phagocytosis and is the same in both cases.

After one week of experiment, the intensity of phagocytic activity in the case of *Staphylococcus aureus* as an object of phagocytosis and of oxygen-dependent bactericidal activity increased (Table 7.7). In the case of *Latex* as an object of phagocytosis, most parameters of phagocytic activity also increased (Table 7.8).

After two weeks of experiment, the intensity of phagocytic activity in the case of *Staphylococcus aureus* as an object of phagocytosis and of oxygen-dependent bactericidal activity significantly increased (Table 7.9, Figs. 7.12–7.14). Most of the results are statistically significant with $p < 0.05$ compared to nonactivated water.

Table 7.5. The functional activity of neutrophils and macrophages in the beginning of experiment (object of phagocytosis — *Staphylococcus aureus*).

Group of experimental animals (N — number of mice in group)	Characteristics of functional activity of phagocytes		
	Index of phagocytosys, %	Phagocytic number, standard units	NBT-TEST spontaneous, %
Macrophages			
Control group, N = 15	42.1 ± 2.0	9.2 ± 1.8	32.4 ± 4.8
Group 1, N = 15	49.5 ± 7.1*	13.4 ± 4.2*	30.9 ± 2.1
Group 2, N = 15	54.7 ± 4.1*	7.6 ± 5.3	46.3 ± 1.5*
Neutrophils			
Control group, N = 15	61.5 ± 6.1	13.0 ± 1.2	43.1 ± 4.2
Group 1, N = 15	47.3 ± 4.4*	7.1 ± 1.8*	45.9 ± 3.1
Group 2, N = 15	53.1 ± 2.9*	8.6 ± 2.1*	39.0 ± 5.3

* — marks statistically significant results with $p < 0.05$ compared to control.

Table 7.6. The functional activity of neutrophils and macrophages in the beginning of experiment (object of phagocytosis — *Latex*).

Group of experimental animals (N — number of mice in group)	Characteristics of functional activity of phagocytes	
	Index of phagocytosis, %	Phagocytic number, standard units
Macrophages		
Control group, N = 5	29.1 ± 0.8	10.5 ± 0.6
Group 1, N = 5	39.7 ± 1.4*	9.4 ± 0.8
Group 2, N = 5	42.8 ± 0.9*	13.2 ± 0.9
Neutrophils		
Control group, N = 5	32.3 ± 1.0	8.3 ± 0.2
Group 1, N = 5	48.6 ± 1.2*	11.5 ± 0.4*
Group 2, N = 5	51.1 ± 1.4*	14.0 ± 0.5*

* — marks statistically significant results with $p < 0.05$ compared to control.

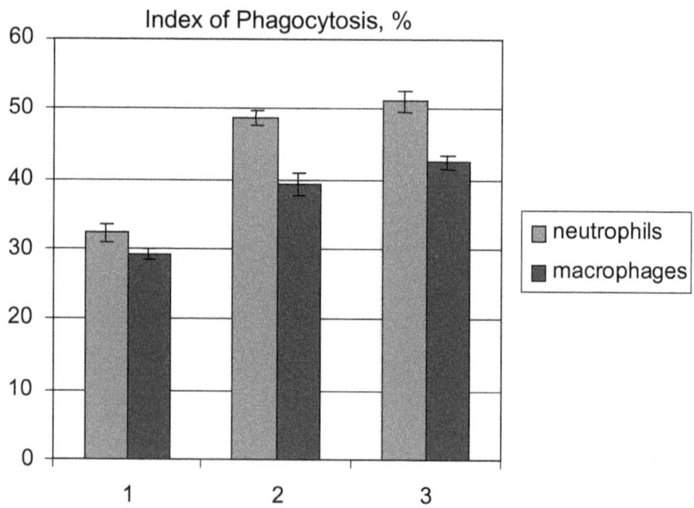

Figure 7.10. Index of phagocytosis in the beginning of experiment (object of phagocytosis — *Latex*): 1 — Control group; 2 — Group No. 1 which consumed MRET water (preventive for four weeks); 3 — Group No. 2 which consumed MRET water (preventive for two weeks).

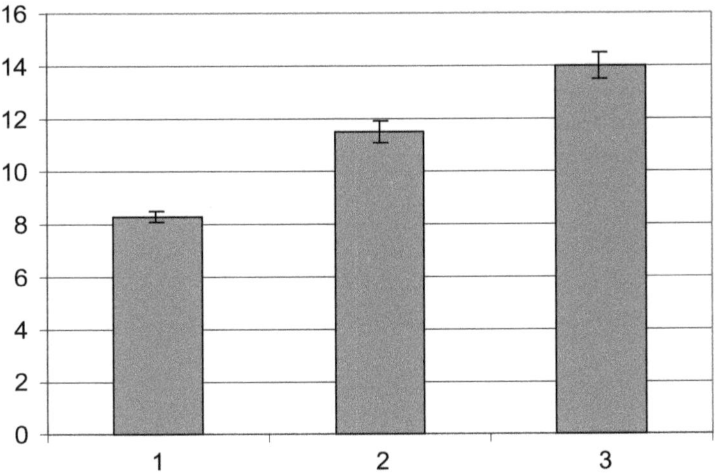

Figure 7.11. Phagocytic number of neutrophils in the beginning of experiment (object of phagocytosis — *Latex*): 1 — Control group; 2 — Group No. 1 which consumed MRET water (preventive for four weeks); 3 — Group No. 2 which consumed MRET water (preventive for two weeks).

Table 7.7. The functional activity of neutrophils and macrophages in one week of experiment (object of phagocytosis — *Staphylococcus aureus*).

Group of experimental animals (N — number of mice in group)	Characteristics of functional activity of phagocytes		
	Index of phagocytosis, %	Phagocytic number, standard units	NBT-TEST spontaneous, %
Macrophages			
Control group, N = 5	34.5 ± 2.7	7.0 ± 3.8	25.4 ± 2.0
Group 1, N = 5	51.3 ± 2.1*	11.4 ± 6.2	31.5 ± 6.1
Group 2, N = 5	63.7 ± 4.6*	8.5 ± 3.3	28.8 ± 3.5
Neutrophils			
Control group, N = 5	53.4 ± 5.1	5.3 ± 2.2	21.5 ± 6.2
Group 1, N = 5	71.8 ± 4.1	6.4 ± 3.6	33.6 ± 5.1
Group 2, N = 5	64.1 ± 2.0*	7.7 ± 5.2	38.5 ± 2.2

* — marks statistically significant results with $p < 0.05$ compared to control.

Table 7.8. The functional activity of neutrophils and macrophages in one week of experiment (object of phagocytosis — *Latex*).

Group of experimental animals (N — number of mice in group)	Characteristics of functional activity of phagocytes	
	Index of phagocytosis, %	Phagocytic number, standard units
Macrophages		
Control group, N = 15	38.2 ± 4.2	11.7 ± 2.8
Group 1, N = 15	41.3 ± 5.1	14.4 ± 1.2*
Group 2, N = 15	45.7 ± 3.8*	10.2 ± 3.3
Neutrophils		
Control group, N = 15	40.3 ± 4.3	9.0 ± 2.2
Group 1, N = 15	62.3 ± 7.4*	12.1 ± 3.8
Group 2, N = 15	55.9 ± 5.3*	15.6 ± 7.1*

* — marks statistically significant results with $p < 0.05$ compared to control.

Table 7.9. The functional activity of neutrophils and macrophages in two weeks of experiment (object of phagocytosis — *Staphylococcus aureus*).

Group of experimental animals (N — number of mice in group)	Characteristics of functional activity of phagocytes		
	Index of phagocytosis, %	Phagocytic number, standard units	NBT-TEST spontaneous, %
Macrophages			
Control group, N = 7	31.1 ± 2.9	9.1 ± 5.8	26.6 ± 2.6
Group 1, N = 7	54.5 ± 8.1*	13.5 ± 9.2	39.1 ± 6.4*
Group 2, N = 7	53.7 ± 1.6*	17.5 ± 1.3*	49.0 ± 6.5*
Neutrophils			
Control group, N = 7	47.4 ± 7.1	8.2 ± 3.2	25.0 ± 5.2
Group 1, N = 7	69.8 ± 3.7*	14.2 ± 1.6*	46.2 ± 9.1*
Group 2, N = 7	66.0 ± 2.8*	13.1 ± 4.2	48.5 ± 1.2*

* — marks statistically significant results with $p < 0.05$ compared to control.

These experiments confirmed the increase of effective potential of phagocytes which constitutes one of the main factors of natural protection of the body against infections, and initiates immune response.

The analysis of data in the beginning of experiments leads to the conclusion that significant intensification of phagocytic and bactericidal activity

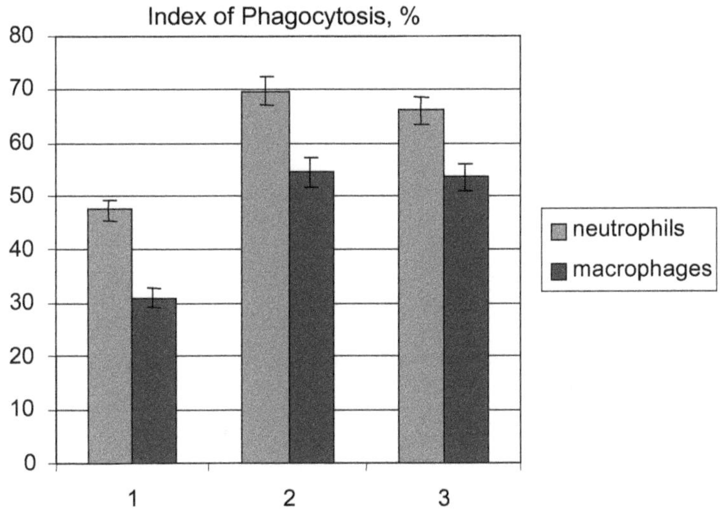

Figure 7.12. Index of phagocytosis of neutrophils and macrophages in two weeks of experiment (object of phagocytosis — *Staphylococcus aureus*): 1 — Control group; 2 — group No. 1 (preventive for four weeks; MRET water); 3 — group No. 2 (preventive for two weeks; MRET water).

of macrophages and neutraphils of mice which consumed MRET water begins only after 24 hours following the intra-peritoneal inoculation of *Staphylococcus* culture. At the end of two weeks of experiment, the mean values of studied parameters in both groups of mice which consumed MRET water substantially increased in comparison to the control group. The differences in mean values of the parameters of functional activity of phagocytes of groups of mice consuming MRET water compared to the control group of mice which were fed with nonactivated water were statistically significant with $p < 0.05$ (for the Index of Phagocytosis and NBT test). These results confirm the significant intensification of phagocytic and bactericidal activity and of immune-system response following the consumption of MRET water.

The differences in mean values of studied parameters for the groups of mice which consumed MRET water were statistically insignificant compared to each other, which confirms the similarity of the level of beneficial effect of MRET water in both groups. This fact also confirms that regular consumption of MRET water provides health benefits in a rather short time (two weeks in case of the animal mice model).

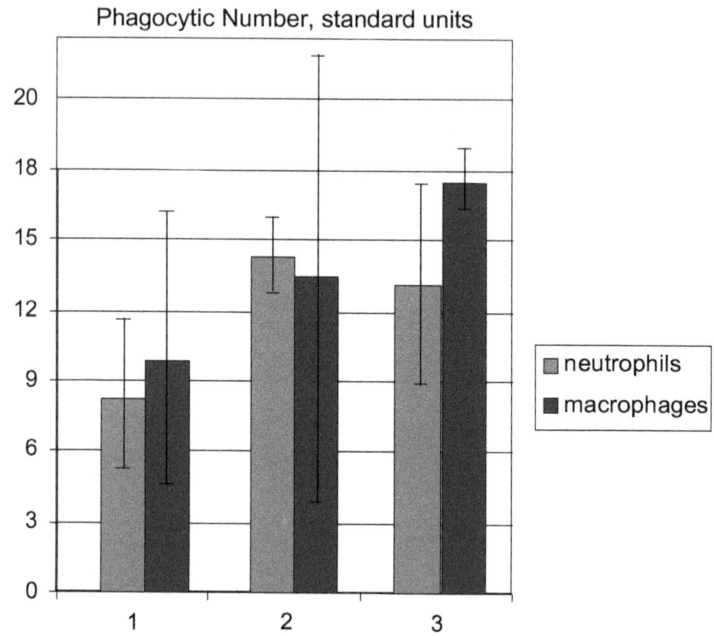

Figure 7.13. Phagocytic number of neutrophils and macrophags in two weeks of experiment (object of phagocytosis — *Staphylococcus aureus*): 1 — Control group; 2 — group No. 1 (preventive for 4 weeks; MRET water); 3 — group No. 2 (preventive for 2 weeks; MRET water).

After two weeks of experiment in the case of *Latex* as an object of phagocytosis, most parameters of phagocytic activity also increased (Table 7.10), but much less than in the case of *Staphylococcus aureus* as an object of phagocytosis. It can be explained as a result of the fact that during the first two weeks of experiment, the phagocytes "learned" to counteract *Staphylococcus aureus* microorganisms, and were not "accustomed" to destroy *Latex* culture.

7.3.6. *Conclusions to the section*

(1) The consumption of MRET Activated Water significantly enhances the factors of natural resistance of the body which constitutes the first line of protection of an organism against the penetration and reproduction of pathogenic microorganisms.

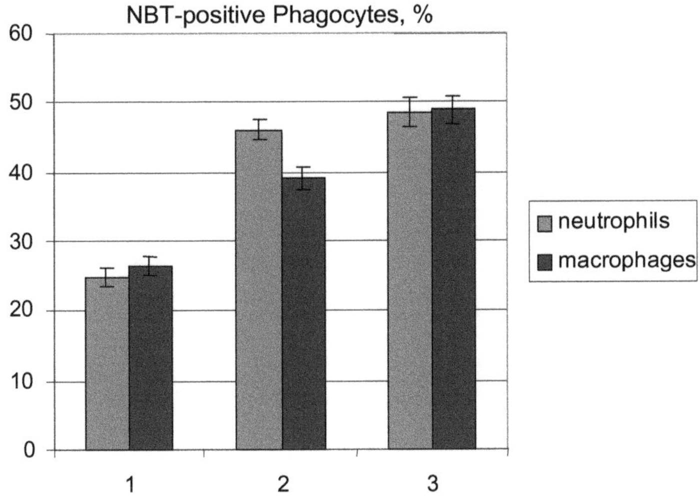

Figure 7.14. The oxygen-depended bactericidal activity (NBT test) of neutrophils and macrophages in two weeks of experiment: 1 — Control group; 2 — group No. 1 (preventive for four weeks; MRET water); 3 — group No. 2 (preventive for two weeks; MRET water).

Table 7.10. The functional activity of neutrophils and macrophages in two weeks of experiment (object of phagocytosis — *Latex*).

Group of experimental animals (N — number of mice in group)	Characteristics of functional activity of phagocytes	
	Index of phagocytosis, %	Phagocytic number, standard units
Macrophages		
Control group, N = 7	41.5 ± 3.6	11.1 ± 4.1
Group 1, N = 7	46.3 ± 7.1	9.8 ± 1.1
Group 2, N = 7	56.8 ± 3.6*	14.3 ± 2.3
Neutrophils		
Control group, N = 7	51.4 ± 2.1	9.6 ± 1.2
Group 1, N = 7	53.5 ± 2.4	15.2 ± 3.1*
Group 2, N = 7	59.0 ± 1.6*	12.9 ± 4.3

* — marks statistically significant results with $p < 0.05$ compared to control.

The analysis of data in the beginning of experiment leads to the conclusion that significant changes in all studied parameters of mice which consumed MRET water (decrease of pathogen colonies in homogenate of kidneys, increase of the weight and the cellularity of lymphoid organs, intensification of the phagocytic and bactericidal activity of macrophages and neutraphils) begin only after 24 hours following the inoculation of *Staphylococcus* culture. In other words, the consumption of MRET water increases the potential of immune capacities of the body to counteract the infections without any changes in the vital parameters of immune organs and functions prior to the penetration of infectious pathogens in the body.

At the end of two weeks of experiment, the mean values of studied parameters in both groups of mice which consumed MRET water (preventive for four and two weeks respectively) significantly increased in comparison to the control group. The differences in mean values of the studied parameters of the groups of mice consuming MRET water compared to the control group of mice were statistically significant with $p < 0.05$ (for most of the parameters). These results confirm the significant intensification of phagocytic activity and of immune system response following the consumption of MRET water.

The differences in mean values of studied parameters for the groups of mice which consumed MRET water, when compared to each other, were statistically insignificant. This confirms the similarity of the level of the beneficial effect of MRET water in both groups. This fact also confirms that regular consumption of MRET water provides health benefits in rather short period of time (two weeks in case of the animal mice model).

(2) During the infection period in both groups of mice which consumed MRET water, there was observed the significant increase of the cellularity (quantity of cells) and the weight of the spleen and lymph nodes as well as the insignificant increase of the cellularity and the weight of thymus. These results confirm the fact that MRET water intensifies the response of immune system in animals which are subjected to *Staphylococcus* infection.

(3) MRET water stimulated the phagocytic capacities of neutrophils of the peripheral blood and peritoneal macrophages. It increased their phagocytic activity (intensity of engulfing alien microorganisms) and stimulated the hyper-activation of their oxygen-dependent

bactericidal activity, particularly the increase of quantity of NBT-positive phagocytes. These results confirm the increase of effective potential of phagocytes, which constitute one of the main factors of natural protection of an organism against infections and are essential for the initiation of immune response.

(4) The consumption of MRET water has significant bactericidal effect that was confirmed by substantial decrease of *Staphylococcus* CFU (colony forming units) in homogenate of kidneys of mice. The consumption of MRET water also reduced the death rate from 30% (control group) to 0% (MRET groups) during the first nine days of experiment.

(5) The development of the local acute inflammation was significantly inhibited in the case of preventive consumption of MRET Activated Water by animals.

7.4. The Effect of MRET Activation on the Process of Growth of Staphylococcal Culture in Nutrient Medium

7.4.1. *Materials and methods of examinations*

The following part of examinations is related to the study of the effect of MRET activation process on the growth and development of *Staphylococcus aureus Wood-46* culture *in vitro* in nutrient medium. The bacterial cultures were grown on meat-pepton agar (MPA) with different initial concentration of culture cells. They were introduced to MPA in the form of suspensions and the nutrient medium with culture was MRET activated for the different periods of time (activation for 15 min, 30 min, 45 min, and 60 min respectively) following the requirements of sterility.

Petri dishes with the activated medium and culture were covered with glass caps (aerobic environment) and placed in the thermostat for cultivation at a temperature of 37°C for 18–24 hours. After that, the morphological and tinctorial properties of cultures were observed and the concentrations of colonies grown on MPA surface were calculated. Then, the bacteriostatic activity of MRET activated nutrient medium was defined with the help of appropriate statistical calculations.

The *Index of Bacteriostatic Activity* (IBA) is defined as a coefficient of the inhibition of growth and reproduction of pathogens in bacteriostatic medium, particularly in MRET activated nutrient medium. It is calculated as reduction of the concentration of colonies in MRET activated medium related to the control samples not exposed to the activation:

$$\mathrm{IBA} = \frac{(N_{\mathrm{control}} - N_{\mathrm{act}})}{N_{\mathrm{control}}}$$

where N is the concentration of colonies.

In order to verify the sterility of experiments, Petri dishes with nutrient medium (MPA) without staphylococcal culture were exposed to the process of activation and then kept in the thermostat. No colonies of culture were observed, which confirms the sterility of environment.

7.4.2. *Results and discussion*

Following the investigation, the direct correlations between the time of activation (t_{act}), the initial concentration of culture cells (N_0) and the quantity of colonies grown on MRET activated medium were observed. The results are presented below in the form of a series of photos of Petri dishes with the colonies grown on MPA surfaces, and the diagrams based on the data of these experiments (Figs. 7.15–7.20).

In the process of investigation, the effect of MRET activation on the growth of staphylococcal culture at rather small initial concentration of culture cells was analyzed. The data corresponding to higher initial concentrations $N_0 > 10^3$ cells/ml were not analyzed due to the difficulties related to calculation of very high values of concentrations of colonies, despite the fact of the high bacteriostatic activity of MRET activated nutrient medium in the case of high initial concentrations.

The highly significant bacteriostatic effect of 92–93% was observed after MRET activation for 30 min for cultures with initial concentration $N_0 = 10^3$ cells/ml (Figs. 7.15 and 7.16); the effect is 70–90% with initial concentration of $N_0 = 10^2$ cells/ml (Figs. 7.17 and 7.18). In the case of cultures with low initial concentration, $N_0 = 10$ cells/ml, the bacteriostatic activity in 15-min activated nutrient medium exceeded 93%, and in 30-min activated nutrient medium, 100% inhibition of *staphylococcal colonies* was observed (Figs. 7.19 and 7.20).

The last result confirms that the bacteriostatic effect of nutrient medium based on MRET Activated Water is caused by the effect of MRET water environment on each pathogenic cell. It is possible to assume that there is a zone of blocking of the bacteriostatic activity around each pathogenic cell (the germ of the future colony), where such activity is the most efficient.

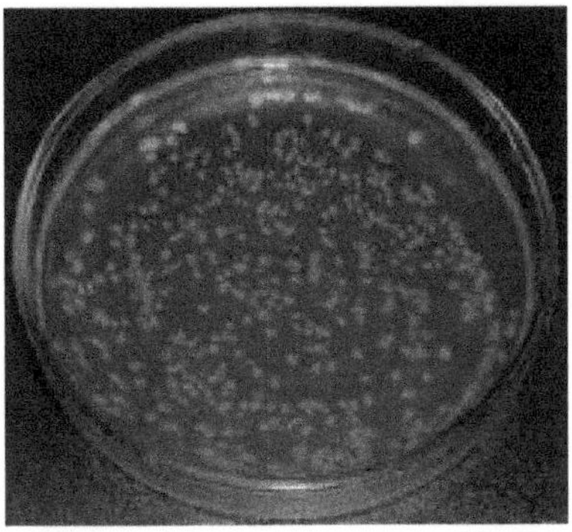

$N_0 = 10^3$ cell/ml, Control

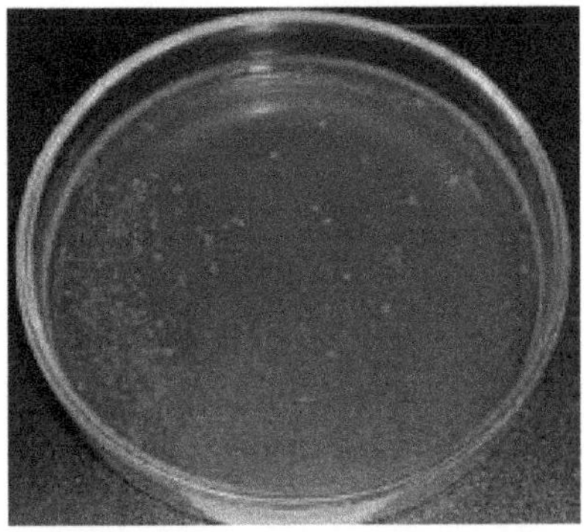

$N_0 = 10^3$ cell/ml, $t_{act} = 15$ min

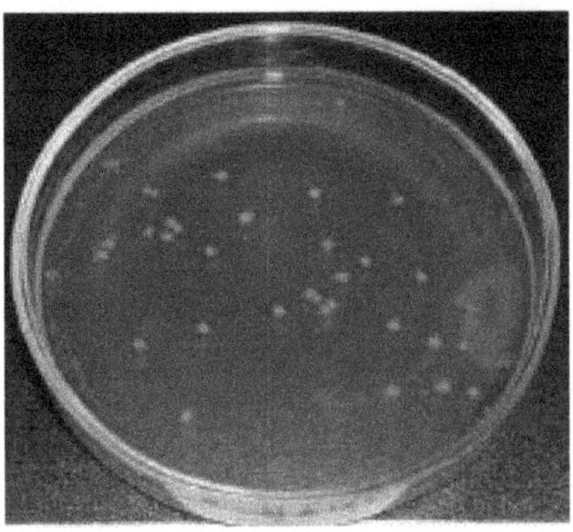

$N_0 = 10^3$ cell/ml, $t_{act} = 30$ min

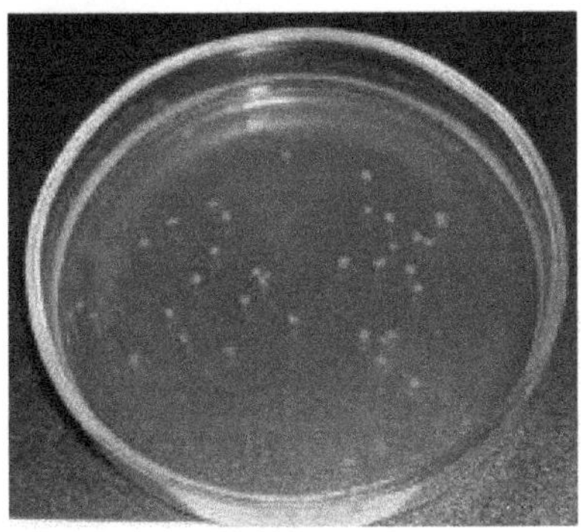

$N_0 = 10^3$ cell/ml, $t_{act} = 45$ min

$N_0 = 10^3$ cell/ml, $t_{act} = 60$ min

Figure 7.15. The effect of time duration of MRET activation on the inhibition of growth of *Staphylococcus aureus Wood-46* culture with initial concentration of $N_0 = 10^3$ cells/ml.

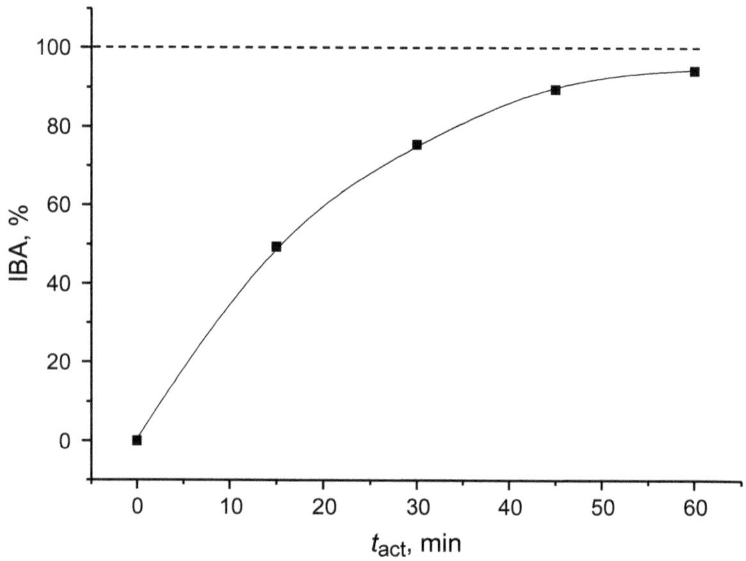

Figure 7.16. The effect of time duration of MRET activation on the inhibition of growth of *Staphylococcus aureus Wood-46* culture with initial concentration of $N_0 = 10^3$ cells/ml. IBA — Index of Bacteriostatic Activity (reduction of the number of colonies related to the control samples not exposed to activation).

In the case that such zones do not overlap, the bacteriostatic activity of MRET activated medium is the most efficient. When there are several colonies in such zone, the bacteriostatic activity is less efficient. Such assumption can explain the dependence of the bacteriostatic effect of MRET water–based medium on the initial concentration of pathogenic cells.

The photos of Petri dishes with the colonies grown on MPA surfaces and the diagrams based on the data of these experiments are shown in Figs. 7.15–7.20:

7.4.3. *Conclusions to the section*

(1) MRET Activated Water–based nutrient medium with suspended staphylococcal culture leads to the origination of the high bacteriostatic activity of such nutrient medium, which depends on the time duration of activation and the initial concentration of culture cells.
(2) The bacteriostatic activity increases following the increase of time of activation (the time of activation up to 60 min was studied).

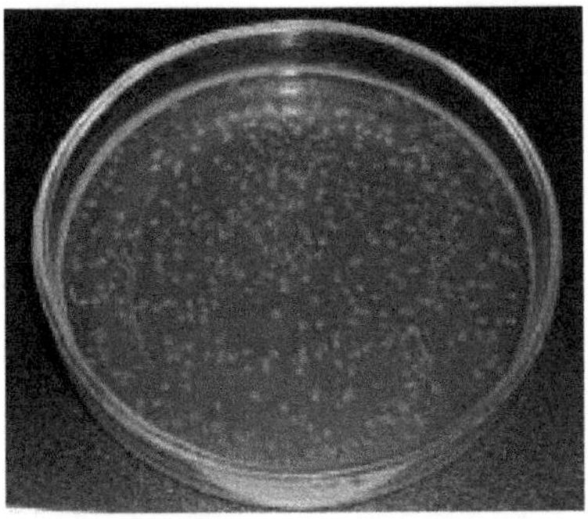

$N_0 = 10^2$ cell/ml, Control

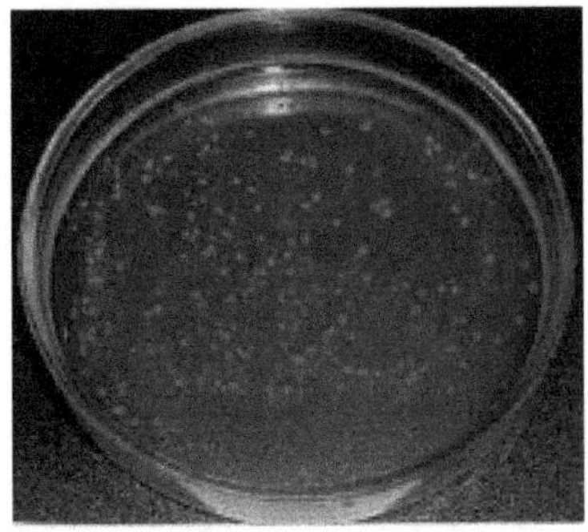

$N_0 = 10^2$ cell/ml, $t_{act} = 15$ min

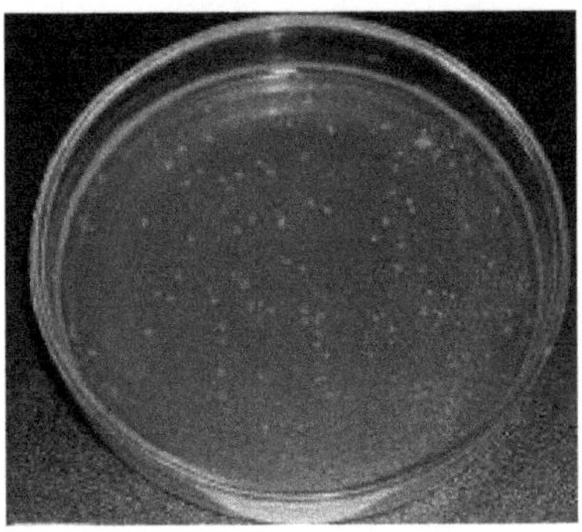

$N_0 = 10^2$ cell/ml, $t_{act} = 30$ min

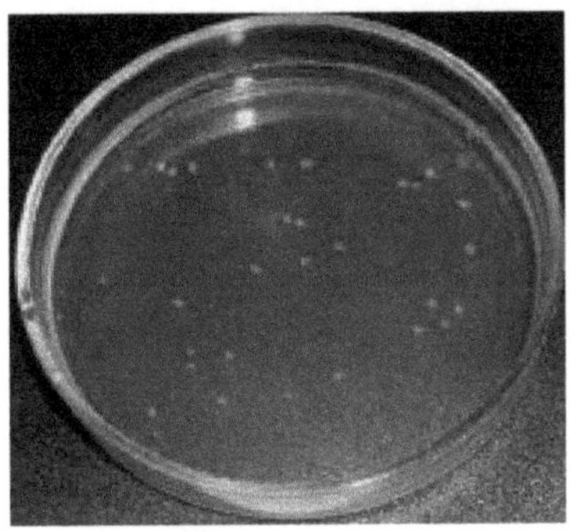

$N_0 = 10^2$ cell/ml, $t_{act} = 45$ min

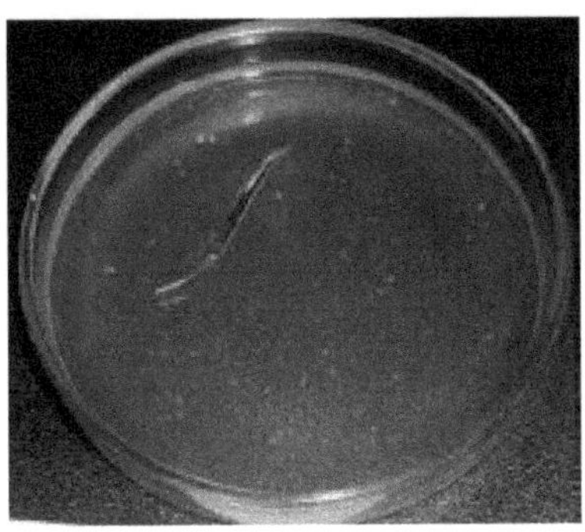

$N_0 = 10^2$ cell/ml, $t_{act} = 60$ min

Figure 7.17. The effect of time duration of MRET activation on the inhibition of growth of *Staphylococcus aureus Wood-46* culture with initial concentration of $N_0 = 10^2$ cells/ml.

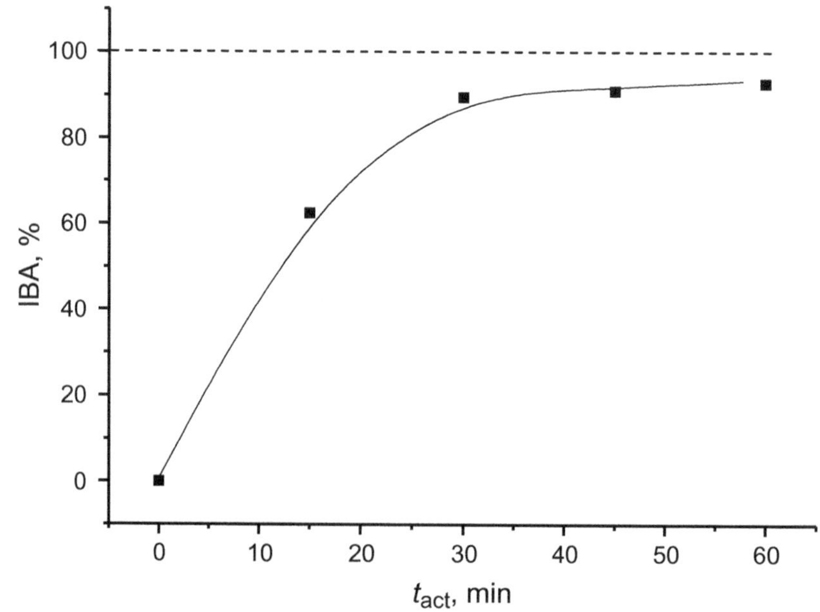

Figure 7.18. The effect of time duration of MRET activation on the inhibition of growth of *Staphylococcus aureus Wood-46* culture with initial concentration of $N_0 = 10^2$ cells/ml. IBA — Index of Bacteriostatic Activity.

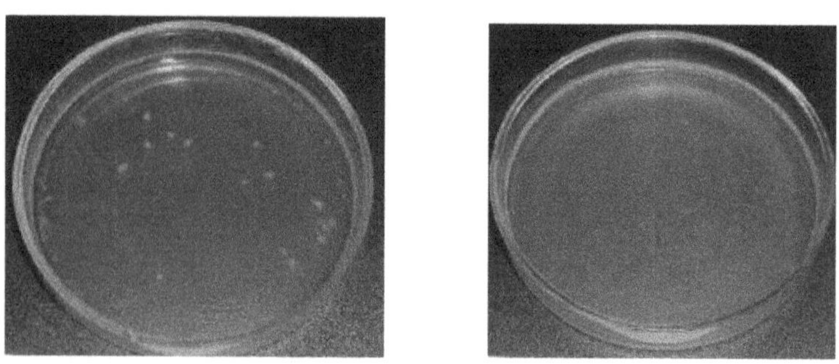

$N_0 = 10$ cell/ml, Control $N_0 = 10^1$ cell/ml, $t_{act} = 15$ min

Figure 7.19. The effect of time duration of MRET activation on the inhibition of growth of *Staphylococcus aureus Wood-46* culture with initial concentration of $N_0 = 10$ cells/ml.

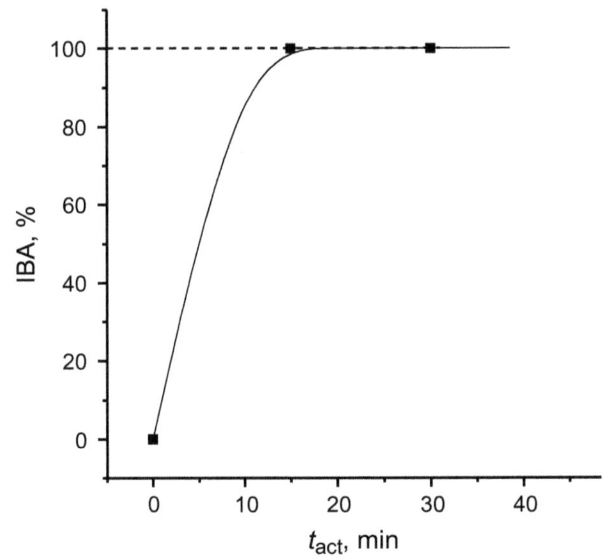

Figure 7.20. The effect of time duration of MRET activation on the inhibition of growth of *Staphylococcus aureus Wood-46* culture with initial concentration of culture $N_0 = 10$ cells/ml. IBA — Index of Bacteriostatic Activity.

(3) The efficacy of bacteriostatic activity increases following the decrease of initial concentration of the suspension of staphylococcal culture. The process of MRET activation is most effective for culture suspensions with the concentration not more than 10^3 cells/ml.
(4) The results of the second part of investigation provide the evidence regarding the high efficacy of MRET activation on the inhibition of growth of colonies and reproduction of staphylococcal microorganisms *in vitro*.

CHAPTER 8

The Possible Mechanisms of Effects of Activated Water on Biological Systems

8.1. General Regularities of the Action of MRET Activated Water on Biological Objects

The results of direct experimental studies of plants, microorganisms, and animals testify that the MRET Activated Water obtained in the optimum way renders a very strong influence on various biological objects.

Below, we give some most significant examples of such an action.

The MRET Activated Water under study

- inhibits the growth and the amount of colonies of pathogenic microbiological cultures;
- modifies very significantly the influence of various antibiotics on microbiological cultures;
- renders a very strong influence on the size and form of cells of microbiological cultures upon their division;
- changes the reductase activity;
- affects the germination rate of seeds of the plants of vegetable crops;
- affects the growth of a phytomass and other parameters of vegetable crops;
- modifies very significantly and inhibits the growth of callus tissue;
- inhibits the growth of oncological cells in the modes of prophylaxis and therapy, decreases the volume of ascitic liquid and the number of infected cells in tumors, which together leads to the significant increase of the lifetime of infected animals; and
- stimulates the antitumoral cytotoxic activity of murine lymphocytes and renders a stimulating influence on the cytotoxic potential of cell-killers.

In some cases, the action of activated water turns out to be so strong that it can be compared by efficiency with extremely-inhibiting factors such as a concentrated acid, strong antibiotics, or specialized chemotherapy. In this case, activated water, contrary to the standard preparations of chemotherapy, renders no negative side action on other organs of a living organism.

Let us separate those principal features which join these phenomena:

- They do not affect the genetic characteristics of living organisms. In all the analyzed cases, no reliable case of a change in specific characteristics of organisms was registered. That is, MRET Activated Water is safe for the action on the genetic apparatus.
- These effects concern only the quantitative characteristics of the development of biological systems (the acceleration or inhibition of the growth of cells and biomass).
- The influence of activated water depends on the time interval from the time moment of the activation, and the effect decreases with the increase of this time interval, which corresponds to the presence of the memory of activated water.
- The influence becomes much weaker, when there is even a small dilution of activated water by the addition of analogous, but nonactivated water.
- There exists a very sharply-pronounced dependence of the efficiency of the action of activated water on the duration of activation, and the optimum efficiency will be different for different types of the action. In all the studied cases, the antitumoral action of water activated for 30 min is the most optimum.

It could be expected *a priori* that activated water must have some anomalous chemical properties for the manifestation of such strong influence. However, the comprehensive studies showed that water after the activation has the same chemical composition and almost the same hydrogen index (pH), contains the same small amount of free radicals, and has no induced radioactivity. But some of its physical characteristics have varied after the activation. In particular, the conductivity, dielectric permittivity, optical density, and viscosity have become different.

It is extremely important to determine how such changes can affect the character of the interrelation of water and the elements of a living system, as well as the very functioning of a living organism.

This is one of the central problems! The further perspective of the application of activated water as a very powerful tool of life-protecting biotechnologies depends on both the degree of its comprehension and the reliability of its solution.

We understand that it is impossible now to find a full final solution to this problem. This will be made in the future. A living organism is a too-complicated, multilevel, multifunctional system. At the same time, it is very important to find those main links which could give a satisfactory answer to the posed global question about the mechanism of action of activated water.

It is necessary to indicate the possible mechanisms which ensure, with a significant probability, those radical changes which are observed in numerous experiments and are induced by the action of activated water on dissimilar biological systems such as microbiological objects, higher plants, phytogenous callus tissue, oncological cells, etc.

8.2. Possible Superficial Viscosity-Based Mechanism of the Influence of MRET Activated Water on the Division of Cells

In our opinion, the most part of the above-presented specific features of the action of activated water can be explained on the basis of the successive analysis of the influence of such water on the process of division of cells.

The division of a cell plays the decisive role in the development of any biological object. It is well known that the division of cells of multicellular organisms is the basis of both the sexual multiplication and the individual development (ontogenesis). The cell division process involves several basic stages and is terminated by the division of the cell content into two equal parts. Each part is identical to the initial cell (Fig. 8.1).

It is necessary to note that such a process by itself is not unique in nature. From the qualitative side, the division of a cell is very similar to nuclear fission of a heavy nucleus. In addition to the external similarity, there exists a profound analogy related to the reasons of the division. In both cases, the efficiency of this process is determined by the balance of acting forces.

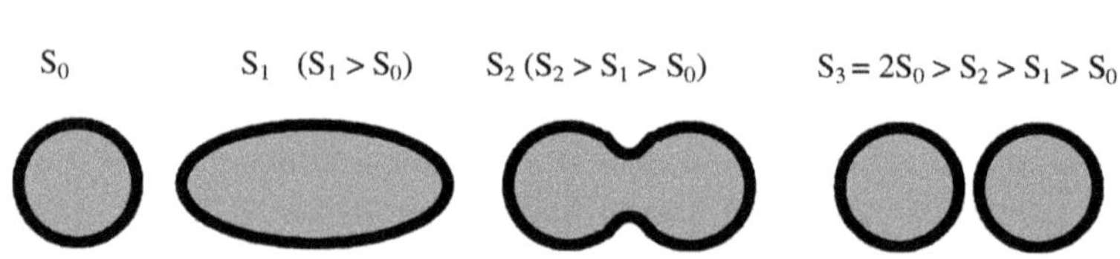

Figure 8.1. Scheme of the division of a cell in normal (nonactivated) water. The values of the area of the outer membrane surface of the cell and its form in different stages of the division are presented.

In the process of division, certain conditions must be satisfied.

The division of a cell is a complicated, ordered, and multistage process. It is related to the transfer of the genetic information written on DNA. In the stage preceding to the topological division, components of the nucleus, chromosome, and cytoplasm must be divided in half between two daughter cells. These are the internal stages of the development which are running in the volume of a cell. The direction and the character of these stages are realized according to the information written on DNA. Since the intracellular stages of the development run in the close vicinity of the genetic archive (DNA) and are realized with the help of a powerful controlling action from the side of enzymes, the external action on the processes of division of cells is very limited in these stages. It is quite natural, because the stages of the development related to DNA replication are the most important link in the process of transfer of the undistorted genetic information. These processes belong to the zone where the biochemistry of enzymes is unconditionally major.

However, the final stages of the division of cells are, to a great extent, under control of purely physical phenomena. This is conditioned by the fact that the division is determined by the balance of the superficial and bulk energy of a cell.

In the final stage of the division (when the copying of genetic information is completed), the synthesis of membranes begins to play the defining role. An increase of the membrane area leads to the appearance of distinctive spatial "folds" (sections of a corrugated surface). These folds are directed inward a cell. By continuing to grow inward, these folds are joined and form two topologically-separated closed cytoplasmic membranes. On the surface of the membrane, a cellular shell begins to grow. When the layers of this shell cover the whole surface of a newly formed cellular membrane, two formed cells can be separated from each other in space. This is the general pattern of the process of division.

At once we note that the newly formed cells can be separated only if it is advantageous from the viewpoint of the energy gain.

What forces act on a cell in the stage of its division?

Omitting the consideration of the well-known mechanism of development of a cell in the stages preceding its division, we note that, from the viewpoint of thermodynamics, the process of development of a cell is related to the transport of micro- and macro-elements in a cell across the bioplasmatic membrane enveloping a cell. In the cell, these elements are continuously transformed in proteins, nucleic acids, and lipids. The ballast elements, which are separated in a cell and are not used in biological synthesis, are removed from a cell across the same membrane. The exchange rate of substances across a membrane is very great. A growing bacteria cell can synthesize up to 40,000 amino acids of a single type per second. At the same time interval, up to 150,000 peptide bonds are created in the scope of a cell. Under the favorable conditions of the optimized withdrawal of metabolites from the cell volume across a membrane, a single cell can assimilate the amount of a substance which exceeds the cell weight at the beginning of a cell cycle by 50 times (Setlow, 1962)!

In each stage of the development, the internal pressure in the scope of a cell is balanced by the external pressure, being the sum of the pressure of the environment and the pressure of the membrane related to the forces of its surface tension. The typical values of this surface tension lie in a wide range from $\sigma = 3 \times 10^3$ erg/cm^2 to $\sigma = 1$ erg/cm^2 (Volkenshtein, 1981). The membranes of cells of different types are characterized by different surface tensions. In particular, $\sigma = 5$ erg/cm^2 for ameba (Setlow, 1962). It is worth noting that these "tabular" values of the surface tension are constants only for a specific isolated cell.

If a cell is placed in the liquid medium, the value of σ is changed. Such a result follows from the obvious fact that the appearance of the forces of surface tension reflects the difference in the character of the interaction between atoms and molecules in the volume of the unbound medium, and that of the interaction of analogous atoms and molecules which are positioned on the surface with atoms and molecules of the medium that are outside of the membrane.

For the sake of simplicity, we may assert that all components of the forces are completely compensated for atoms in the volume of a homogeneous body. Namely, the force of mutual attraction or repulsion acting from one side will always be balanced by an analogous force acting from the other side. At the same time, every atom of a specific object located on its boundary with vacuum is subjected to the action of a single noncompensated component of the force of attraction of the adjacent atom located normally to the surface in the bulk. The totality of such forces defines the pressure force compressing the given object and tending to decrease its surface. Based on such an interpretation, the surface tension defines a value of the additional energy which must be spent in order to increase the body area by 1 cm^2.

The situation is changed, if the other medium is near the surface of this object. In this case, the atoms located on the surface undergo the action of the other component, which is opposite in direction, of the force of attraction of the atoms of the environment. It is obvious that, in this case, the direction of the force of surface tension is determined by the balance (difference) of these forces. In particular, if both forces are equal in magnitude, then the surface tension is absent. It is possible to observe such a situation where the interaction force of surface atoms with atoms of the external medium (the "external liquid" in the case of cells) will be greater than that with "own" atom-neighbors. In this case, the surface tension changes its sign, and it should be named more correctly as the force of "surface stretch". In such a situation, an object immersed in a liquid begins to grow in size.

The above-discussed case of a cell in the stage of division has its own specificity. The matter is that a cell is surrounded by a plasmatic membrane, whose thickness (on the scale of a mid-size cell of the order of 1–10 μm) is very small and does not exceed $\Delta R \approx 60$–100 Å. Under the membrane surface, the liquid water-containing medium is placed. This is the cytoplasm in bacteria cells.

In this system, two complexes of forces act on a membrane. The forces from the side of the external medium tend to decrease the surface tension, and the forces from the side of the internal content of the cell tend to increase the surface tension. If the surface of the membrane were completely plane, then the force of surface tension of the whole membrane would be equal to zero in the case of the identical media on both sides of the membrane. On a concave membrane, the surface tension (for the same external and internal media) is always directed to the curvature center, which leads to the compression of the closed membrane.

Let us consider this situation in more details using the idealized example of a spherical cell with the outer radius R and the area of the external and internal surfaces $S_{out} = 4\pi R^2$ and $S_{in} = 4\pi(R - \Delta R)^2$, respectively.

The forces of surface tension acting on these surfaces are

$$F_{out} = -\frac{d(S_{out}\sigma)}{dR} = -8\pi R\sigma, \quad F_{in} = 8\pi(R - \Delta R)\sigma. \tag{8.1}$$

The resulting force F and the mean pressure on the surface of the membrane P are

$$F = F_{out} + F_{in} = -8\pi\Delta R\sigma, \tag{8.2}$$

$$P = -2\sigma\Delta R/R^2. \tag{8.3}$$

The negative sign in these relations corresponds to the fact that the pressure and the force of surface tension are directed to the center of the cell. The compression is limited by the increase in intracellular pressure.

The other situation will be in the case where different liquid media which interact differently with the surface of the membrane are present outside and inside of a cell. In this case, the surface tension of the membrane will be different: σ_{out} and σ_{in}, respectively, and

$$F_{\text{out}} = -\frac{d(S_{\text{out}}\sigma)}{dR} = -8\pi R\sigma_{\text{out}}, \quad F_{\text{in}} = 8\pi(R - \Delta R)\sigma_{\text{in}}. \tag{8.4}$$

In this case, the resulting force acting on the surface of a cell is equal to

$$F = -8\pi R\sigma_{\text{out}}\left\{1 - \left(1 - \frac{\Delta R}{R}\right)\frac{\sigma_{\text{in}}}{\sigma_{\text{out}}}\right\}. \tag{8.5}$$

The final scenario of the evolution of a cell depends on specific values of σ_{out} and σ_{in}.

If

$$\sigma_{\text{out}}/\sigma_{\text{in}} < (1 - \Delta R/R), \tag{8.6}$$

then the total force will be positive ($F > 0$), and this corresponds to the stretch of a cell.

Otherwise, where

$$\sigma_{\text{out}}/\sigma_{\text{in}} > (1 - \Delta R/R), \tag{8.7}$$

the total force will be negative ($F < 0$), and this corresponds to the compression of a cell.

It is easy to make sure that, for any ratio between σ_{out} and σ_{in}, the stretch or compression will result eventually in the stabilization and the formation of a stable cell with the equilibrium radius

$$R_0 = \Delta R/(1 - \sigma_{\text{out}}/\sigma_{\text{in}}). \tag{8.8}$$

It is expedient to determine how the activation of water affects the coefficient of surface tension.

The performed experiments, of which the results were presented in Chap. 3, show that the activation causes a very significant decrease of the viscosity coefficient measured by the interaction of water with the surface of the testing cylinder. If we take into account that the viscosity coefficient defines the character and the efficiency of molecular bonds on the boundary separating water from the surface contacting with it, then a decrease of this coefficient indicates unambiguously that the activation of water is associated with a weakening of the forces of attraction to other objects on the external boundary of the volume of activated water. It is obvious that such an effect allows one to conclude at once that the activation of water leads to an increase of the coefficient of surface tension:

$$\sigma_{\text{out}} \to \sigma^*_{\text{out}}, \qquad \sigma^*_{\text{out}} > \sigma_{\text{out}}. \tag{8.9}$$

Starting from relation (8.8), we find that, in this case, the equilibrium radius of the cell decreases by a value of

$$\delta R^*_0 = -\Delta R(1 - \sigma^*_{\text{out}}/\sigma_{\text{out}})$$

and becomes

$$R^*_0 = \Delta R/(1 - \sigma^*_{\text{out}}/\sigma_{\text{in}}). \tag{8.10}$$

It is natural to consider that if the equilibrium radius of a cell R^*_0 were to become less than the optimum radius R_0 for the normal functioning of all intracellular elements, then such a cell would not be viable. Moreover, the very process of division of a normal cell in activated water can turn out to be very decelerated. In particular, the last stage of the growth of a forming daughter cell to the optimum size, at which the full separation of cells occurs, can turn out to be impossible in activated water. That is, the cells being separated by their physiology and metabolism turn out to be topologically joined in a single undivided structure.

This process is presented symbolically in Fig. 8.2.

It is easy to make sure that a difference of the form of a real cell from the idealized spherical form does not change the conclusion about the influence of the characteristics of water on the division of cells.

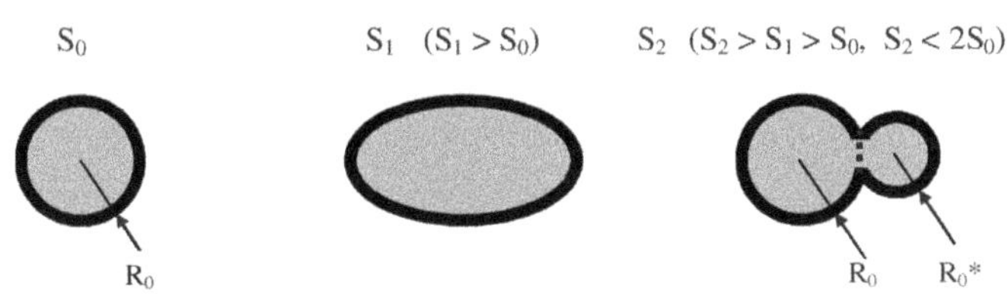

Figure 8.2. Possible scheme of the division of an initial cell in activated water. The values of the area of the outer membrane surface of the cell and its form in different stages of the division are presented. R_0 and R^*_0 are the equilibrium radii of the cell in nonactivated and activated water, respectively.

We now make the simple estimates, for example, on the basis of relations (8.6) and (8.7), in which we replace the reciprocal radius of the surface of a spherical cell $1/R$ by the mean reciprocal radius of the surface of a nonspherical closed cell:

$$\langle 1/R\rangle = \frac{1}{4\pi}\int_0^{\pi}\int_0^{2\pi}(1/r(\theta,\varphi))\sin\theta d\theta d\varphi. \tag{8.11}$$

Then, relations (8.6) and (8.7) take the form

$$\sigma_{\text{out}}/\sigma_{\text{in}} < (1-\Delta R\langle 1/R\rangle), \tag{8.6a}$$

$$\sigma_{\text{out}}/\sigma_{\text{in}} > (1-\Delta R\langle 1/R\rangle). \tag{8.7b}$$

In the case where a pair of cells are not separated, the mean value of $\langle 1/R\rangle$ for each of the unseparated cells will be greater than that for completely separated cells (as well as for the initial maternal cell), which proves the possibility of the termination of the process of separation of cells at the attainment of the optimum value of $\langle 1/R\rangle_{\text{opt}}$, i.e. in the stage where the size of the cell will be less than that in the case of nonactivated water.

This case is presented in Fig. 8.3.

As a convincing, though indirect, confirmation of the above-considered scenario, we recall the experiments performed with callus tissue described in Chap. 4. It was determined that the dilution of activated water by a small amount of nonactivated water, which is analogous by its chemical composition, leads to a very sharp decrease of the effect of inhibition of the growth of such tissue, rather than to a monotonous one.

We may advance several assumptions for this effect. If we take into account that activated water in itself has no toxic properties and, as a

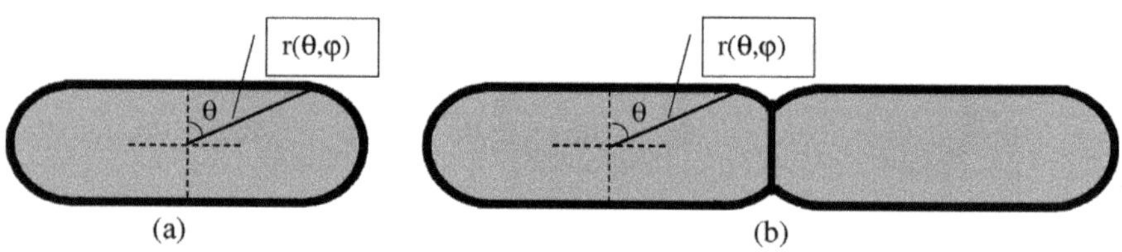

Figure 8.3. Possible scheme of the incomplete division of a nonspherical cell in activated water. The mean radius of each nonseparated cell (b) is less than that of the initial maternal cell (a). The angles θ and φ correspond to the spherical coordinate system with the origin coinciding with the geometric center of each cell.

chemical agent, renders no direct inhibiting action, we can suppose that a small admixture of ordinary nonactivated water changes significantly the character of the influence of activated water on the surface of a cell.

Based on the above-presented conception of the influence of activated water on the total surface energy of separating cells, we can assume that the inhibition of the growth of callus tissue in the medium with a great concentration of MRET Activated Water and the failure of the separation of cells of the microbiological culture of *Escherichia coli* on meat-peptone agar under aerobic conditions occur immediately for the same reason. This leads to the natural conclusion that there exists a definite threshold for the surface energy, such that its crossing makes the division of cells unfavorable in energy and, hence, improbable. In this case, though small changes in the relative concentration of activated water leads to small changes in the surface energy, if the surface energy turns out below the threshold, the division becomes again possible.

Of course, the above-considered mechanism of the influence of activated water on the division of cells is a model and omits many questions which concern the specificity of the vital activity of specific organisms. At the same time, it is obvious that the significant influence of activated water on the division of cells and, as a consequence, on the growth of micro- and macro-organisms is a universal one, and must manifest itself in any dividing biological systems.

Such a result allows us to substantiate a possible scenario of the running of many anomalous processes which occur in biological objects in the presence of MRET Activated Water (in particular, the inhibition of the growth and the number of colonies of pathogenic microbiological cultures, the appearance of anomalous nonseparated cells. inhibition of the growth of callus tissue, inhibition of the growth of cancer cells, etc.).

It is also obvious that, by virtue of the universality of this mechanism, quite real is such a situation where the presence of one definite sort of activated water (obtained from a certain duration of activation) leads to the deceleration of the division and the inhibition of the growth of certain types of cells, but we will observe the acceleration of the division for other cells and, possibly, for activated water of other sort.

We note one more aspect of the influence of the inhibition of the division of cells on their vital activity. It is related to that the actively used part of the free bioplasmatic membrane (i.e. in contact with the intercellular liquid) enveloping the cell decreases significantly in partially-separated cells. This can induce the deceleration of the process of withdrawal of ballast elements formed in a cell in the process of metabolism and, eventually, the self-poisoning of a cell.

It is natural that all these questions require further profound study.

8.3. Electrodynamic Dispersive and Viscosity-Related Mechanical Principles of the Influence of Activated Water on the Vital Activity of Biological Objects

An alternative mechanism of the influence of activated water on biological objects can be related to the peculiarities of the influence of such water on the interaction of biological molecules and microorganisms separated by a large distance in the bulk of a living organism.

As known, such an interaction is conditioned by two main processes: the electrostatic attraction and repulsion of the distributed charges of these objects and the dispersive force of the van der Waals interaction between them.

Both types of interaction depend significantly on the properties of water, which is the basis of the intracellular and intercellular water-salt liquid media.

In particular, the electrostatic interaction energy can be determined on the basis of the Debye–Hückel theory, which accounts for the self-consistent screening of the field of the charges of macromolecules by ions of the intracellular liquid, which is a weakly ionized plasma.

In order to find such screened field, it is necessary to solve the linearized Poisson–Boltzmann equation for the electrostatic field potential φ

$$\nabla^2\varphi = g^2\varphi. \tag{8.12}$$

Here, $g = \sqrt{\frac{8\pi ne^2}{\varepsilon_{W0}k_bT}}$ is the Debye constant numerically equal to the reciprocal Debye screening radius of charges in a plasma, $g = 1/\Lambda$, n is the concentration of charge carriers of each sign, and ε_{W0} is the orientational part of the dielectric permittivity of the intracellular liquid at a low frequency.

For a neutral solution, the total charge of the system is equal to zero.

The solution of this equation is the screened electrostatic field potential

$$\varphi(r) = \frac{q}{r\varepsilon_{W0}}e^{-r/\Lambda}. \tag{8.13}$$

For a neutral medium with pH = 7 and $n_{H^+} = n_{OH^-} \approx 6\times10^{15}$ cm^{-3} at room temperature $T = 373$ K, the screening radius $\Lambda \approx 0.1$ μm.

As a result, the total energy of the electrostatic interaction between all N_j charges $q_{j\alpha}$ situated on the object j and all N_k charges $q_{k\beta}$ on the object k is as follows:

$$V_{jk}^{Q} = \sum_{\alpha=1}^{N_j} \sum_{\beta=1}^{N_k} \frac{q_{j\alpha} q_{k\beta}}{r_{j\alpha,k\beta} \varepsilon_{W0}(0)} e^{-r_{j\alpha,k\beta}/\Lambda}. \tag{8.14}$$

Depending on a specific distribution of charges, this energy can correspond to the attraction, as well as the repulsion (Vysotskii *et al.*, 2005).

The other type of interaction between the same objects is determined by the dispersion van der Waals forces. The interaction energy, in the case of extended bodies including a great number of atoms and molecules, can be written as

$$V_{jk}^{\mathrm{VDW}}(r) \approx -\frac{27\hbar}{16\pi^3} \frac{V_k V_j}{r^6} \int_0^{\omega_{\max}} \times \frac{\left(\varepsilon_j(i\omega) - \varepsilon_W(i\omega)\right)\left(\varepsilon_k(i\omega) - \varepsilon_W(i\omega)\right)}{\left(\varepsilon_j(i\omega) + 2\varepsilon_W(i\omega)\right)\left(\varepsilon_k(i\omega) + 2\varepsilon_W(i\omega)\right)} d\omega. \tag{8.15}$$

The quantity $V_{jk}^{\mathrm{VDW}}(r)$ depends on the distance r between the surfaces of these objects, the spectrum of the total dielectric permittivity of water $\varepsilon_W(\omega)$, and the corresponding spectra of the dielectric permittivities $\varepsilon_j(\omega)$ and $\varepsilon_k(\omega)$ of the interacting objects (Pinchuk and Vysotskii, 2001).

Here, $\omega_{\max} = 2\pi c/r$ is the maximum frequency of the fluctuating electromagnetic field which should be accounted in the calculation of the van der Waals interaction energy between two bodies (objects) with volumes V_k and V_j.

The fundamental reason for the appearance of the van der Waals forces is a change in the energy of the systems of quantum oscillators (atoms, molecules) on their convergence. The nature of the appearance of these forces is related to the specificity of quantum electrodynamics. In the modern interpretation of quantum electrodynamics, the electromagnetic field is considered as a collection of mutually independent electromagnetic modes. Each mode is a distinctive harmonic oscillator, in which the continuous interconversion of the electric and magnetic components of the field occurs. Like any oscillator, the minimum energy of a separate mode corresponds to zero oscillations, is nonzero, and depends on the mode frequency. The reason for the appearance of these zero oscillations is related to the uncertainty relation. The total energy of all modes of the field depends on the number and the structure of modes. Upon any change of the spatial configuration of a separate part of the space, the structure of the fluctuating electromagnetic field in this region and its total energy are changed.

Thus, the total bulk energy density of the field depends on the electromagnetic properties of interacting objects in the whole range of frequencies. The presence of the maximum frequency $\omega_{\max}$ is conditioned by the influence of the effects of retardation of electromagnetic waves.

It is seen from relations (8.14) and (8.15) that the final character of the interaction between any bodies, its sign, and the intensity depend on the spectrum of the dielectric permittivities of these bodies and the water-salt medium in the region between them. They also depend on the distance between bodies. Typical, for example, is the situation where $\varepsilon_j(\omega) > \varepsilon_W(\omega)$ and $\varepsilon_k(\omega) > \varepsilon_W(\omega)$ in some part of the spectrum and $\varepsilon_j(\omega) > \varepsilon_W(\omega)$ and $\varepsilon_k(\omega) < \varepsilon_W(\omega)$ or $\varepsilon_j(\omega) < \varepsilon_W(\omega)$ and $\varepsilon_k(\omega) > \varepsilon_W(\omega)$ in other parts. This leads to that the integrand in Eq. (8.15) becomes an alternating function of the frequency. Accordingly, the interaction corresponds to the attraction of bodies in one region of frequencies and to their repulsion in the other one. The resulting force is determined by the algebraic sum of all alternating contributions from different parts of the electromagnetic spectrum.

This allows us to conclude that the controlled change in the dispersion characteristics of the water-salt medium separating the interacting objects gives the possibility to influence the sign and the intensity of the interaction between bodies. In particular, a change of the dielectric permittivity of water can stimulate the mutual attraction of, for example, viruses and cells, but can also favor their mutual repulsion at large distances.

We note that such specific features (the possibility for both attraction and repulsion) are inherent only in the total van der Waals interaction for two microbodies positioned in a medium with the dispersion of the dielectric permittivity.

Contrary to that, the first term of the frequently-used simplified relation for the van der Waals energy (the so-called "6–12" interaction)

$$V^{\mathrm{VDW}}(r) = -\frac{A}{r^6} + \frac{B}{r^{12}} \tag{8.16}$$

is a partial case of the general relation (8.15) and follows directly from it in the absence of a medium (liquid) between microbodies [in this case, $\varepsilon_W(\omega) = 1$]. Relation (8.15) yields

$$A = \frac{27\hbar}{16\pi^3} V_j V_k \int_0^{\omega_{\max}} \frac{\left(\varepsilon_j(i\omega) - 1\right)\left(\varepsilon_k(i\omega) - 1\right)}{\left(\varepsilon_j(i\omega) + 2\right)\left(\varepsilon_k(i\omega) + 2\right)} d\omega > 0. \tag{8.17}$$

It follows from Eqs. (8.16) and (8.17) that the interaction energy [Eq. (8.16)] is always negative in vacuum at large distances (at $r > \sqrt[6]{B/A}$), which corresponds to the attraction.

In the presence of a medium between the interacting microbodies (in the presence of the aqueous medium for a biological system), these objects can either attract one another or repel. It all depends on the electrodynamic properties of the microbodies and the characteristics of the water-salt medium.

The proper account of the dispersive properties of water allows one to substantiate specific effects such as the mutual recognition for two biological micro-objects at a comparatively large distance [objects with some values of $\varepsilon_j(\omega)$ and $\varepsilon_k(\omega)$ will attract one another, and objects with other values will repel one another]. For example, the process of recognition of an extraneous pathogenic virus by a leukocyte at a large distance can occur in such a way. Moreover, if the different sections of the virus surface have different chemical compositions and are characterized by different dielectric permittivities $\varepsilon_k(\omega)$, then such a situation is possible at a definite dielectric permittivity of water where some part of the virus will attract to, for example, a leukocyte, whereas the other part will repel from it. It is natural that such a character of the interaction will lead to the possibility of a spatial turn of the virus and to a sharp change of the efficiency of the interaction between a virus and a leukocyte, or between a virus and a cell.

In particular, such a mechanism allows us to substantiate one of the possible reasons for the increase of the efficiency of the modulating action of MRET Activated Water on the antitumoral cytotoxic activity of murine lymphocytes and on the cytotoxic potential of cell-killers.

It is also possible that the same mechanism can explain the great influence of activated water on the efficiency of action of various antibiotics on specific microbiological cultures, which was observed in the process of experiments.

Indeed, if it turns out that for a certain type of antibiotics and a specific microbiological culture, the ratio of their dielectric permittivities ensures that the resulting energy [Eq. (8.15)] $V_{jk}^{\mathrm{VDW}}(r) < 0$, then such micro-objects can approach each other to a small distance, which corresponds to the zone of inhibition.

At the same time, it can turn out that $V_{jk}^{\mathrm{VDW}}(r) > 0$ for the other type of antibiotics and the same specific microbiological culture. Such a result leads at once to the impossibility for the micro-objects under consideration to approach each other, which corresponds to a great size of the zone of inhibition.

The analogous assertion can be made for a change in the reductase activity discovered in the experiments.

It is expedient to note that there exist many other potentially possible mechanisms which can ensure the action of activated water on biological systems.

First of all, we mention the mechanisms which are directly related to the immediate influence of the anomalously-low viscosity of optimally activated water. The meaning of this factor becomes obvious, if we account for the decisive role of water in the processes of salt exchange and ion transport in membranes. Moreover, such a mechanism is global, and its action is revealed in every organ of a living organism. In particular, for animals and human, the use of activated water can influence the blood system.

One more aspect of the influence of the anomalously-low viscosity of activated water can be a reason for the stimulation of the cytotoxic activity of lymphocytes possessing the natural killing ability which was described in Sec. 6.4.

The above-considered mechanisms of the activation of water allow one to give a qualitative explanation for a number of effects related to the specificity of the use of such water.

We may assume that the clathrate structure of the "memory" of water is of great importance for the processes where a diverse modification of the vital activity happens. In particular, the high-temperature stability of a human organism leads automatically to that all its water content must have a fixed number of filled microcavities corresponding to the normal temperature (Tables 1.1 and 1.2). If activated water is introduced in organism, this leads to a change of the parameters of the aqueous medium. The calculations indicate that water activated in a certain way at the normal temperature of a man can be preserved for 24 hours, which is quite sufficient for the therapeutic action of such water to be realized.

If, for example, water was preliminarily rapidly heated, these contains an excess of amorphous water and many unfilled microcavities in the volume of the clathrate frame. Such water will possess a lower bulk density, which can significantly decrease the load on the heart and other organs of human beings. Its viscosity willx be significantly less, which can significantly facilitate the transport of salts in organism. A change in the viscosity of water also affects the process of enzyme-free self-reparation of double radiation-induced breaks of DNA (Pinchuk and Vysotskii, 2001; Vysotskii *et al.*, 2002; Vysotskii *et al.*, 2005). A significant change in the dielectric permittivity of activated water in the UV region of the spectrum leads to the same effect.

Activated water obtained, for example, by means of fast cooling, reveals the deficit of amorphous water and the excess of filled microcavities. The same composition is characteristic of water preliminarily subjected to long-term strong compression or water taken from mountainous springs (water molecules will be "pressed" in microcavities and have no time to leave them after the termination of the pressure action). Such water has the enhanced density and viscosity. In the process of relaxation of such water, isolated H_2O molecules will leave microcavities and pass into the volume of amorphous water. These molecules can neutralize free radicals via the free bonds.

The results presented above give quite a satisfactory explanation to the recipe well-known in medical practice: in order to preserve the activated state of water obtained in the initial hot state, it should be very rapidly cooled to a lower temperature (for example, to the temperature of a living organism). This result follows directly from the above-presented analysis of the processes of direct and reversible relaxations running in the course of the preservation of water memory at the expense of the different relaxation rates with the establishment of a thermodynamic equilibrium in two bound systems: the system of clathrate microcavities and the system of quasiamorphous water. In the case of excess concentration of water molecules in the volume of the clathrate frame (this occurs upon the cooling of water), the process of relaxation and, respectively, the process of loss of the information occur much faster than those in the process of heating of water. The last situation is characterized by the excess of empty microcavities in the same clathrate frame.

Summarizing the brief analysis of the possible mechanisms of action of MRET Activated Water on biological objects, it is worth noting that this field is, by essence, *terra incognita* and needs further profound studies.

CHAPTER 9

Conclusions and Recommendations

The performed detailed studies of the physicomolecular characteristics of water activated on the basis of the MRET technology have evidently shown that such water has a number of special or even anomalous properties relative to similar nonactivated water with the same chemical structure.

It is worth noting the following most essential features of MRET Activated Water:

- It is characterized by a very great change (the reduction by five and more times) of the dielectric permittivity and the conductivity in the fields of low and ultralow frequencies;
- a very significant reduction of both the viscosity and the coefficient of friction is registered in the regions of small pressure acting on the water and of small speed of its movement;
- the temporal nonmonotone behavior of the hydrogen index (pH) is revealed, and this change can be characterized by sharp maxima or minima for a certain type of activated water, which are manifested in many days after the completion of the activation of water, and have the form corresponding to phase transitions;
- anomalous properties of activated water are preserved for a large time interval which can last many hours and days;
- the duration of preservation of the anomalous properties of activated water depends strongly on its temperature (it increases as the temperature drops and decreases as the temperature grows).

These properties can be used for the identification of MRET Activated Water and for the explanation of some effects of the anomalous influence of such water on biological objects.

Activated water changes some very important biochemical and biophysical processes in living systems. It renders essential influence on the processes of cell division and ionic transport, and on the interaction between biological macromolecules, cells, viruses, leukocytes, etc.

The results of studies allow us to make some preliminary conclusions about features of the influence of activated water on biological objects, as well as about the probable use of such influence.

The first and most significant result of our studies is the unequivocal and proved conclusion that water activated on the basis of MRET technology influences essentially the metabolism and internal stability of microorganisms, plants, and animals. The reliability of such a conclusion is proved by the fact that a series of experiments have shown the essential influence of activated water (or a medium on the basis of activated water) on the following six levels of organization of the living matter:

- 1st level — microbial cultures *Escherichia coli* and *Staphylococcus aureus*;
- 2nd level — microbial syntrophic associations including the maximum number of species and physiological groups of microorganisms, being in the state of symbiosis and natural synergism;
- 3rd level — culture of higher plants *in vitro* (callus cells);
- 4th level — full-value higher plants *in vitro* (sterile plants on the agarized nutrient medium) and *in vivo* (plants in soil);
- 5th level — animal cells *in vitro* (as-separated tumoral cells of the ascitic form of inoculated *Ehrlich carcinoma* separated from abdominal cavities of mice);
- 6th level — animals with inoculated cancer cells of different lines (*Ehrlich carcinoma* and *Sarcoma 37*).

The second significant result is the fact that the action of activated water on living organisms is ambiguous. The ambiguity is shown by the temporal parameter (there exists the maximum effective duration for the activation of water or a nutrient medium prepared from activated water) and by the effect of action (a stronger or weaker influence on living organisms leading to a positive, neutral, or negative effect).

We note that the interpretation of the effect as positive or negative is not absolute, because it depends on the conjectural use of this effect. In particular, the effect of inhibition of the growth of a pathogenic culture is certainly positive for its use as a bacteriostatic means or one inhibiting growths, though it can be considered as negative for special types of biotechnology.

We now enumerate the most significant results obtained from studying the action of activated water on living organisms in a generalized form, and will estimate the possibility of their application in the areas concerning health protection, optimization of agricultural production, and commercial targets. Furthermore, it is expedient to formulate some recommendations for a possible use of the obtained results, as well as for their subsequent optimization.

(1) Water activated for 30 min enhances the reductase activity of microbial syntrophic associations under anaerobic conditions. Such water entering the organism of warm-blooded animals and human (with food and drink) can render essential influence on the metabolic processes of the microflora of macroorganisms.

In what follows, we present some examples of the presumable influence of such a type of activated water on different living objects.

In a human organism, water activated in the optimum way can increase the metabolic activity of symbiotic microbial associations of intestine: producers of vitamins (e.g. B_{12}), lactic acid bacteria, etc.

In the organisms of cows and other herbivorous animals, the intake of optimally activated water can help to regulate the complex of symbiotic microorganisms digesting cellulose and, on the whole, can probably increase essentially the production efficiency of milk and meat. This conclusion has simple logic substantiation. We recall that the symbiotic microflora of the digestive tract of herbivorous animals (cows and others) is a typical representative of syntrophic associations. In the scope of digestive tract, anaerobic conditions are realized, under which, as shown in Sec. 5.5, the reductase activity increases. An increase of the metabolic activity of associations of digestive tract of herbivorous animals is accompanied by an increase of the rate of digestion of cellulose and, hence, results in the increase of milk yields and the acceleration of weight gain of animals.

In the organisms of poultry (e.g. hens, ducks), the use of optimally activated water can raise the grain assimilation speed owing to the activation of the metabolism of microorganisms of intestine and, as a result, it can increase the weight-gaining speed and raise egg-laying qualities.

(2) Microflora of gastroenteric tract of human is also a syntrophic association. The composition of microorganisms useful to a man includes the symbiotic producers of vitamins, lactic acid bacteria, and other microorganisms. The activation of water entering a human organism can render essential influence on the biochemical activity of the mentioned symbiotic microorganisms. This yields the obvious necessity of studying the action of the activation of water on symbionts, which are the producers of biologically active substances. The importance of the optimization and balance of the functioning of the microflora of human intestine, with the help of water

activated in the most optimum way, is theoretically comparable with that of the application of the most effective biological preparations such as yogurt, special lactic-acid fermented product, etc.

(3) The separate and rather scientific-commercial perspective is the influence of activated water (for various modes of its activation) on the efficiency of the synthesis of biologically active substances using industrial microorganisms. The activation of water influences the metabolism of microorganisms. It is obvious that various modes of activation can result in the essential increase of the synthesis rate of valuable products. For example, it is possible to significantly increase the yield of citric acid by acting on microbiological culture *Aspergillus niger* or to attain a high yield of the synthesis of various antibiotics with culture *Streptomyces*, etc. Hence, MRET Activated Water presents the basis for a significant enhancement of the efficiency of industrial biotechnology.

(4) In our studies, we have found the essential dependence of the efficiency of the action of MRET Activated Water on microbiological cultures and their associations in the presence of a sufficient amount of free oxygen (i.e. in the conditions under which the growth of microbiological cultures takes place: aerobic or anaerobic).

In particular, we observed a very strong increase of the reductase activity of microbiological associations grown under anaerobic conditions.

At the same time, during the growth of the pure *Escherichia coli* culture under the same anaerobic conditions, no change of reductase activity was registered for all investigated types of activated water.

On the other hand, whereas a small reduction of the reductase activity was registered for the pure culture of *Escherichia coli* under aerobic conditions during first hours of the growth in water activated for 30 min, we observed an increase of this activity under the same conditions in water activated for 60 min.

These examples show the complicated and ambiguous character of the action of MRET Activated Water on the processes of metabolism in microorganisms in the presence or absence of a sufficient amount of free oxygen. For this reason, the final recommendations for a practical application of activated water in applied biotechnology involving the processes of metabolism in various microorganisms should be given with regard for the whole complex of conditions. It is also obvious that, in order to draw a more certain conclusion on the general tendencies of the action of such water, it is necessary to carry out more precise experiments with the other types of pure cultures.

(5) MRET Activated Water influences essentially the stability of a typical representative of conditionally pathogenic microflora, the culture of *Escherichia coli*, to the action of different types of antibiotics. The action of activated water turned out ambiguous. For various modes of activation, we observed both an increase of the stability of the culture (a decrease of the sensitivity) and a decrease of the stability (an increase of the sensitivity) with respect to various antibiotics. For example, water activated for 30 min enhances the stability to chloromycetin, but water activated for 60 min, on the contrary, reduces this stability and increases the sensitivity.

It is necessary to emphasize especially that the influence of activated water on the efficiency of action of antibiotics can be very strong.

For example, in the use of water activated for 30 min, the stability of *Escherichia coli* culture to chloromycetin increases in comparison with the control by 19 times! In the use of water activated for 60 min, the stability to kanamycin and cephalexin increases by 12–13 times.

In contrast to this, with the use of the same type of activated water, the sensitivity of *Escherichia coli* culture to the action of ampicillin increases in comparison with that under the influence of nonactivated water by 2.3 times.

These rather surprising and even paradoxical results have been obtained from one representative of conditionally pathogenic microflora — *Escherichia coli*. It is obvious that the same studies should be carried out with other pathogenic cultures. In addition, since the action of activated water depends very significantly on the duration of activation, it is important to carry out similar studies with samples of water activated with smaller discrete steps in time (for example, for the duration of activation equal to 20 min, 25 min, 30 min, 35 min, and 40 min).

The determination and use of the modes of activation resulting in the maximal increase of the sensitivity or stability of specific microbiological cultures to antibiotics can lead to the overturn in clinical medical microbiology and a significant reduction of the dosage of antibiotics, and can sharply weaken the negative side effects accompanying the use of antibiotics.

(6) Activated water renders bactericidal (or bacteriostatic) action on *Escherichia coli* culture growing under aerobic conditions. It is reasonable to expect a similar influence for other pathogenic cultures (for example, activated water should suppress the development of conditionally pathogenic bacteria *Citrobacter*, etc.).

The corresponding optimization of the modes of activation of aqueous solutions can be used effectively for the therapy of *E. coli* enteritis and other similar infectious diseases caused by enterobacteria.

In addition to purely medical aspects of such an action of MRET Activated Water, it is necessary to emphasize the potential of a very wide spectrum of its possible applications. For example, it is possible to produce and store great amount of pure water, in which the development of microorganisms is extremely inhibited. This will be of great importance in solving the problems of hygiene, especially in countries with adverse epidemiological conditions. The same applications are possible in the case of big-scale natural disasters (flood, earthquake, etc.).

(7) The activation of water suppresses very significantly (by tens and hundreds of times) the development of a callus culture of plant cells (in particular, *Solanum rickii*) in the presence of a stimulator of uncontrollable nonspecific growth. A callus culture of cells can serve as a representative model for studying the inhibition of the uncontrollable growth of the cells of epidermis in the case of psoriasis by activated water. There are weighty arguments to believe that such water can be very effective nonmedicamentous means for the treatment of psoriasis without inherent negative side effects. This effect should be comprehensively investigated in clinical practice in a wide range of changes of the duration of activation. The preliminary experiments confirm its efficiency on such a treatment.

(8) With respect to the efficiency of action of MRET Activated Water on callus tissue we revealed a very sharp dependence on the degree of its dilution by similar, but nonactivated water. In particular, the reduction of the relative concentration of water activated for 30 min from 80% up to 72% in the cultural medium led to the decrease of the coefficient of inhibition of the growth of callus tissue by 20 times. Such an effect can be applied to clinical medical pharmacology and chemotherapy. In particular, let us consider the situation of the preliminary injection or the saturation of certain organs of a living organism with ordinary water. In this case, the sharply pronounced selectivity or differentiation of the action of activated water or various pharmaceutical preparations (including antibiotics) on different organs can take place.

(9) The application of optimally activated water results in very significant positive effect on the prophylaxis and treatment of the tumoral process of *Ehrlich carcinoma* and *Sarcoma 37* in animals.

Water activated for 30 min in prophylactic action (water consumed before the start of the oncological process) inhibits the growth rate of

Ehrlich carcinoma by two times and reduces the number of cancer cells in the volume of a tumor by more than four times. Such water also increases the lifetime of mice infected by *Ehrlich carcinoma* by 61.7%. The inhibition of tumor growth, which is similar in efficiency and increase of the lifetime, was also observed with the intake of water activated for 15 min and 45 min.

The same type of activated water inhibits the development of a tumor (cancer cells) of *Sarcoma 37* by 67% and reduces the number of active oncological cells in the volume of a tumor by three times. In this case, the lifetime of sick animals grew by 50%.

Experiments have shown that water activated for 30 min is optimum for the therapeutic mode of treatment (water consumed after the beginning of the oncological process). The efficiency of the therapeutic action of activated water turned out two to three times worse than that in the case of the prophylactic use of this water.

Summing up, we note that as prepared activated water with the duration of activation of 30 min, which is used for prophylaxis, is a very effective means for the growth inhibition of tumors caused by cells of *Ehrlich carcinoma*. The efficiency of the prophylactic action of such water approaches the efficiency of chemotherapy but, unlike chemotherapy, the application of activated water does not result in negative side effects.

(10) Water activated for 60 min renders a much weaker positive influence on the tumoral process when used prophylactically (as compared with the action of water activated for 30 min), and no appreciable (statistically reliable) action on the therapeutic use was observed.

(11) The long-term (in the interval of 15–45 days) storage of water activated for 30 min and stored after that in a cooler does not result in the loss of antitumoral activity, though reduces it a little. Such water is still an effective antitumoral means and ensures the inhibition of the growth of tumors and thus increases the lifetime. During the investigated period of storage of such water, its antitumoral efficiency was reduced approximately twice. This result allows us to predict the opportunity to use not only as prepared water, but also water after a long-term storage.

(12) The application of water activated in the optimum way results in an increase of the cytotoxic activity of lymphocytes and natural cell-killers developed in the organisms of animals. Water activated for 30 min for prophylactic use for 21 days before the inoculation of tumor cells increases the index of cytotoxic activity of lymphocytes by 20%. For prophylactic application, such water can be used to increase the natural immunity. It is found out that, within the limits of the studied time-interval, the

immunostimulating properties of such water increase with the duration of prophylactic intake. In particular, when the duration of intake of water increased from 14 to 21 days, the index of cytotoxic activity grew from 6 to 20%.

On the other hand, we note that water activated for 15 min, 45 min, and 60 min and used for 21 days has no noticeable immunostimulating properties (a change of the index of cytotoxic activity is within the limits of statistical error).

These results testify to a very important value of the duration of activation of water before its use for the purpose of prophylaxis. It is obvious that carrying out additional detailed studies with a smaller increase of the duration of activation within the limits of the interval from 15 to 45 min would provide useful information (for example, for the duration of activation equal to 20 min, 25 min, 30 min, 35 min, or 40 min with a step of five min; or even two min).

In addition, it is desirable to carry out more detailed studies of the dependence of the prophylactic action of activated water on the time of its intake. It is also necessary to clarify the question whether the constant intake of such water is the optimum method of prophylaxis.

Such studies will allow us to find the optimum duration of activation, at which the effect of the prophylactic antitumoral action of activated water will be maximum.

(13) The use of activated water stimulates the development of many plants. In particular, the use of water activated for 60 min results in the acceleration of the germination of seeds of radish, peas, and cabbage (representatives of crucifers) by 80–200% on the third day after sowing in soil. It is obvious that this phenomenon can be efficiently used commercially, for example, for the accelerated cultivation of the sprouts of valuable ornamental plants or vegetable crops.

(14) The use of water activated for 60 min leads to the acceleration of the growth and an increase of the parameters of some vegetable culture such as cabbage, pumpkin, string bean, peas, and radish. The positive effect is manifested, first of all, in the increase of the biomass growth, i.e. an increase of the productivity of agricultural culture.

Summing up, we note that, for a correction of the range and the efficiency of the action of activated water on living organisms, it is necessary to expand the specific and taxonomic assortments of objects under study. For example, it is desirable to investigate the influence of the activation not only on objects such as bacteria of intestinal group, but also on yeast, actinomyces, micromyces, facultative-anaerobic and obligatory-anaerobic microorganisms. The same concerns higher plants — it is also necessary to expand the composition of investigated objects. The detailed research of the influence of activated water on the processes of metabolism of various animals is extremely desirable as well.

Very important is the further study of the action of activated water on the specific features of the process and the specificity of treatment of various oncological diseases. The study of influence of the duration of the application of optimally-activated water on the optimization of the immune system of organism, including the influence on the optimization of the cytotoxic activity of lymphocytes and natural cell-killers, is also of great importance. There is ground to consider that the increase of the index of cytotoxic activity by 20% discovered in our experiments is not limiting and can be significantly enhanced.

The same wishes can be stated concerning the opportunity to use activated water for the therapy of other human diseases.

(15) The consumption of MRET Activated Water significantly enhances the factors of natural resistance of the body which constitute the first line of protection of an organism against the penetration and reproduction of pathogenic microorganisms.

The analysis leads to the conclusion that significant changes in all studied parameters of mice which consumed MRET water (decrease of pathogen colonies in homogenate of kidneys, increase of the weight and the cellularity of lymphoid organs, intensification of the phagocytic and bactericidal activity of macrophages and neutrophils) begin only after 24 hours following the inoculation of *Staphylococcus* culture. So, the consumption of MRET water increases the potential of immune capacities of the body to counteract the infections without any changes in the vital parameters of immune organs and functions prior to the penetration of infectious pathogens in the body.

Experimental results confirm the significant intensification of phagocytic activity and of immune system response following the consumption of MRET water.

MRET water stimulated the phagocytic capacities of neutrophils of the peripheral blood and peritoneal macrophages. It increased their phagocytic activity (intensity of engulfing of alien microorganisms) and stimulated the hyper-activation of their oxygen-dependent bactericidal activity, particularly the increase of quantity of NBT-positive phagocytes. These results confirm the increase of effective potential of phagocytes, which constitute one of the main factors of natural protection of an organism against infections and are essential for the initiation of immune response.

(16) MRET activation of the water-based nutrient medium with suspended *Staphylococcus* culture leads to the origination of very high bacteriostatic activity of such nutrient medium which depends on the time duration of activation and the initial concentration of culture cells. The bacteriostatic activity increases following the increase of time of activation and decrease of initial concentration of the suspension of staphylococcal culture!

The results of investigation provide the evidence regarding the high efficacy of MRET activation on the inhibition of growth of colonies and reproduction of staphylococcal microorganisms *in vitro.*

References

Bernal J. D. and Fowler R. H. (1933) *J. Chem. Phys.* 1, 513.

Carion H. P. (1978). *Applied Optics.* 17, 3192.

Chaplin M. (2005). Water dielectric and microwave radiation, London South Bank University, http://www.lsbu.ac.uk/water/microwave.html.

Chau Z. J. (1996). *The New Construction Method of Concrete*, The Publishing House of Chinese Architectural Industry, Beijing, pp. 401–407.

Davenas E. *et al.* (1988). *Nature* 333, 816–818.

Drachev V. P., Bragg W. D., Podolskiy V. A. *et al.* (2001). Large local optical activity in fractal aggregates of nanoparticles, *J. Opt. Soc. Am. B* 18(12), 1896–1903.

Drost-Hansen W. (1971). *Chemistry of the Cell Surface, Part B*, Academic Press, New York, p. 1.

Drost-Hansen W. and Singleton J. L. (1991). Our aqueous heritage: evidence for vicicnal water in cells, *Fundamentals of Medical Cell Biology*, Vol. 3A, Chap. 5, JAJ Press Inc.

Gapochka L. D., Koroliov A. F. and Sukhorukov A. P. (1994). *Moscow State University Physics Bulletin*, Vol. 36, pp. 71–75.

Klassen V. I. (1973). *Water and Magnet*, Nauka, Moscow.

Ling G. N. (2003). A new theoretical foundation for the polarized-oriented multilayer theory of cell water and for inanimate systems demonstrating long-range dynamic structuring of water molecules, *Physiol. Chem. Phys. Med. NMR*, USA, 35, 91–130.

Maksimenko V. V. (1999). Localization of light in fractal systems, *Conference on Biophotons*, International Institute of Biophysics.

Marin-Almazo M. *et al.* (2004). Cobalt-based superparamagnetic nanorings, *Nano Lett.*, 4(8), 1365–1371.

Meier W. and Finkelmann H. (1993). Piezoelectricity of cholesteric elastomers. Influence of the helicoidal pitch on the piezoelectric coefficient, *J. Macromol.*, 26, 1811–1817.

Murashige T. and Skoog F. (1962). *Physiol. Plantarum* 15, 191–195.

Nemethy G. and Seheraga H. A. (1962). *J. Chem. Phys.* 36, 3382.

Ong K. C. G. (2004). *The Effect of Activated Water on the Properties of Concrete*, National University of Singapore Press.

Patsis G. P. and Glezos N. (1999). Molecular dynamics simulation of gel formation and acid diffusion in negative tone chemically amplified resists, *Microelec. Engin.* 46, 359.

Pauling E. (1959). *Hydrogen Bonding*, ed. D. Hadzi, Pergamon Press, London.

Pinchuk A. O. and Vysotskii V. I. (2001). Long-range intermolecular interaction between broken DNA fragments, *Phys. Rev. E* 63, 31904–31910.

Samoilov O. Ya. (1957). *Structure of Water Solutions and Hydration of Ions*, USSR Academy of Sciences, Moscow.

Serov I. N. (2003). Aires ecological converter, *Aires New Medical Technologies Foundation*, St. Petersburg.

Setlow R. B. and Pollard E. C. (1962). *Molecular Biophysics*, Addison-Wesley, Reading, Massachusetts, USA.

Smirnov I. V. (2003). The effect of a specially modified electromagnetic field on the molecular structure of liquid water, *Explore Mag.* 13(1).

Smirnov I. V. Method and device for producing activated liquids and method for use thereof, February 2000, US Patent No. 6.002.479.

Sobry R., Rassel Y., Fontaine F. *et al.* (1991). Structural SAXS study of epoxy-acrylate interpenetrating polymer networks. Fractal geometry and spherical domain sizes, *J. Appl. Crystallogr.* 24, 692–701.

Villee C. A. (1967). *Biology*, W.B. Saunders Company, Philadelphia and London.

Volkenshtain M. V. (1981). *Biophysics*, Nauka, Moscow, Chap. 10.2.

Vysotskii V. I. and Kornilova A. A. (2004). The physical basis of long-term water memory effect, *Moscow State University Physics Bulletin*, 59(3), 58–62.

Vysotskii V. I., Pinchuk A. O., Kornilova A. A. and Samoylenko I. I. (2002). *Radiation Physics and Chemistry*, Vol. 65, p. 487.

Vysotskii V., Olishevsky S., Yanish Yu. and Kornilova A. (2006). Investigation of physical properties of MRET activated water and its successful application for prophylaxis and treatment of oncology, *World Congress on Medical Physics and Biomedical Engineering 2006 (WC 2006)*, Seoul, Korea, IFMBE Proceedings, 14, 1788–1791.

Vysotskii V., Tashyrev A., Tashyreva A. and Kornilova A. (2006). The biophysical model and experimental observation of strong inhibition activity of water activated with the help of MRET process, *World Congress on Medical Physics and*

Biomedical Engineering 2006 (WC 2006), Seoul, Korea, IFMBE Proceedings, 14, 3093–3096.

Vysotskii V. I., Smirnov I. V. and Kornilova A. A. (2005). *Introduction to the Biophysics of Activated Water*, Universal Publishers, USA.

Zatsepina G. P. (1998). Physical properties and structure of water, Moscow State University Publishing House.

Zenin S. V. (1999) *Structured State of Water as a Basis of Control of Behavior and Safety of Living Systems*, Habilitation thesis, Moscow State University, Moscow, Russia.

Index

www.ingramcontent.com/pod-product-compliance
Lightning Source LLC
LaVergne TN
LVHW051057210926
848486LV00002B/132